HVAC INSTANT ANSWERS

HVAC INSTANT ANSWERS

Peter Curtiss
Newton Breth

McGRAW-HILL

New York Chicago San Francisco Lisbon London
Madrid Mexico City Milan New Delhi San Juan
Seoul Singapore Sydney Toronto

2 3 4 5 6 7 8 9 0 AGM/AGM 0 9 8 7 6 5 4 3 2

ISBN 0-07-138701-3

The sponsoring editor for this book was Larry S. Hager and the production supervisor was Pamela Pelton. It was set in Stone Sans by Lone Wolf Enterprises, Ltd.

Printed and bound by Quebecor/Martinsburg.

This book is printed on recycled, acid-free paper containing a minimum of 50% recycled, de-inked fiber.

CONTENTS

ACKNOWLEDGMENTS

Many of the photographs in this book were taken on the Boulder campus of the University of Colorado. The Facilities Management department at the University graciously gave permission to access the various mechanical spaces and to use the resulting images. In doing so, they have helped provide a clarity to the descriptions that would have been impossible otherwise.

The final manuscript was reviewed and edited by Stuart Waterbury and Heidi Crimmin. Their comments and suggestions were vital in making a book that is hopefully easy to read and use. Thanks also to Elizabeth Gehring, who contributed the section on condensing and non-condensing furnaces.

Finally, the authors dedicate this book to their wives, Arlene Breth and Heidi Crimmin, who provided immeasurable support and kept them going and in good health during the writing process.

ABOUT THE AUTHORS

Peter Curtiss is a consulting engineer in Boulder, Colorado. He received his Ph.D. in Civil Engineering at the University of Colorado, where he was one the first to investigate the use of neural networks in the control and optimization of large HVAC systems.

Newton Breth has worked as an Instrument Technician at the University of Colorado since 1977. Prior to that, he spent time in the U.S. Army Signal Corps and the Denver Water Board. He is certified by the Instrument Society of America and holds several U.S. patents related to automatic control devices.

chapter

INTRODUCTION AND OVERVIEW

This book is aimed at the field technician who has the of task solving problems rapidly and efficiently. It is the unfortunate truth that most facility management teams are understaffed. The technicians spend a lot of time satisfying occupant comfort complaints and often do not have the time to spend tracking down the root of many problems.

This guide is meant to provide both theoretical background and principles of operation of equipment, while at the same time giving the reader a practical means for quickly solving problems. The basic philosophy of this guidebook is to work backwards from the effects of a problem, hopefully eliminating many false leads along the way. We have tried to reach a compromise between solutions that are too simplistic or too complicated. It is emphasized that more often than not the cause of a control or equipment problem may not be in the obvious place. The key point is to identify what is causing the building occupants to express concern.

HOW BUILDINGS WORK

Every building is different. Even those that appear similar from the outside will certainly have differences in their structure, their use, and the quirks of their behaviors. Like people, no two are alike. In fact, the idea of a building as a person has some merit: most buildings have a need to take in fresh air and expel contaminated air. Many buildings provide a constant circulation of fluids around the building, and practically all buildings must maintain an internal temperature setpoint for the comfort of the occupants. We even now refer to problems as if we were describing people: a structure may have *sick building syndrome*. Some buildings are naturally relaxing to the occupants, others are disconcerting for one reason or another.

To the HVAC engineers and technicians, the goal is to make the building a comfortable and safe environment and to do so at a minimum cost for energy and maintenance. This is a noble, yet occasionally difficult task. As the building size increases and the function becomes more varied, the complexity of the HVAC systems can grow. Eventually, it can get to the point where a technician may feel like he is constantly just putting out small fires rather than getting a handle on the entire building.

Figure 1.1 shows a cross-section of a typical residence. Here the HVAC is quite simple: a furnace provides heating in the winter and an air-conditioning unit provides cooling in the summer. Just about the only required maintenance is the annual replacement of the furnace filter and an occasional inspection of the air conditioning unit. Even then, the occupants of many houses pay almost no attention to these systems. This is to be expected, since for most people the residential HVAC is more or less invisible until something goes wrong.

In contrast, Figure 1.2 shows a cross section of a typical office building. Here there are many more components to the HVAC system. A mechanical room in the basement sports a boiler, chiller, and central air-handling unit. A series of pipes and ductwork allow the water and air to flow through the building to equipment where it may help keep the building at the desired indoor temperature and humidity. Each occupant will have a different tolerance for heat and cold and humidity, and it is best to try to maintain the building at the setpoint conditions to avoid occupant complaints. In some cases, the physical layout of the building or the original HVAC design can make it very difficult to provide quick solutions to comfort problems. However, the sum of annual salaries in a commercial building is often several hundred times the sum of annual energy costs, so the argument can be made that occupant comfort is worth the investment in HVAC repair and upgrades. Also, in commercial retail buildings the volume of sales is related to the comfort of the shoppers. Some large market chains accept relatively large energy and lighting bills precisely because these add to the ability to get merchandise out the door.

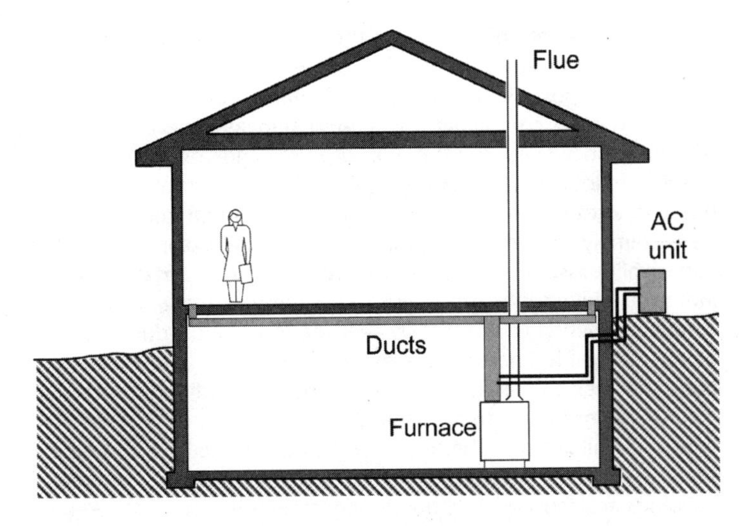

FIGURE 1.1 Typical residential HVAC system components.

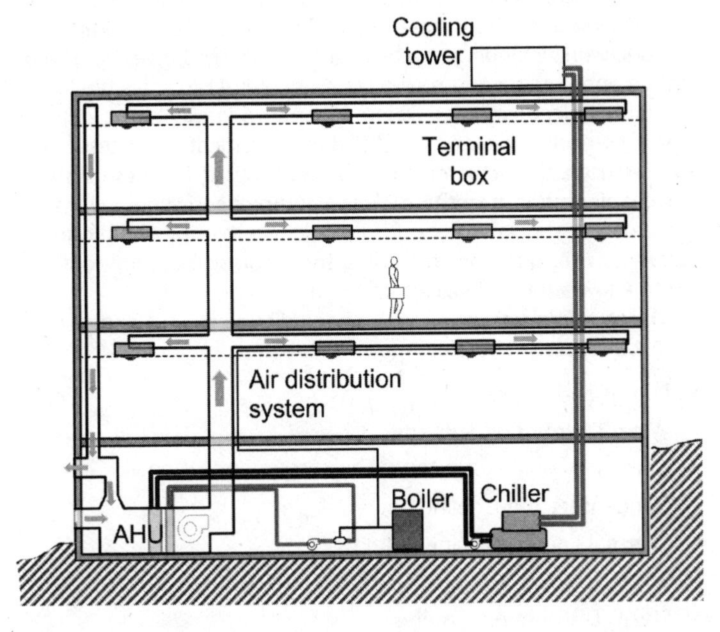

FIGURE 1.2 Typical commercial building HVAC system components.

A third category of building is the industrial site as illustrated in Figure 1.3. These kinds of buildings come in all shapes and sizes and it is almost impossible to make generalizations about what even constitutes an industrial building. The figure here assumes a large manufacturing space with rooftop units used to condition the space. But in some manufacturing buildings, especially those that create electronic devices or use chemical reactions, the indoor conditions can be extremely important to produce a high yield of products that pass quality assurance tests off the assembly line. Certain zones of industrial buildings may require exact temperature and humidity conditions. When one considers that the value of products coming off the assembly line may be measured in millions of dollars per hour, the need to maintain the environmental conditions is obvious.

It is also important to note that the line between commercial and industrial buildings can be thin. Many industrial facilities also have a significant amount of attached office space, while some commercial buildings are combined with light industrial use.

Chapters 2 through 4 in this book provide general information on the basic materials used in building HVAC systems. This information is not critical to the quick solution of HVAC problems but is included here as a reference for the technician who wants to dive deeper into the details of a specific issue. Chapter 2 describes the properties of air and ductwork, including psychrometrics. Chapter 3 gives basic information about water and piping, and Chapter 4 discusses the fundamentals of electricity and wiring. Chapter 5 presents the different types of sensors used for control and measurement of processes, and also discusses the actuators and controllers used for these systems. The chapter is not meant to replace a comprehensive guide to pneumatic or electronic control—it is included so that the technician can quickly answer questions regarding these components without having to know a lot of detail about them.

The remaining chapters in the book discuss various components of HVAC systems. These components are

- Pumps and Valves
- Water Distribution Systems
- Chillers
- Condensers and Cooling Towers
- Thermal Energy Storage Systems
- Boilers
- Steam Distribution Systems
- Fans and Dampers

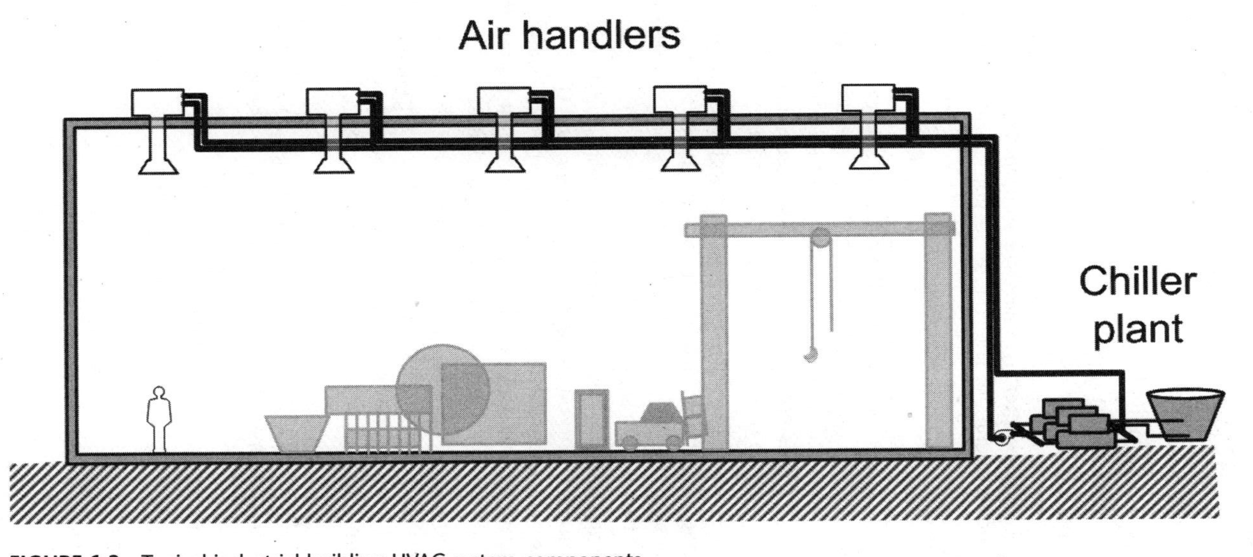

FIGURE 1.3 Typical industrial building HVAC system components.

- Air Handling Units
- Air Distribution Systems
- Zone Terminal Systems
- Evaporative Cooling
- Residential Systems

Each of these chapters contains five sections describing the component and its operation. The chapter sections appear in each chapter in the same order so that you can quickly find the information you are looking for.

Principles of Operation presents the theoretical background and function of the component. This is not an exhaustive coverage of every single different type of system, but is instead an introduction to the essential elements that the technician would need in order to quickly identify the component.

Controls describes the different control algorithms that would be expected with this particular component. The positive and negatives of many of the different control algorithms are discussed.

Safety gives tips on how to ensure the personal safety of the individual working on these systems. Since many HVAC processes incorporate high voltages, dangerous pressures and temperatures, and moving machinery, there must be considerable emphasis on the preservation of life and limb.

Troubleshooting provides a series of explanations and flow charts on how to approach problems involving the component. The method takes a top-down approach where we try to solve the problem simply at first and then move to more sophisticated solutions as necessary.

Practical Considerations gives the reader an idea about how to make the equipment operate more dependably and at lower cost.

chapter 2

AIR AND DUCTS

The flow of fresh air through a building is essential to providing a safe and comfortable environment to the occupants. Maintaining the proper airflow through a building provides a constant supply of oxygen, removes indoor pollutants, and helps keep the temperature and humidity at comfortable levels

Air is typically moved through commercial and industrial buildings by a series of fans and ductworks. This chapter discusses the properties of air and the mechanics of ductwork. See Chapters 13, 14, and 15 for more details on fans and air distribution systems.

AIR

At normal temperatures and pressures, air can be treated like a fluid. Clean, dry air at sea level is composed of about 78 percent nitrogen, 21 percent oxygen, 1 percent argon, 0.033 percent carbon dioxide, and the remaining small fraction is composed of a number of trace gases. Often, air is referred to at the so-called "standard temperature and pressure" or STP. This simply means air at 32°F at sea level. The pressure of the air at these conditions is called 1 atmosphere and is equivalent to the following:

- 14.696 psi
- 33.899 feet of water
- 2116.2 lbs per square foot
- 101.325 kilopascals
- 29.921 inches of mercury
- 10.332 meters of water
- 406.782 inches of water
- 760 mm of mercury

At standard temperature and pressure, the density of air is 0.0807 lbs/ft^3. At the air warms up, the density decreases. This is why warm air rises. As the altitude increases, the density also decreases, which is why combustion equipment runs less efficiently in places like Denver. Table 2.1 shows how the density of air changes as a function of temperature and relative humidity for cities at sea level and at 5000 feet.

It takes a certain amount of heat to raise the temperature of air. This amount of heat needed to warm a given amount of air is called the specific heat. For most HVAC applications, the specific heat of air at one atmosphere can be taken as 0.24 Btu/lb·°F, meaning it takes 0.24 BTU to raise the temperature of one pound of air by one degree Fahrenheit.

Moving air can produce surprisingly strong forces on a body or a door. The force of an airstream is given by pressure = ½ ρ v^2 where ρ is the air density and v is the air velocity. To find the pressure caused by a moving airstream at sea level, use the equation:

$$\text{Lbs / square foot} = (\text{feet / min})^2 \div 3{,}093{,}000 \qquad (2.1)$$

Figure 2.1 shows the velocity pressure of an air stream at different velocities. The charts in this figure can be used to determine the pressure exerted on an object in the air stream. For example, at an air velocity of 2500 feet per minute (about 28 mph) the air exerts a pressure of about two pounds per square foot. This may not sound like much, but on a standard 3 foot x 6 foot door the resulting pressure would be almost 40 pounds.

Psychrometrics

We never experience truly dry air. The ambient air around us is a mixture of air and water vapor. The standard explanation is that the

TABLE 2.1 Air Density Lookup Tables

Values in lb/ft^3 for sea level

	Relative humidity									
	10%	20%	30%	40%	50%	60%	70%	80%	90%	100%
0°F	0.086	0.086	0.086	0.086	0.086	0.086	0.086	0.086	0.086	0.086
10°F	0.084	0.084	0.084	0.084	0.084	0.084	0.084	0.084	0.084	0.084
20°F	0.083	0.083	0.083	0.083	0.083	0.083	0.082	0.082	0.082	0.082
30°F	0.081	0.081	0.081	0.081	0.081	0.081	0.081	0.081	0.081	0.081
40°F	0.079	0.079	0.079	0.079	0.079	0.079	0.079	0.079	0.079	0.079
50°F	0.078	0.078	0.078	0.077	0.077	0.077	0.077	0.077	0.077	0.077
60°F	0.076	0.076	0.076	0.076	0.076	0.076	0.075	0.075	0.075	0.075
70°F	0.075	0.075	0.074	0.074	0.074	0.074	0.074	0.073	0.073	0.073
80°F	0.073	0.073	0.073	0.072	0.072	0.072	0.072	0.071	0.071	0.071
90°F	0.072	0.071	0.071	0.071	0.070	0.070	0.070	0.069	0.069	0.069
100°F	0.070	0.070	0.069	0.069	0.069	0.068	0.068	0.067	0.067	0.066
110°F	0.069	0.068	0.068	0.067	0.067	0.066	0.065	0.065	0.064	0.064
120°F	0.068	0.067	0.066	0.065	0.064	0.064	0.063	0.062	0.061	0.061

Values in lb/ft^3 for 5000 feet elevation

	Relative humidity									
	10%	20%	30%	40%	50%	60%	70%	80%	90%	100%
0°F	0.072	0.072	0.072	0.072	0.072	0.072	0.072	0.072	0.072	0.072
10°F	0.070	0.070	0.070	0.070	0.070	0.070	0.070	0.070	0.070	0.070
20°F	0.069	0.069	0.069	0.069	0.069	0.069	0.069	0.069	0.069	0.069
30°F	0.067	0.067	0.067	0.067	0.067	0.067	0.067	0.067	0.067	0.067
40°F	0.066	0.066	0.066	0.066	0.066	0.066	0.066	0.066	0.065	0.065
50°F	0.065	0.065	0.064	0.064	0.064	0.064	0.064	0.064	0.064	0.064
60°F	0.063	0.063	0.063	0.063	0.063	0.063	0.063	0.062	0.062	0.062
70°F	0.062	0.062	0.062	0.062	0.061	0.061	0.061	0.061	0.061	0.060
80°F	0.061	0.061	0.060	0.060	0.060	0.060	0.059	0.059	0.059	0.059
90°F	0.060	0.059	0.059	0.059	0.058	0.058	0.058	0.057	0.057	0.057
100°F	0.059	0.058	0.058	0.057	0.057	0.056	0.056	0.055	0.055	0.054
110°F	0.057	0.057	0.056	0.056	0.055	0.054	0.054	0.053	0.052	0.052
120°F	0.056	0.055	0.055	0.054	0.053	0.052	0.051	0.051	0.050	0.049

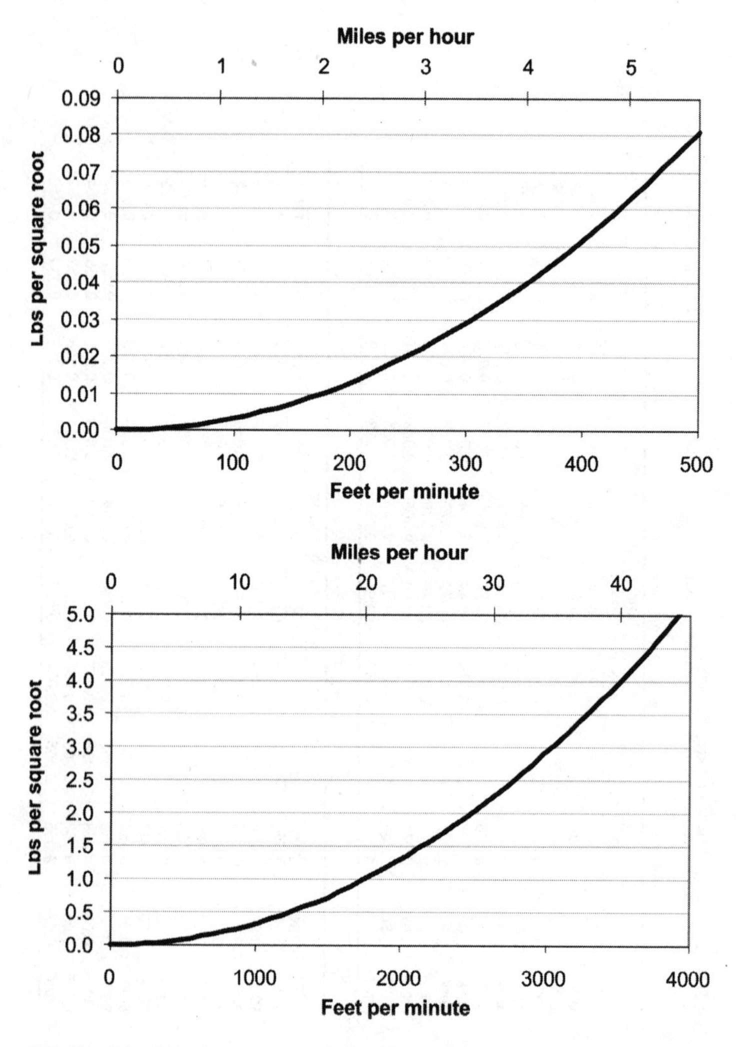

FIGURE 2.1 Velocity pressure of air at low velocities (top) and high velocities (bottom).

air "absorbs" the water vapor, although this is not really true; the air and water vapor exist together but are not linked. For a given temperature and pressure, the water vapor would exist even if there were no air. However, the absorption analogy is convenient because it fits the observed behavior. The total atmospheric pressure is the sum of the partial pressure of the water vapor, the partial pressure of the air, and the partial pressure of all the other gases present. As the temperature of the air increases, the partial pressure of the water vapor increases.

The amount of water vapor mixed with the air makes a big difference in the comfort level of people. Too little water vapor and people complain because it is too dry—their sinuses and eyes dry out. Too much water vapor and it is considered too muggy. It does not take much water to make people uncomfortable. For example, at 75°F only about one fiftieth of an ounce of water vapor is enough to saturate a cubic foot of air. That is, to have the maximum amount of water vapor that can exist at that temperature. This is equivalent to 100% relative humidity. As the temperature goes up, it takes more water to saturate the air and at lower temperatures it takes less water. Figure 2.2 shows how the saturation amount changes as the ambient air temperature changes. If you start with saturated air at a high temperature and then lower the temperature, the water condenses out of the air.

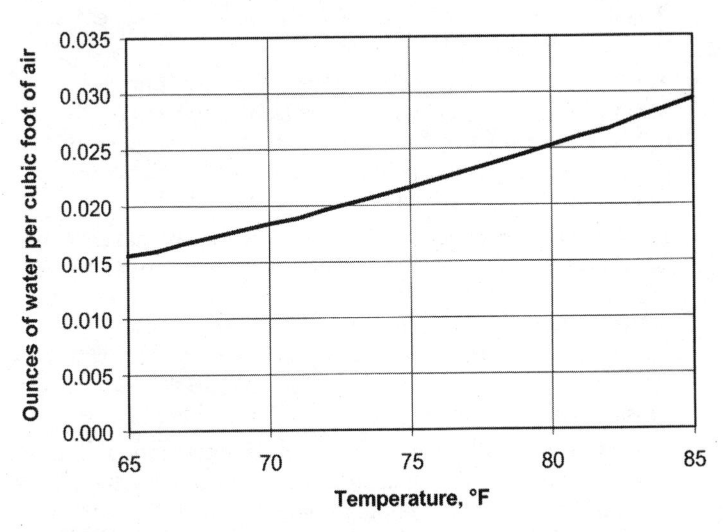

FIGURE 2.2 Amount of water needed to saturate room-temperature air at sea level.

The study of moist air is called *psychrometrics*, and the *psychrometric chart* (Figure 2.3) is often used to quickly find the various properties of moist air. The psychrometric chart has the air temperature on the horizontal axis and the amount of water vapor in the air on the vertical axis. The values on these charts are described here.

- The *humidity ratio* is the mass of water per mass of dry air; for example, lb. H_20 per lb. dry air. This is sometimes expressed as grains of water per pound of dry air (there are 7000 grains per pound.) Figure 2.4 shows how the humidity ratio varies on the psychrometric chart, and Table 2.2 gives the humidity ratio for sea level and high altitude cities.
- The *specific humidity* is the ratio of the water vapor to the total mass of the moist air sample. This is very similar to the humidity ratio except that the reference is the total mass of the air sample with units lb H_20 per lb of air + water vapor mixture.
- The *relative humidity* is the amount of water vapor in the air divided by the amount of water in the air at saturation at the same temperature and pressure. Figure 2.5 shows how the relative humidity changes in the psychrometric chart.
- The *dew point* temperature is the temperature at saturation for a given humidity ratio and a given pressure. If moist air encounters a surface at the dew point, water begins to condense out of the air. This is how a cooling coil is able to remove water from the supply air stream to a building.
- As water evaporates, it cools off. This is the basic principle behind evaporative cooling. The *wet bulb* temperature is the temperature caused by the evaporation of water at a given pressure. Figure 2.6 shows how the wet bulb temperature changes in the psychrometric chart.
- The *enthalpy* of the moist air is the sum of the enthalpies of the air and the water vapor. The enthalpy, with units of Btu per pound of air, is used to determine how much energy needs to be added or removed from air to heat or cool the air. Figure 2.7 shows how the enthalpy changes in the psychrometric chart. Note that the lines of constant enthalpy are parallel to those of constant wet-bulb, so take care not to confuse the two.

The psychrometric chart is used when designing and troubleshooting HVAC systems that heat, cool, or humidify the air. By identifying the starting and ending points of a process on the psychrometric chart, you can determine where problems might be occurring. For example, Figure 2.8 shows example psychrometric

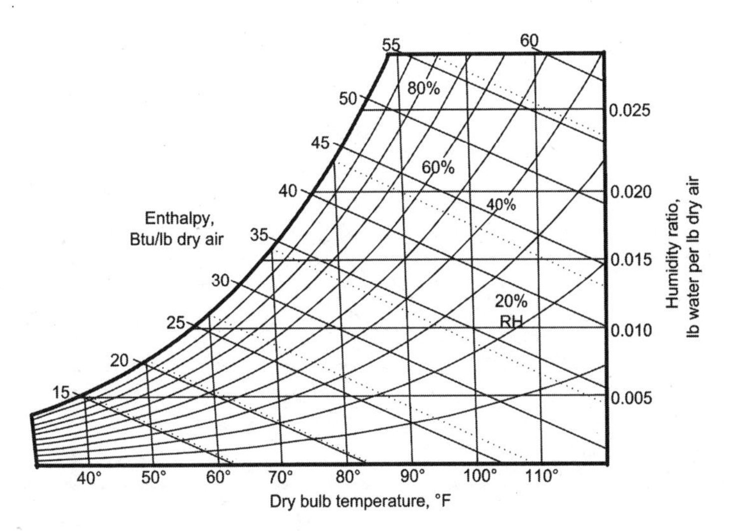

FIGURE 2.3 Psychrometric chart for sea level.

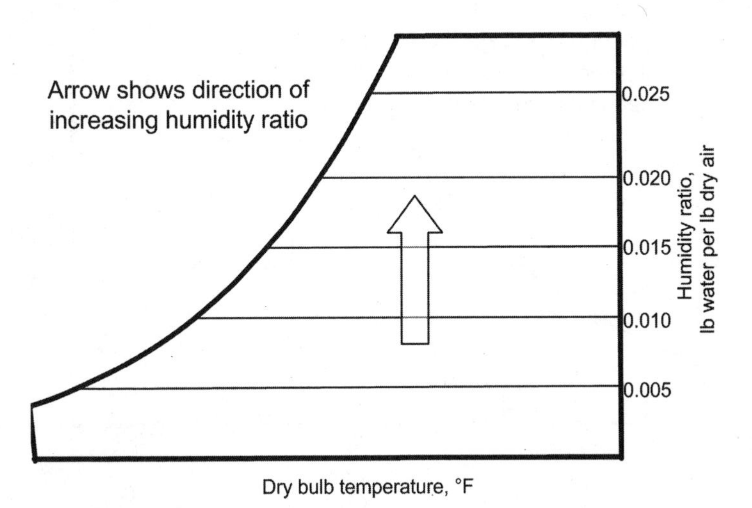

FIGURE 2.4 Psychrometric chart showing lines of constant humidity ratio.

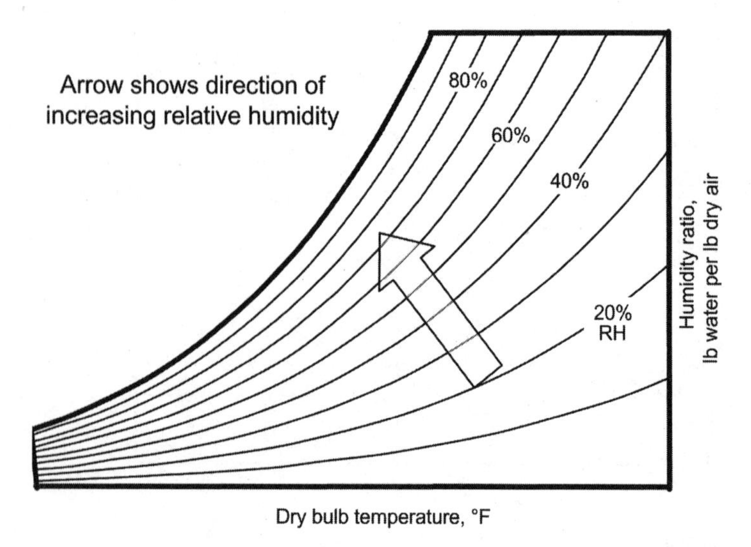

FIGURE 2.5 Psychrometric chart showing lines of constant relative humidity.

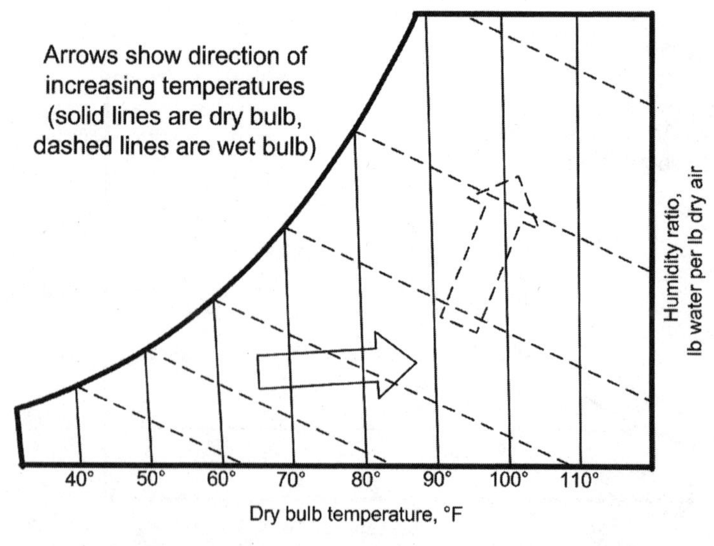

FIGURE 2.6 Psychrometric chart showing lines of constant dry-bulb and wet-bulb temperatures.

TABLE 2.2 Humidity Ratio Look-up Tables

Values in lb water per lb dry air for sea level

Dry bulb temperature	Relative humidity									
	10%	20%	30%	40%	50%	60%	70%	80%	90%	100%
0°F	0.0001	0.0002	0.0002	0.0003	0.0004	0.0005	0.0005	0.0006	0.0007	0.0008
10°F	0.0001	0.0003	0.0004	0.0005	0.0007	0.0008	0.0009	0.0010	0.0012	0.0013
20°F	0.0002	0.0004	0.0006	0.0009	0.0011	0.0013	0.0015	0.0017	0.0019	0.0021
30°F	0.0003	0.0007	0.0010	0.0014	0.0017	0.0021	0.0024	0.0027	0.0031	0.0034
40°F	0.0005	0.0010	0.0015	0.0021	0.0026	0.0031	0.0036	0.0041	0.0047	0.0052
50°F	0.0008	0.0015	0.0023	0.0030	0.0038	0.0046	0.0053	0.0061	0.0069	0.0076
60°F	0.0011	0.0022	0.0033	0.0044	0.0055	0.0066	0.0077	0.0088	0.0099	0.0110
70°F	0.0015	0.0031	0.0046	0.0062	0.0078	0.0094	0.0110	0.0125	0.0142	0.0158
80°F	0.0022	0.0043	0.0065	0.0087	0.0109	0.0132	0.0154	0.0177	0.0199	0.0222
90°F	0.0030	0.0060	0.0090	0.0121	0.0151	0.0183	0.0214	0.0246	0.0278	0.0311
100°F	0.0040	0.0081	0.0123	0.0165	0.0208	0.0251	0.0295	0.0339	0.0384	0.0430
110°F	0.0054	0.0110	0.0166	0.0224	0.0282	0.0342	0.0403	0.0464	0.0527	0.0592
120°F	0.0073	0.0147	0.0223	0.0301	0.0381	0.0462	0.0546	0.0632	0.0720	0.0811

Values in lb water per lb dry air for 5000 feet elevation

Dry bulb temperature	Relative humidity									
	10%	20%	30%	40%	50%	60%	70%	80%	90%	100%
0°F	0.0001	0.0002	0.0003	0.0004	0.0005	0.0006	0.0007	0.0008	0.0008	0.0009
10°F	0.0002	0.0003	0.0005	0.0006	0.0008	0.0009	0.0011	0.0013	0.0014	0.0016
20°F	0.0003	0.0005	0.0008	0.0010	0.0013	0.0015	0.0018	0.0021	0.0023	0.0026
30°F	0.0004	0.0008	0.0012	0.0016	0.0021	0.0025	0.0029	0.0033	0.0037	0.0041
40°F	0.0006	0.0012	0.0019	0.0025	0.0031	0.0037	0.0044	0.0050	0.0056	0.0063
50°F	0.0009	0.0018	0.0027	0.0036	0.0046	0.0055	0.0064	0.0073	0.0083	0.0092
60°F	0.0013	0.0026	0.0039	0.0053	0.0066	0.0079	0.0093	0.0106	0.0120	0.0133
70°F	0.0019	0.0037	0.0056	0.0075	0.0094	0.0113	0.0132	0.0151	0.0171	0.0190
80°F	0.0026	0.0052	0.0078	0.0105	0.0132	0.0159	0.0186	0.0214	0.0241	0.0269
90°F	0.0036	0.0072	0.0108	0.0145	0.0183	0.0221	0.0259	0.0298	0.0337	0.0377
100°F	0.0049	0.0098	0.0148	0.0200	0.0251	0.0304	0.0358	0.0412	0.0468	0.0524
110°F	0.0066	0.0133	0.0201	0.0271	0.0342	0.0416	0.0490	0.0567	0.0645	0.0725
120°F	0.0087	0.0177	0.0270	0.0365	0.0463	0.0564	0.0668	0.0775	0.0886	0.1001

TABLE 2.3 Dew Point Lookup Tables

Values for sea level

Dry bulb temperature	Relative humidity									
	10%	**20%**	**30%**	**40%**	**50%**	**60%**	**70%**	**80%**	**90%**	**100%**
0°F	-40°F	-29°F	-22°F	-17°F	-13°F	-9°F	-7°F	-4°F	-2°F	0°F
10°F	-32°F	-20°F	-13°F	-7°F	-3°F	0°F	3°F	6°F	8°F	10°F
20°F	-23°F	-11°F	-4°F	2°F	6°F	10°F	13°F	16°F	18°F	20°F
30°F	-15°F	-2°F	5°F	11°F	16°F	19°F	22°F	25°F	28°F	30°F
40°F	-6°F	5°F	13°F	19°F	24°F	28°F	31°F	34°F	37°F	40°F
50°F	0°F	12°F	21°F	27°F	32°F	37°F	41°F	44°F	47°F	50°F
60°F	6°F	20°F	29°F	36°F	41°F	46°F	50°F	54°F	57°F	60°F
70°F	13°F	28°F	37°F	44°F	50°F	55°F	59°F	63°F	67°F	70°F
80°F	20°F	35°F	46°F	53°F	59°F	65°F	69°F	73°F	77°F	80°F
90°F	27°F	43°F	54°F	62°F	69°F	74°F	79°F	83°F	87°F	90°F
100°F	34°F	51°F	63°F	71°F	78°F	84°F	88°F	93°F	97°F	100°F
110°F	41°F	60°F	71°F	80°F	87°F	93°F	98°F	103°F	107°F	110°F
120°F	48°F	68°F	80°F	89°F	96°F	102°F	108°F	112°F	117°F	120°F

Values for 5000 feet elevation

Dry bulb temperature	Relative humidity									
	10%	**20%**	**30%**	**40%**	**50%**	**60%**	**70%**	**80%**	**90%**	**100%**
0°F	-40°F	-29°F	-22°F	-17°F	-13°F	-9°F	-7°F	-4°F	-2°F	0°F
10°F	-32°F	-20°F	-13°F	-7°F	-3°F	0°F	3°F	6°F	8°F	10°F
20°F	-23°F	-11°F	-4°F	2°F	6°F	10°F	13°F	16°F	18°F	20°F
30°F	-15°F	-2°F	5°F	11°F	16°F	19°F	22°F	25°F	28°F	30°F
40°F	-6°F	5°F	13°F	19°F	24°F	28°F	31°F	34°F	37°F	40°F
50°F	0°F	12°F	21°F	27°F	32°F	37°F	41°F	44°F	47°F	50°F
60°F	6°F	20°F	29°F	36°F	41°F	46°F	50°F	54°F	57°F	60°F
70°F	13°F	28°F	37°F	44°F	50°F	55°F	59°F	63°F	67°F	70°F
80°F	20°F	35°F	46°F	53°F	59°F	65°F	69°F	73°F	77°F	80°F
90°F	27°F	43°F	54°F	62°F	69°F	74°F	79°F	83°F	87°F	90°F
100°F	34°F	51°F	63°F	71°F	78°F	84°F	88°F	93°F	97°F	100°F
110°F	41°F	60°F	71°F	80°F	87°F	93°F	98°F	103°F	107°F	110°F
120°F	48°F	68°F	80°F	89°F	96°F	102°F	108°F	112°F	117°F	120°F

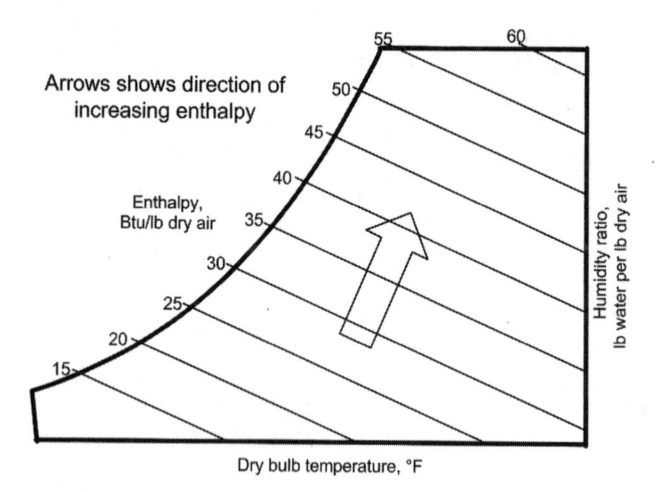

FIGURE 2.7 Psychrometric chart showing lines of constant enthalpy.

TABLE 2.4 Wet Bulb Temperature Look-up Tables

Values for sea level

Dry bulb temperature	Relative humidity 10%	20%	30%	40%	50%	60%	70%	80%	90%	100%
0°F	-3°F	-2°F	-2°F	-2°F	-2°F	-1°F	-1°F	-1°F	0°F	0°F
10°F	6°F	6°F	7°F	7°F	8°F	8°F	9°F	9°F	10°F	10°F
20°F	14°F	14°F	15°F	16°F	17°F	17°F	18°F	19°F	19°F	20°F
30°F	21°F	22°F	23°F	24°F	25°F	26°F	27°F	28°F	29°F	30°F
40°F	28°F	30°F	31°F	32°F	34°F	35°F	36°F	38°F	39°F	40°F
50°F	35°F	37°F	38°F	40°F	42°F	44°F	45°F	47°F	48°F	50°F
60°F	41°F	43°F	46°F	48°F	50°F	52°F	54°F	56°F	58°F	60°F
70°F	47°F	50°F	53°F	56°F	58°F	61°F	63°F	66°F	68°F	70°F
80°F	52°F	56°F	60°F	63°F	67°F	70°F	72°F	75°F	78°F	80°F
90°F	58°F	63°F	67°F	71°F	75°F	78°F	82°F	85°F	87°F	90°F
100°F	63°F	69°F	74°F	79°F	83°F	87°F	91°F	94°F	97°F	100°F
110°F	68°F	75°F	81°F	87°F	92°F	96°F	100°F	103°F	107°F	110°F
120°F	74°F	82°F	89°F	95°F	100°F	105°F	109°F	113°F	117°F	120°F

Values for 5000 feet elevation

Dry bulb temperature	Relative humidity 10%	20%	30%	40%	50%	60%	70%	80%	90%	100%
0°F	-3°F	-3°F	-2°F	-2°F	-2°F	-1°F	-1°F	-1°F	0°F	0°F
10°F	5°F	6°F	6°F	7°F	7°F	8°F	8°F	9°F	9°F	10°F
20°F	13°F	14°F	15°F	15°F	16°F	17°F	18°F	18°F	19°F	20°F
30°F	20°F	21°F	22°F	24°F	25°F	26°F	27°F	28°F	29°F	30°F
40°F	27°F	28°F	30°F	31°F	33°F	34°F	36°F	37°F	39°F	40°F
50°F	33°F	35°F	37°F	39°F	41°F	43°F	45°F	47°F	48°F	50°F
60°F	39°F	42°F	44°F	47°F	49°F	52°F	54°F	56°F	58°F	60°F
70°F	45°F	48°F	52°F	55°F	58°F	60°F	63°F	65°F	68°F	70°F
80°F	50°F	55°F	59°F	62°F	66°F	69°F	72°F	75°F	78°F	80°F
90°F	56°F	61°F	66°F	70°F	74°F	78°F	81°F	84°F	87°F	90°F
100°F	61°F	67°F	73°F	78°F	83°F	87°F	90°F	94°F	97°F	100°F
110°F	66°F	74°F	80°F	86°F	91°F	95°F	100°F	103°F	107°F	110°F
120°F	71°F	80°F	88°F	94°F	99°F	104°F	109°F	113°F	117°F	120°F

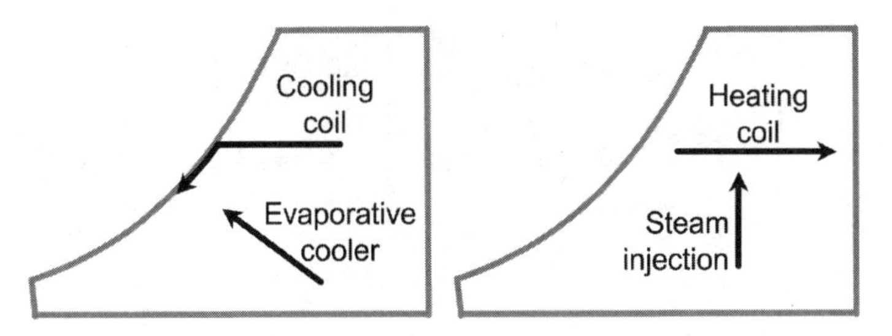

FIGURE 2.8 Example cooling (left) and heating (right) psychrometric processes.

TABLE 2.5 Enthalpy Look-up Tables

Values in Btu/lb for sea level

Dry bulb temperature	Relative humidity									
	10%	20%	30%	40%	50%	60%	70%	80%	90%	100%
0°F	0.042	0.125	0.208	0.291	0.375	0.458	0.541	0.624	0.708	0.791
10°F	2.486	2.625	2.765	2.904	3.044	3.184	3.323	3.463	3.603	3.743
20°F	4.963	5.192	5.421	5.650	5.880	6.109	6.339	6.568	6.798	7.028
30°F	7.490	7.858	8.227	8.596	8.965	9.335	9.705	10.08	10.45	10.82
40°F	10.07	10.62	11.18	11.74	12.30	12.86	13.42	13.98	14.55	15.11
50°F	12.72	13.53	14.36	15.18	16.00	16.83	17.66	18.49	19.32	20.16
60°F	15.47	16.65	17.84	19.04	20.23	21.44	22.64	23.85	25.07	26.29
70°F	18.36	20.05	21.75	23.45	25.17	26.89	28.63	30.37	32.12	33.88
80°F	21.42	23.80	26.19	28.60	31.03	33.48	35.94	38.42	40.91	43.43
90°F	24.72	28.02	31.35	34.72	38.11	41.54	45.01	48.51	52.04	55.61
100°F	28.31	32.84	37.43	42.07	46.79	51.56	56.40	61.31	66.28	71.32
110°F	32.27	38.41	44.67	51.04	57.53	64.13	70.86	77.71	84.70	91.81
120°F	36.69	44.95	53.42	62.09	70.97	80.07	89.40	98.96	108.8	118.8

Values in Btu/lb for 5000 ft elevation

Dry bulb temperature	Relative humidity									
	10%	20%	30%	40%	50%	60%	70%	80%	90%	100%
0°F	0.059	0.159	0.259	0.359	0.459	0.559	0.659	0.759	0.859	0.959
10°F	2.514	2.682	2.849	3.017	3.185	3.353	3.521	3.689	3.857	4.025
20°F	5.009	5.284	5.560	5.835	6.111	6.387	6.663	6.939	7.216	7.493
30°F	7.564	8.007	8.450	8.894	9.338	9.783	10.23	10.67	11.12	11.57
40°F	10.18	10.85	11.52	12.19	12.86	13.54	14.22	14.89	15.57	16.25
50°F	12.88	13.87	14.85	15.84	16.84	17.83	18.83	19.84	20.84	21.85
60°F	15.71	17.13	18.56	20.00	21.45	22.90	24.35	25.82	27.29	28.76
70°F	18.70	20.73	22.78	24.84	26.91	28.99	31.09	33.20	35.32	37.45
80°F	21.90	24.76	27.65	30.56	33.50	36.46	39.44	42.45	45.49	48.56
90°F	25.38	29.36	33.38	37.45	41.57	45.74	49.96	54.22	58.54	62.92
100°F	29.22	34.68	40.23	45.87	51.60	57.42	63.34	69.36	75.48	81.70
110°F	33.50	40.92	48.51	56.26	64.19	72.29	80.57	89.04	97.71	106.6
120°F	38.34	48.34	58.64	69.23	80.14	91.38	103.0	114.9	127.2	140.0

"paths" for cooling processes and heating processes. In this figure, the temperature decrease across the cooling coil is shown, including the moisture removal once water begins condensing from the air. If your supply air is too dry you may be able to use psychrometrics to understand why so much water is being removed from the air stream. The evaporative cooling process path shows how the air temperature goes down as more water is evaporated into the air stream. Psychrometrics are used when investigating occupant comfort problems. Figure 2.9 shows the acceptable range of temperatures and humidities for the summer months, while Figure 2.10 shows the range for winter. The summer temperature range is higher than the winter because people are not as heavily dressed.

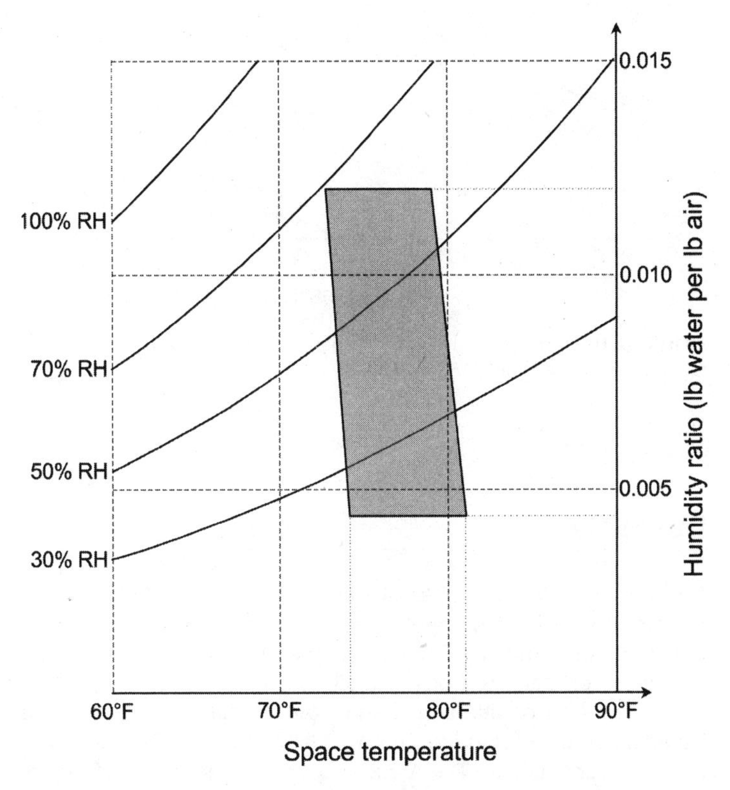

FIGURE 2.9 Acceptable ranges of temperature and humidity for buildings in summer. *(Adapted from ASHRAE, 1981)*

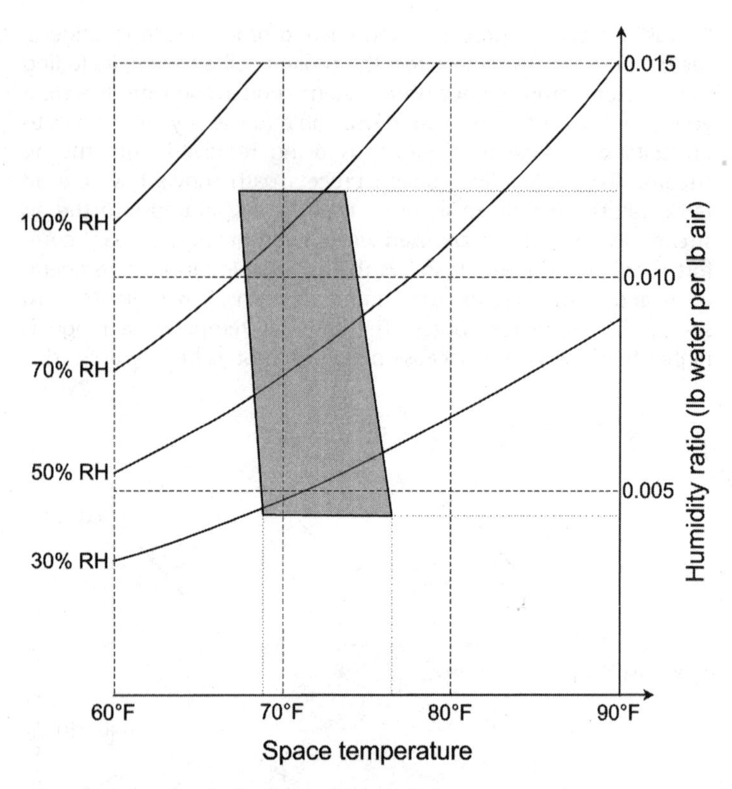

FIGURE 2.10 Acceptable ranges of temperature and humidity for buildings in winter. *(Adapted from ASHRAE, 1981)*

DUCTS

Ducts are aluminum or fiberglass conduits that transport cold and hot air into the building, remove stale air, or eliminate smoke from buildings. In commercial buildings, the air distribution system is composed of many sections of ductwork and dampers. The Sheet Metal and Air Conditioning Contractors' National Association has issued a number of standards for residential, commercial, and industrial duct construction. The ASHRAE Systems Handbook (2000) provides an excellent summary of the different duct standards and construction techniques.

Ducts come in a variety of shapes and sizes, along with different fittings. Figure 2.11 shows some different types of duct elbows. These are typically manufactured in 22.5°, 45°, and 90° turns and come in both rectangular and circular sections. The difference in the elbows comes mostly from the manufacturing technique and materials used. Figure 2.12 shows mitered rectangular elbows. These sections often do not present as smooth a turning radius for the air as do the circular elbows. For this reason, turning vanes are often incorporated into the elbows.

Ducts can be connected by a variety of means. Flanges, friction fits, duct tape, sheet metal screws, and metal clips are all used, sometimes together. If you are having problems taking duct sections apart, check to make sure that there are no hidden screws or clips holding the sections together. Any ductwork that is attached to

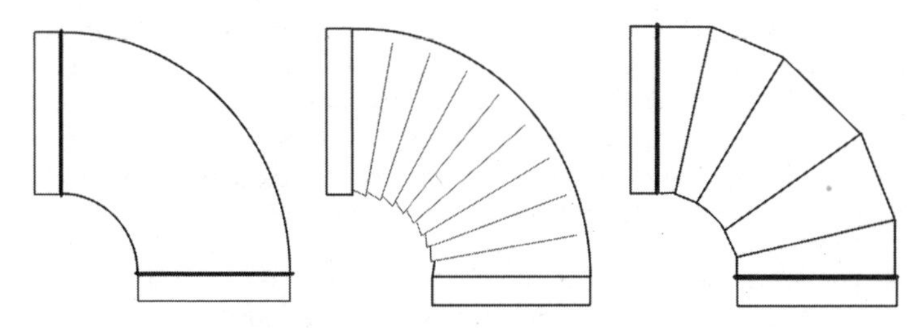

FIGURE 2.11 Die-stamped duct elbow (left), pleated duct elbow (middle), and five-piece gore elbow (right).

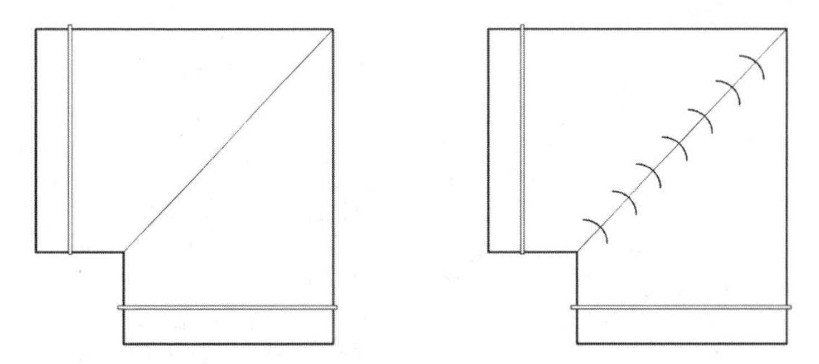

FIGURE 2.12 (Left) Mitered duct elbow, simple duct section; (right) elbow with turning vanes.

moving or vibrating equipment (such as fan units) should use a flexible duct connector. This isolates the ductwork from the vibration source and can help prevent subsequent damage to duct connections and sensors mounted on the duct, as well as reduce noise transmitted from the fans to the zones.

When using electronic sensors to measure air temperatures, flow, and humidity, it can be useful to use wires to electrically connect ductwork sections that are separated by flexible connections. It is possible for different duct sections to be at surprisingly different electronic potentials with respect to ground. If the sensor has a chassis ground reference that is in good electrical contact with the duct, it is possible to set up a current loop through the sensor leads and add noise to your readings.

Sometimes it is necessary to get inside ductwork for repairs, sensor installation, or cleaning. A number of companies manufacture special doors that can be installed in holes cut into the side of ductwork. If you use these types of doors, make sure that the door is in a relatively accessible location. Remember that those who might have to rescue you must also pass through this opening. Also be sure to place the door such that it does in fact provide access to the spot in the duct you want to get to. Despite their appearance in action movies, the inside of ductwork can be a very dangerous place, so there are a few rules to follow:

- The inside of ducts is completely dark. Be sure to bring a work lamp and preferably a backup flashlight. Turning vanes and other duct obstructions can be located where you least expect them, as can vertical duct take-offs that may drop down for many feet. These are the kinds of things you want to know about before you encounter them.
- *Always* let someone know when you are entering the duct and inform them when you are out again. They should be instructed to look for you if you do not report back within a reasonable amount of time.
- Screws, sharp edges, and other protrusions exist everywhere in ducts. Often these are hidden by duct insulation. Wear head and eye protection if possible. Pay close attention to what is around you in the ductwork so you don't accidentally put all your weight down onto the sharp end of a nail or screw.
- Remember that moving air can exert a strong pressure, to say nothing of any drag forces. A worker in a duct can have the equivalent area of 5 to 10 square feet, and even a low velocity air stream can knock you over. If you are anywhere near a verti-

fastfacts

Ducts should be sized so that the air velocity is around 500 feet per minute through the duct.

cal duct take-off, you might end up taking a bad fall. Always be prepared for sudden changes in the duct air velocity.

- *Never* approach an operating fan from inside the duct. The turbulence near a fan intake can be surprisingly strong. Also, fans can create a significant amount of noise in the confined spaces of a duct. Wear ear protection if you plan to be in a particularly noisy area longer than ten minutes.

Due to the constant presence of moist air in ducts, some air distribution systems can develop serious problems due to the growth of molds, fungi, and microorganisms. Sick building syndrome and Legionnaire's disease are two of the better-known issues that can arise from dirty or contaminated ducts. It is therefore important to periodically inspect ductwork and filters for dirt, excessive humidity or standing water, and any kind of bacterial or organic growth. Both the National Air Duct Cleaners Association and the North American Insulation Contractors Association offer guidelines on how to properly clean commercial duct systems.

Duct air leakage wastes energy, makes duct pressure control difficult, and can lead to occupant comfort complaints. Begin your troubleshooting of duct systems by initially checking the duct for unnecessary leakage. A few strips from a roll of duct tape are cheap and can help solve and prevent problems.

AIR COMPRESSORS

Air compressors are used in HVAC systems for the operation of pneumatic sensors and actuators (see Chapter 5 for more information about pneumatic control systems). The air supply in a compressed air system must be clean and free from oil, dirt, and water. Figure 2.13 shows a simple schematic of the major components in a compressed air system. A compressor keeps a pressurized storage tank at some pressure usually between 50 and 150 psi. The air then passes through a dryer and some filters before it arrives at a pressure regulation valve (PRV). The PRV maintains a downstream pressure of about 18 to 25 psi that is used by the sensors and actuators. Figure 2.14 shows what the compressor and storage tank look like in an actual installation.

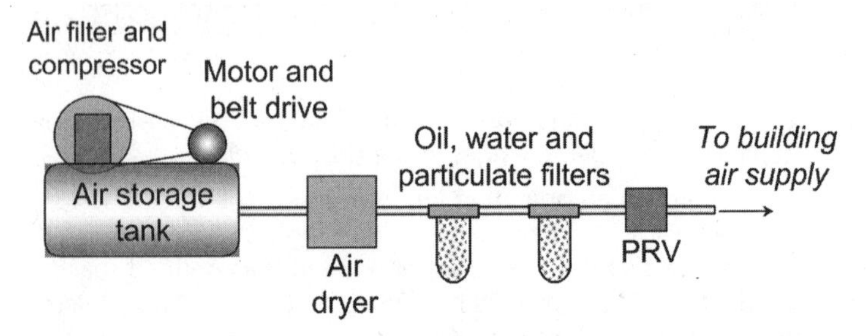

FIGURE 2.13 Typical compressed air supply system.

FIGURE 2.14 Combined air compressor and storage tank.

The compressor motor is sized to match the amount of air and the amount of compression required. Table 2.6 shows the theoretical horsepower required by an air compressor for a given outlet pressure and air use rate. These numbers should be increased by about 15 percent for high altitude applications.

The storage tank is sized so that the motor and compressor do not cycle too much. A compressor that is running all the time indicates that there is a leak somewhere in the system, quite possibly upstream of the PRV. In practice, the motor should cycle no more than about 10 times per hour. It is also important to periodically check the pressure relief valve on the storage tank to make sure it is working properly.

If the air compressor is used to provide pressurized air to a relative large system (for example, multiple buildings), a large standalone compressor may be used in conjunction with a separate storage tank. Figure 2.15 shows such a storage tank, which stands around 12 feet tall and is about 5 feet around. Care must be taken around tanks this large so that they do not get damaged by forklifts, welding torches, etc. Failure of a tank of this size and pressure can put a big damper on your day.

Once the air has been compressed, it must be dried and cleaned. The air dryer works by either increasing the air pressure (to force the water vapor out of the air), by decreasing the air temperature (to condense water vapor), or through desiccant absorption. No matter what the drying method, there is usually a water drain coming from the air storage tank, the dryer, and the filters. Check this line periodically to make sure it is draining properly. It is much cheaper and takes a lot less time to ensure that your air supply is clean than it is to replace fouled pneumatic relays and sensors.

TABLE 2.6 Theoretical Air Compressor Horsepower

		CFM											
		5	**10**	**20**	**40**	**60**	**80**	**100**	**150**	**200**	**300**	**400**	**500**
PSI	**5**	0.10	0.20	0.39	0.8	1.2	1.6	2.0	2.9	3.9	5.9	7.8	10
	10	0.18	0.36	0.7	1.4	2.2	2.9	3.6	5.4	7.2	11	14	18
	15	0.25	0.50	1.0	2.0	3.0	4.0	5.0	7.5	10	15	20	25
	20	0.31	0.6	1.2	2.5	3.7	5.0	6.2	9.4	12	19	25	31
	30	0.42	0.8	1.7	3.4	5.0	6.7	8.4	13	17	25	34	42
	40	0.51	1.0	2.0	4.1	6.1	8.2	10	15	20	31	41	51
	60	0.66	1.3	2.6	5.3	7.9	11	13	20	26	40	53	66
	80	0.79	1.6	3.1	6.3	9.4	13	16	24	31	47	63	79
	100	0.89	1.8	3.6	7.2	11	14	18	27	36	54	72	89
	120	1.0	2.0	4.0	7.9	12	16	20	30	40	59	79	99
	140	1.1	2.1	4.3	8.6	13	17	21	32	43	64	86	107
	160	1.2	2.3	4.6	9.2	14	18	23	35	46	69	92	115
	180	1.2	2.4	4.9	10	15	20	24	37	49	73	98	122
	200	1.3	2.6	5.1	10	15	21	26	39	51	77	103	129

FIGURE 2.15 100 psi compressed air tank.

Compressor oil can become entrained in the compressed air and travel into the pipes, where it is removed by the filters and dryers. Most local codes now require that compressed air systems have equipment in place that prevents this compressor oil from entering the water drains. Figure 2.16 shows a water/oil separator that is used to float the oil so that it can be removed before the water is released. Note also that this device sits on an oil catch pad. This, too, is often required by code.

After leaving the compressed air tank, the air travels to controllers, actuators, and other compressed air processes throughout the building. In cases where the compressor serves a campus of buildings, there can be compressed air lines that travel relatively large distances between buildings. Copper tubing and pipe in the ¼ inch to ½ inch range is used for this. Obviously, a leak in the copper tubing results in a failure of the downstream devices. If you have pneumatic controls that don't seem to be behaving properly, one of the first things to check is the air pressure.

FIGURE 2.16 Water/oil separator.

Also keep in mind that long runs of compressed air tubing usually share a space with other utilities such as electrical lines, steam pipes, and water distribution systems. You often find that the pipes and wires have not exactly been laid out with maintenance in mind. Scrambling over high voltage cables to fix leaky compressed air lines is not fun, so when designing systems try to place the runs such that they are easily accessible.

If the runs are over a hundred feet or so, you should also probably include expansion joints in the tubing. The air temperature in the utility corridor may not always be the same (particularly if steam or water systems are shut down). As the temperature changes, the copper tubing expands or contracts and can kink the pipe or even pull sweated fittings apart. Figure 2.17 shows the expansion characteristics of copper tubing as a function of temperature. For example, if you have a 200-foot length of copper tubing and that decreases in temperature by 20°F, the pipe shrinks by almost half an inch. If the pipe is rigidly fixed to the wall and supports, the shrinkage could easily cause cracks to develop, particularly around elbows and tees. Include expansion joints or fittings as appropriate, even if you don't think it will be a problem. It costs very little in additional tubing and may save a bundle on repair costs.

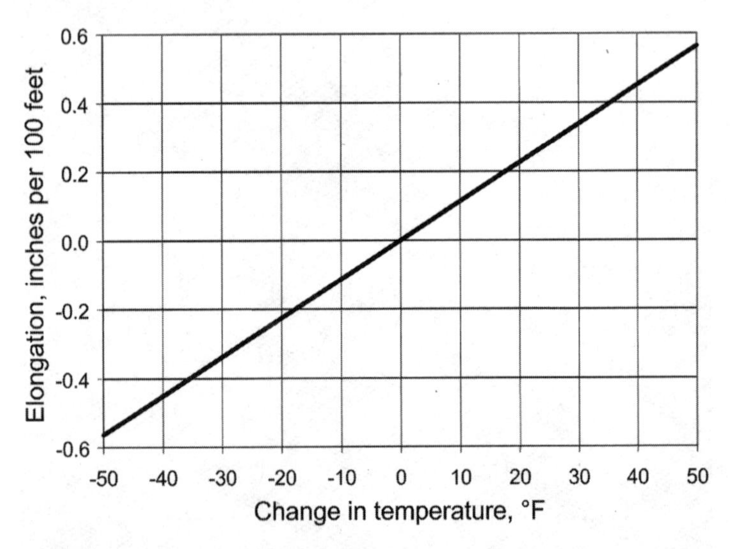

FIGURE 2.17 Elongation of copper tubing as a function of temperature change.

WATER, STEAM, AND PIPES

WATER

Water is used in large buildings to move heating and cooling energy around. For example, water is used to distribute the cooling from a chiller plant to one or more air handlers. Under all normal HVAC conditions, water is incompressible and the only significant property that varies is the water density as shown in Table 3.1.

TABLE 3.1 Properties of Liquid Water

Temperature	Density	
°F	lb/ft^3	lb/gallon
32	63.41	8.48
39	63.42	8.48
50	63.40	8.48
68	63.31	8.46
86	63.15	8.44
122	62.67	8.38
140	62.35	8.34
158	62.01	8.29
176	61.63	8.24
194	61.22	8.18
212	60.78	8.13

Since the properties of water in HVAC applications do not change very much, we can come up with the following general equations that may be useful. The pumping power (in horsepower) for water is given by:

$$\text{Horsepower} = \frac{\text{gallons per minute} \times \text{total head in feet}}{3960}$$

The flow rate of water through a pipe (in gallons per minute) is given by:

$$\text{GPM} = 0.0408 \times (\text{pipe diameter})^2 \times (\text{water velocity})$$

where the *pipe diameter* is in inches and the *water velocity* is in feet per minute. The weight of water in a given section of pipe is found from:

$$\text{Lbs of water} = 0.34 \times \text{pipe length} \times (\text{pipe diameter})^2$$

where the *pipe length* is in feet and the *pipe diameter* is in inches.

STEAM

Like its liquid counterpart, steam is used to move energy around buildings. For example, it is used to transport energy from a boiler to various processes in a building. The properties of steam depend on both its temperature and pressure. Steam is often referred to as being dry or wet. *Dry steam* is 100 percent water vapor—a colorless gas—while *wet steam* contains water droplets. *Saturated steam* is water vapor at the boiling point of water at a given pressure. For example, steam at sea level at 212°F would be saturated. Saturated steam is at the lowest temperature it can be—if the steam were any colder it would condense back into water. *Superheated* steam is just steam that is hotter than the saturation temperature. A pot of boiling water on a stove produces steam that is about 98 percent saturated water vapor and about 2 percent water droplets. In other words, it is both wet and saturated. The *quality* of steam is the percentage of the steam that is dry, so in the boiling water example just given the quality of the steam is 98 percent. The ASHRAE Handbook of Fundamentals (1997) gives complete tables for the properties of steam. Table 3.2 lists some of these properties for saturated steam, while Tables 3.3 and 3.4 list the volume and enthalpy of superheated steam.

PIPES AND TUBING

Pipes are used for transporting air, water, refrigerants and steam throughout buildings. The vast majority of commercial pipe comes in standard sizes with varying wall thickness. The standard size for pipes up to 12 inches thick is given by the inside diameter, whereas for pipes over 12 inches thick the outside diameter is used.

TABLE 3.2 Properties of Saturated Steam

Pressure, psia	Saturation temp., °F	Specific volume, ft^3/lb_m		Enthalpy, Btu/lb_m	
		Liquid	Vapor	Liquid	Vapor
1	101.70	0.016136	333.6	69.74	1105.8
2	126.04	0.016230	173.75	94.02	1116.1
3	141.43	0.016300	118.72	109.39	1122.5
4	152.93	0.016358	90.64	120.89	1127.3
5	162.21	0.016407	73.53	130.17	1131.0
6	170.03	0.016451	61.98	138.00	1134.2
8	192.84	0.016526	47.35	150.84	1139.3
10	193.19	0.016590	38.42	161.23	1143.3
15	213.03	0.016723	26.29	181.19	1150.9
20	227.96	0.016830	20.09	196.26	1156.4
25	240.08	0.016922	16.306	208.52	1160.7
30	250.34	0.017004	13.748	218.93	1164.3
35	259.30	0.017073	11.900	228.04	1167.4
40	267.26	0.017146	10.501	236.16	1170.0
45	274.46	0.017209	9.403	243.51	1172.3
50	281.03	0.017269	8.518	250.24	1174.4
55	287.10	0.017325	7.789	256.46	1176.3
60	292.73	0.017378	7.177	262.25	1178.0
65	298.00	0.017429	6.657	267.67	1179.6
70	302.96	0.017478	6.209	272.79	1181.0
75	307.63	0.017524	5.818	277.61	1182.4
80	312.07	0.017570	5.474	282.21	1183.6
85	316.29	0.017613	5.170	286.58	1184.8
90	320.31	0.017655	4.898	290.76	1185.9
95	324.16	0.017696	4.654	294.76	1186.9
100	327.86	0.017736	4.434	298.61	1187.8
110	334.82	0.017813	4.051	305.88	1189.6
120	341.30	0.017886	3.730	312.67	1191.1
130	347.37	0.017957	3.457	319.04	1192.5
140	353.08	0.018024	3.221	325.05	1193.8
150	358.48	0.018089	3.016	330.75	1194.9

TABLE 3.3 Volume of Superheated Steam

Values in ft^3/lb_m

Temperature, °F	Pressure, psia 1	5	14.7	20
Sat.	333.6	73.53	26.80	20.09
200	392.5	78.15		
240	416.4	83.00	28.00	20.47
280	440.3	81.83	29.69	21.73
320	464.2	92.64	31.36	22.98
360	488.1	97.45	33.02	24.21
400	511.9	102.24	34.67	25.43
440	535.8	107.03	36.31	26.64
500	571.5	114.20	38.77	28.46
600	631.1	126.15	42.86	31.47
700	690.7	138.08	46.93	34.47
800	750.3	150.01	51.00	37.46
1000	869.5	173.86	59.13	43.44
1200	988.6	197.70	67.25	49.41
1400	1107.7	221.54	75.36	55.37
1600			83.47	61.33

TABLE 3.4 Enthalpy of Superheated Steam

Values in Btu/lb_m

Temperature, °F	Pressure, psia 1	5	14.7	20
Sat.	1105.8	1131.0	1150.5	1156.4
200	1150.1	1148.6		
240	1168.3	1167.1	1164.0	1162.3
280	1186.5	1185.5	1183.1	1181.8
320	1204.8	1204.0	1202.1	1201.0
360	1223.2	1222.6	1221.0	1220.1
400	1241.8	1241.2	1239.9	1239.2
440	1260.4	1259.9	1258.8	1258.2
500	1288.5	1288.2	1287.3	1286.8
600	1336.1	1335.8	1335.2	1334.8
700	1384.5	1384.3	1383.8	1383.5
800	1433.7	1433.5	1433.1	1432.9
1000	1534.8	1534.7	1534.5	1534.3
1200	1639.6	1639.5	1639.3	1639.2
1400	1748.1	1748.1	1747.9	1747.9
1600			1860.2	1860.1

Pipes are joined together using either threaded, flanged, or soldered connections. Threaded connections are usually used for pipe less than 4 inches in diameter and low-to-medium pressure. Flanged connections are used for high pressure and large pipes. Soldered connections are generally reserved for copper pipes used in domestic water service. Piping sections can be broken down into the following categories:

- Bushings are used to connect different sized pipes.
- Couplings are used to connect two pieces of pipe of the same size.
- Crosses connect four pipes together, each at right angles to the other.
- Elbows are used to make angle turns, usually at 45° or 90°.
- Nipples are threaded on both ends and are used for making close connections.
- Reducers are used to connect different sized pipes.
- Return bends provide a U-turn for reversing the direction of piping.
- Straight pipe is simply a straight run of piping.
- Tees connect three pipes together with one perpendicular to the other two.
- Unions join pipe sections and are easily dismantled
- Wyes connect three pipes together at various angles.

Iron and Steel Pipes

Iron and steel pipes are used in most HVAC water and steam systems. Older iron pipes use the symbols S (standard), XS (extra strong), and XXS (double extra strong) to designate the wall thickness, but you

fastfacts

For water flow, the piping should be sized so that the water velocity does not exceed six to seven feet per second.

rarely find these designations any more. In modern systems, the pipe wall strength is listed using different schedule numbers. The most common pipe schedules are 10, 20, 30, 40 (standard), 60, 80 (extra strong), 100, 120, 140, and 160. Tables 3.5 and 3.6 list some of the properties of schedule 40 and schedule 80 iron pipe.

TABLE 3.5 Properties of Schedule 40 Steel Pipe

Pipe size, inches	Diameter		Flow area		Pounds per foot	
	Outside	Inside	ft²	in²	Pipe	Water
⅛	0.405	0.269	0.000395	0.05688	0.24	0.025
¼	0.540	0.364	0.000723	0.1041	0.42	0.045
⅜	0.675	0.493	0.001326	0.1909	0.57	0.083
½	0.840	0.622	0.002110	0.3038	0.85	0.13
¾	1.050	0.824	0.003703	0.5332	1.13	0.23
1	1.315	1.049	0.006002	0.8643	1.68	0.37
1¼	1.660	1.380	0.01039	1.496	2.27	0.65
1½	1.900	1.610	0.01414	2.036	2.72	0.88
2	2.375	2.068	0.02330	3.355	3.64	1.45
2½	2.875	2.470	0.03325	4.788	5.78	2.07
3	3.500	3.068	0.05134	7.393	7.58	3.20
3½	4.000	3.548	0.06866	9.887	9.11	4.28
4	4.500	4.026	0.08841	12.73	10.8	5.51
5	5.563	5.047	0.1389	20.00	14.6	8.66
6	6.625	6.065	0.2006	28.89	19.0	12.5
8	8.625	7.981	0.3474	50.03	28.6	21.6
10	10.75	10.02	0.5476	78.85	40.5	34.1
12	12.75	12.00	0.7854	113.1	49.6	48.9

TABLE 3.6 Properties of Schedule 80 Steel Pipe

Pipe size, inches	Diameter		Flow area		Pounds per foot	
	Outside	Inside	ft²	in²	Pipe	Water
⅛	0.405	0.215	0.000252	0.03629	0.31	0.016
¼	0.540	0.302	0.000497	0.07157	0.54	0.031
⅜	0.675	0.423	0.000976	0.1405	0.74	0.061
½	0.840	0.546	0.001626	0.2341	1.09	0.10
¾	1.050	0.742	0.003003	0.4324	1.47	0.19
1	1.315	0.957	0.004995	0.7193	2.17	0.31
1¼	1.660	1.278	0.00891	1.283	3.00	0.56
1½	1.900	1.500	0.01227	1.767	3.63	0.76
2	2.375	1.939	0.02051	2.953	5.02	1.28
2½	2.875	2.323	0.02943	4.238	7.66	1.83
3	3.500	2.900	0.04587	6.605	10.3	2.86
3½	4.000	3.364	0.06172	8.888	12.5	3.84
4	4.500	3.838	0.07984	11.50	14.7	5.01
5	5.563	4.813	0.1263	18.19	20.8	7.87
6	6.625	5.761	0.1810	26.06	28.6	11.3
8	8.625	7.625	0.3171	45.66	43.4	19.8

Copper Tubing

Copper tubing is often used for domestic water, refrigerants, natural gas, and compressed air. There are seven different types of copper tubing commonly available:

- Type ACR tubing is used for refrigerants and natural gas and is available from ⅜ up to 4⅛ inch diameters in 20 foot sections.
- Type G tubing is for natural gas and liquid petroleum and is available in sizes up to 1⅛ inch diameters.
- Types K, L, and M are used for general HVAC purposes, compressed air, domestic hot water, fire protection, and solar heating applications. The principal differences between these types are the wall size and strength (K is thicker than L and L is thicker than M). Properties of these tubing types are given in Tables 3.7, 3.8, and 3.9.
- Type DMV is usually used for draining and venting purposes, although it can also be used for general HVAC applications.
- Type MED is reserved for medical gases like pure oxygen.

When connecting copper tubing for refrigerant systems, you should use cadmium-free silver solder with a minimum of 45% silver. Whenever you are brazing or soldering the pipe you should fill the pipes with nitrogen to prevent oxidation on the inside of the tubing. Once soldered, the system should be pressure tested to about 200 psi to make sure there are no leaks. It is in your best interest to minimize the refrigerant line length where possible. The longer the refrigerant runs between chiller and condenser, the more refrigerant is required to fill the system.

fastfacts

Any risers on the suction line of a refrigeration system should have a U-trap at the bottom to catch liquid refrigerant before it enters the compressor.

TABLE 3.7 Properties of Type K Copper Tubing

Standard size,	Outside diameter	Internal diameter		Flow area	
inches	in	ft	in	ft^2	in^2
1/4	0.375	0.0254	0.305	0.00051	0.0734
3/8	0.500	0.0335	0.402	0.00088	0.1267
1/2	0.625	0.0439	0.527	0.00152	0.2189
5/8	0.750	0.0543	0.652	0.00232	0.3341
3/4	0.875	0.0621	0.745	0.00303	0.4363
1	1.125	0.0829	0.995	0.00540	0.7776
1 1/4	1.375	0.1038	1.246	0.00845	1.217
1 1/2	1.625	0.1234	1.481	0.01196	1.722
2	2.125	0.1633	1.960	0.02093	3.014
2 1/2	2.625	0.2029	2.435	0.03234	4.657
3	3.125	0.2423	2.908	0.04609	6.637
3 1/2	3.625	0.2821	3.385	0.06249	8.999
4	4.125	0.3214	3.857	0.08114	11.68
5	5.125	0.4004	4.805	0.1259	18.13
6	6.125	0.4784	5.741	0.1798	25.89
8	8.125	0.6319	7.583	0.3136	45.16

TABLE 3.8 Properties of Type L Copper Tubing

Standard size,	Outside diameter	Internal diameter		Flow area	
inches	in	ft	in	ft^2	in^2
1/4	0.375	0.0263	0.315	0.00054	0.0778
3/8	0.500	0.0358	0.430	0.00101	0.1454
1/2	0.625	0.0454	0.545	0.00162	0.2333
5/8	0.750	0.0555	0.666	0.00242	0.3485
3/4	0.875	0.0654	0.785	0.00336	0.4838
1	1.125	0.0854	1.025	0.00573	0.8251
1 1/4	1.375	0.1054	1.265	0.00873	1.257
1 1/2	1.625	0.1254	1.505	0.01235	1.778
2	2.125	0.1654	1.985	0.02149	3.095
2 1/2	2.625	0.2054	2.465	0.03314	4.772
3	3.125	0.2454	2.945	0.04730	6.811
3 1/2	3.625	0.2854	3.425	0.06398	9.213
4	4.125	0.3254	3.905	0.08317	11.98
5	5.125	0.4063	4.876	0.1296	18.66
6	6.125	0.4871	5.845	0.1863	26.83
8	8.125	0.6438	7.726	0.3255	46.87

TABLE 3.9 Properties of Type M Copper Tubing

Standard size,	Outside diameter	Internal diameter		Flow area	
inches	in	ft	in	ft^2	in^2
3/8	0.500	0.0375	0.450	0.00110	0.1584
1/2	0.625	0.0474	0.569	0.00177	0.2549
3/4	0.875	0.0676	0.811	0.00359	0.5170
1	1.125	0.0879	1.055	0.00607	0.8741
1 1/4	1.375	0.1076	1.291	0.00909	1.309
1 1/2	1.625	0.1273	1.528	0.01272	1.832
2	2.125	0.1674	2.009	0.02701	3.889
2 1/2	2.625	0.2079	2.495	0.03395	4.889
3	3.125	0.2484	2.981	0.04847	6.980
3 1/2	3.625	0.2883	3.460	0.06523	9.393
4	4.125	0.3279	3.935	0.08445	12.16
5	5.125	0.4089	4.907	0.1313	18.91
6	6.125	0.4901	5.881	0.1886	27.16
8	8.125	0.6488	7.786	0.3306	47.61

Plastic Piping

Plastic piping can be used in cases where you don't have hot water service and wish to minimize the expense of the line. Low-pressure cold water lines and open systems (for example, coil condensate drains) are good applications for plastic piping. There are several different kinds of plastic piping available. PE piping stands for polyethylene and should be used for cold water service only. ABS stands for acrylonitrile-butadiene styrene and is also for cold water service only. PVC is polyvinyl chloride and is used for cold water service only, while PVDC is polyvinyl dichloride and can be used for both cold and hot water service.

Support and Insulation

Pipe supports should be used whenever possible to prevent bending or strain of the pipes and connections. This is particularly important with steam, refrigerant, and high-pressure piping where a leak can be a big problem. The thinner the pipe, the less support it can provide itself and the greater the number of hangers and supports required. Figure 3.1 shows the recommended hanger support distances for horizontal pipe used in different applications. Keep in mind that the supports must themselves be supported using steel threaded rods of the proper size and that the total supported weight is not only the pipe but the fluids in the pipe. This is why the pipe spacing on steam pipes is greater than that for water pipes: the steam does not weigh as much.

Piping must also be supported on risers and wall-mounted sections. Here the concern is not so much the weight of the pipe but rather to prevent the pipe from flexing or torquing the horizontal sections. In particular, make sure both the vertical and horizontal sections leading to a riser are supported (see Figure 3.2).

When copper pipe is used for refrigeration service, insulated horizontal suction lines can be laid, unclamped, on hangers provided that they are insulated from vibrations that may abrade the pipe. Compressor discharge piping must be securely clamped to supports. The spacing of the supports depends on the pipe size; thicker pipe is more rigid and can support longer distances. You should use pipe hanger spacing distances similar to that for copper piping in Figure 3.1.

As the pipe changes temperature during service, it will most likely expand and contract. Figure 3.3 shows how much copper tubing

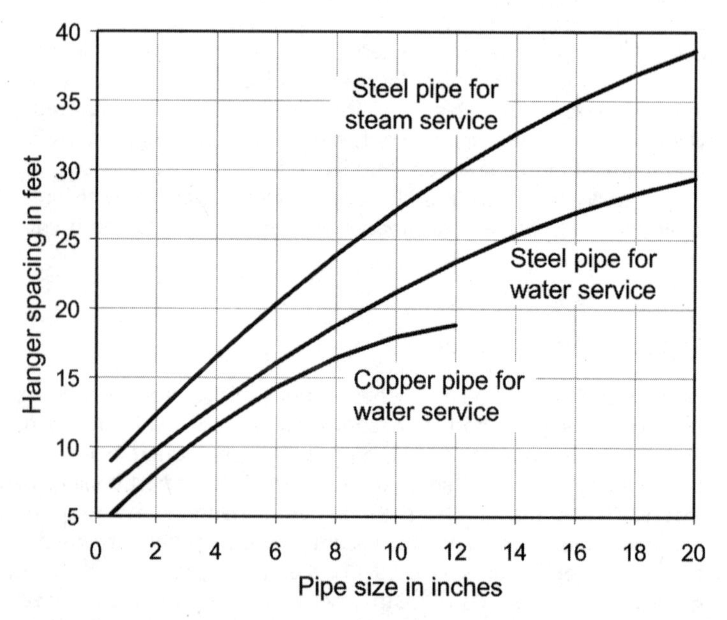

FIGURE 3.1 Recommended horizontal pipe support spacing.

and steel pipe expands as the temperature increases. If there are rigid supports at the end of long runs of the pipe, the expansion may rip the support or put tremendous strain on pipe couplings, elbows, and anywhere else that the expansion may concentrate. If your pipe runs are long enough to provide more than a few tenths of an inch of expansion, you should include Z bends or U bends as shown in Figure 3.4 to help absorb the pipe length change.

FIGURE 3.2 Insulated pipe mounted on wall brackets.

fastfacts

The hanger spacings listed in Figure 3.1 are for just the pipe and do not include valves, tees, or any other point loads. You should include additional hangers if there is additional hardware on the piping system.

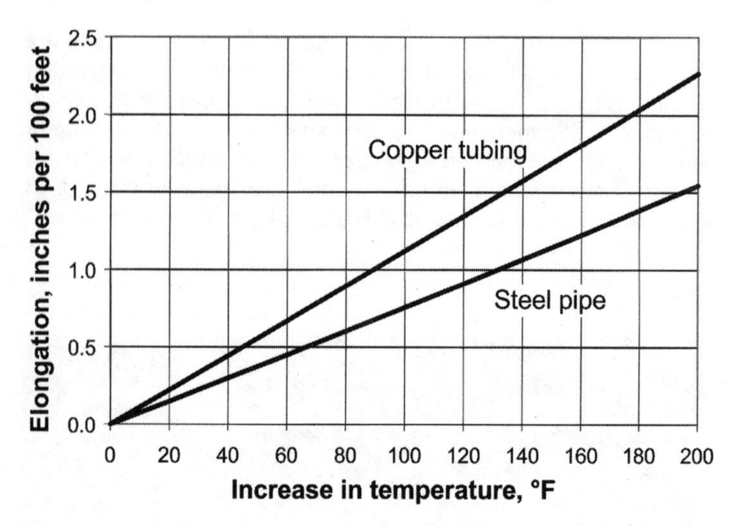

FIGURE 3.3 Expansion of copper tubing and steel pipe on temperature rise.

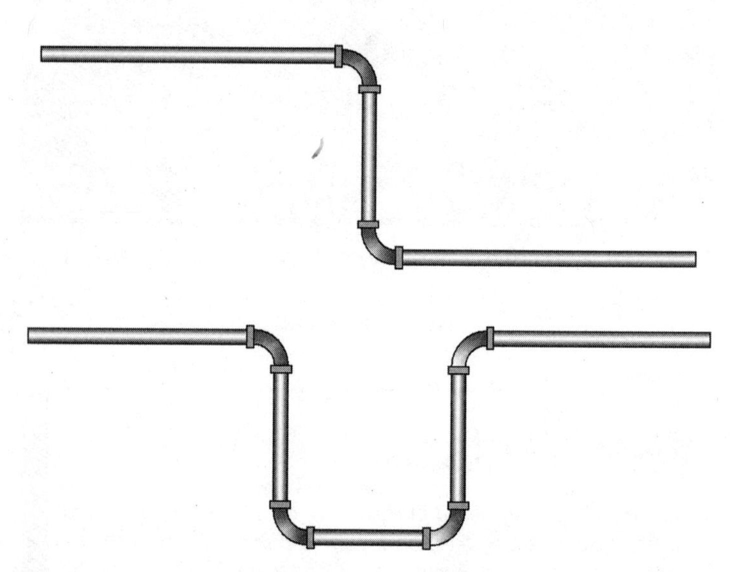

FIGURE 3.4 Z and U bends used to take up expansion and contraction of piping length.

fastfacts

Plastic pipe expands about five times as much as copper piping for a given temperature change. Be very careful when mounting and supporting plastic piping to account for this.

Piping should be insulated to provide both safety and energy conservation. Hot water and steam piping should be insulated to prevent excessive heat loss from the pipe and to prevent the pipe from making the surrounding environment too hot. Uninsulated steam piping can also cause immediate burns should you accidentally touch it. Chilled water piping insulation helps prevent condensation on the outside of the pipe in humid climates. This can damage the pipe and any controls or electronic devices located underneath the pipe. Some example insulation materials are given in Table 3.10. The magnesia insulation is also available in a powdered form that can be mixed with water to form an insulating plaster. This is useful when covering unusual-shaped fittings and unions. Of course, asbestos can't be used as a insulating material any more, but it is pretty common to find asbestos insulation in older buildings.

TABLE 3.10 Different Types of Piping Insulation

Material	Temperature range
Asbestos	up to 1200°F
Calcium Silicate	up to 1200°F
Cellular Glass	up to 800°F
Fiberglass	up to 370°F
Diatomaceous Silica	up to 1850°F
Magnesia	up to 600°F
Plastic Foams	-250 to 250°F

HEAT EXCHANGERS

Heat exchangers transfer heat from one fluid to another without allowing the two to mix. Chilled and hot water coils, steam-to-hot-water converters, and chiller evaporators are all examples of heat exchangers.

In *parallel flow* heat exchangers the fluids both flow in the same direction (see Figure 3.5). The initial heat flow from the hot fluid to the cold fluid is very high because of the large temperature difference between the two. However, as the fluids near the exit, their temperatures become closer and closer and there is less energy exchanged.

The fluids in *counterflow* heat exchangers enter at opposite ends and travel in opposite directions as shown in Figure 3.6. The temperature difference across the length of the heat exchanger remains much more constant than in a similarly-sized parallel-flow heat exchanger. Counterflow heat exchangers are therefore usually more effective at transferring heat than parallel flow heat exchangers.

Another common type of heat exchanger is the *cross flow* type, as illustrated in Figure 3.7. Here the two fluids cross at perpendicular directions, such as in a water-to-air heating or cooling coil (Figure 3.8). Since there is not much surface area between the fluid pipes and the air stream, a series of fins is used to enhance the heat

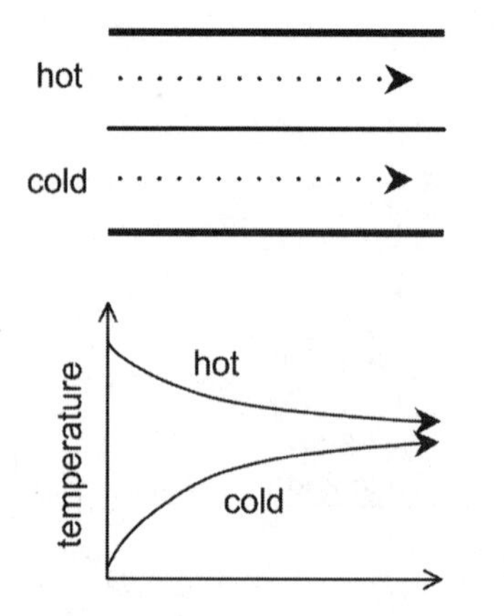

FIGURE 3.5 Fluid flows and temperature profiles in a parallel-flow heat exchanger.

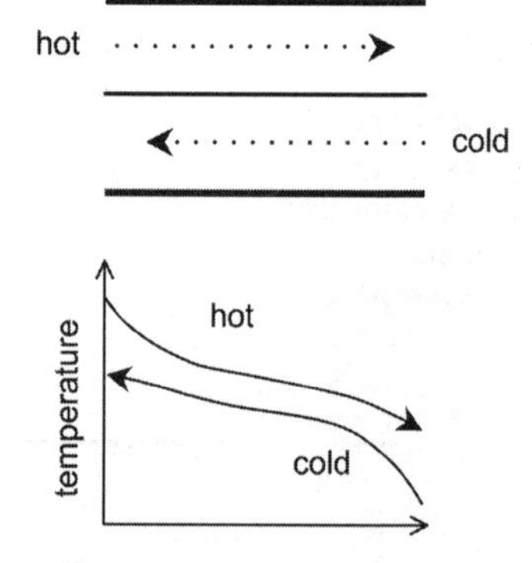

FIGURE 3.6 Fluid flows and temperature profiles in a counterflow heat exchanger.

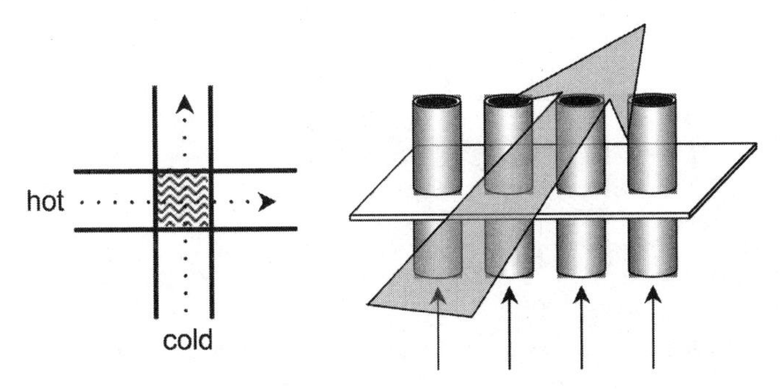

FIGURE 3.7 Fluid flows and example of cross-flow heat exchanger.

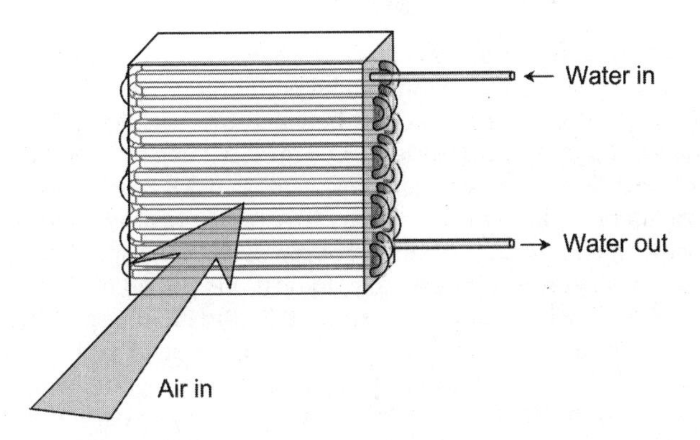

FIGURE 3.8 Water-to-air cross-flow heat exchanger.

transfer. The fins are very thin metal and have a high thermal conductivity. There are typically between 8 and 12 fins per inch in these kinds of heat exchangers and the total fin area can be dozens of square feet even when the face area is relatively small.

Air heating and cooling coils are rated by the maximum heat transfer rate possible at the design conditions. For example, a heating coil might be listed as 100,000 Btu per hour at 10,000 CFM of air and 120°F hot water inlet. Cooling coils are often cited by their cooling capacity in tons of refrigeration.

Another kind of cross-flow heat exchanger is the shell-and-tube heat exchanger as shown in Figure 3.9. These are often used in chiller evaporators, steam-to-hot water converters, and heat-recovery condensers.

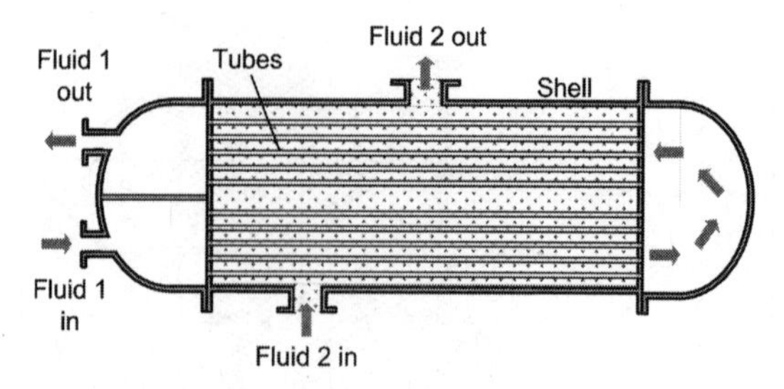

FIGURE 3.9 Cross-section of a shell-and-tube heat exchanger.

Performance of Heat Exchangers

When troubleshooting air-to-air coils or air-to-water coils, it is often useful to know how well the coil should be performing. That is, how much heat can actually be transferred between the two sides. The basic equation for coil heat transfer is $Q = m\ c_p\Delta T$ where Q is the amount of heat flow, m is the mass flow rate, and ΔT is the temperature difference. The term c_p is called the specific heat and indicates how much energy must be added or removed from something to change its temperature. For example, to heat one pound of air by one degree Fahrenheit you must add about 0.24 Btu. So the specific heat of air is around 0.24 Btu/lb·°F. The specific heat of water is about 1.0 Btu/lb·°F. If a heating coil heats an airstream by 20°F and the air is flowing at 10,000 CFM (and assuming an air density of 0.075 lb/ ft^3), the amount of heat added to the air is:

$$\dot{Q} = 10{,}000\frac{ft^3}{min} \times 0.075\frac{lb}{ft^3} \times 0.24\frac{Btu}{lb\cdot{}^\circ F} \times 20\,{}^\circ F = 3600\frac{Btu}{min}$$

or about 216,000 Btu/hr. On the water side, if the coil inlet temperature is 120°F and the water is flowing at 40 GPM, we can use the same equation to find the outlet water temperature. First we have to find the temperature difference of the water,

$$3600\ \frac{Btu}{min} = 40\ \text{GPM} \times 8.33\ \frac{lb}{gal} \times 1.0\frac{Btu}{lb\cdot{}^\circ F} \times \Delta T$$

so the water temperature difference is about 10.8°F and the leaving water temperature would be about 109°F.

This example shows a simple way to determine how much energy is transferred across a coil. This can be compared to the rated coil capacity to see if the coil is working correctly. If you don't have the coil specifications, you can estimate the capacity using the *NTU method.* In this method, the *capacity rates* for each of the two fluid streams is calculated by multiplying the mass flow rate (in pounds per minute) and the specific heat. The capacity rate has the units of energy per time per degree (for example, Btu/hr·°F) and is calculated for the hot and the cold fluids:

$$C_{hot} = \dot{M}_{hot} \cdot c_{p,hot}$$

$$C_{cold} = \dot{M}_{cold} \cdot c_{p,cold}$$

The *minimum capacity rate* C_{min} is the smaller of the two capacity rates and is the limiting factor, as is the temperature difference between the inlet temperatures of the hot and cold fluids. The *maximum heat transfer rate* is then given by

$$\dot{Q}_{max} = C_{min} \times \left(T_{hot,in} - T_{cold,in}\right)$$

This is the largest heat transfer rate possible in the heat exchanger. There will also be certain inefficiencies in the process (nothing is perfect, after all) so we have to define the *heat exchanger effectiveness,*

$$\varepsilon = \frac{\dot{Q}_{actual}}{\dot{Q}_{max}}$$

If we know the effectiveness of a heat exchanger, we can then determine the actual heat transfer,

$$\dot{Q}_{actual} = \varepsilon \cdot \dot{Q}_{max}$$

The effectiveness is calculated using the two capacity rates, the coil type (e.g., parallel, counterflow, or crossflow) and the *number of transfer units* (NTU). The number of transfer units is defined as the product of the total heat transfer area and the surface convection coefficient divided by the minimum capacity rate:

$$NTU \equiv \frac{AU}{C_{min}}$$

The relationship between NTU and effectiveness is plotted in Figures 3.10 and 3.11 for cross flow and shell-and-tube heat exchangers.

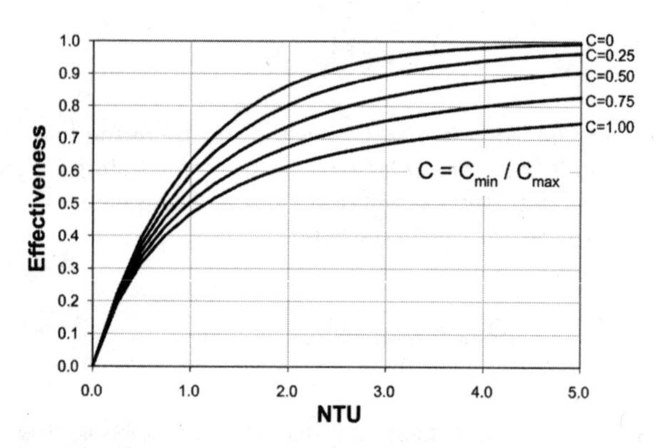

FIGURE 3.10 Effectiveness for cross flow HX.

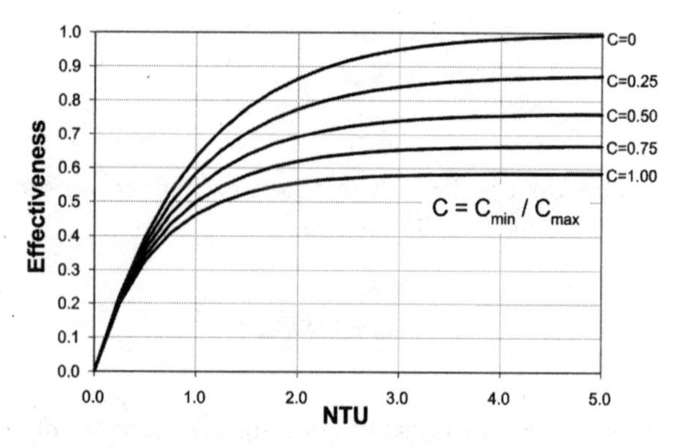

FIGURE 3.11 Effectiveness for shell-and-tube HX.

chapter 4

ELECTRICITY AND WIRING

HVAC systems use high-voltage systems for running motors, compressors and other big equipment. These devices operate at voltages anywhere from 120 V up to 480 V and even higher. On the other end of the scale, many electronic control circuits use lower voltages in the 12 to 24 V range for running actuators, sensors, and controllers. This chapter discusses some of the basics of electronic circuits and equipment.

Electricity is often compared to water. While working with the two together can be a dangerous combination, the analogy between the two is quite workable. Just as water flows through a pipe, you can think of electrons flowing through wires. Water current is measured in gallons per minute, while current flow is measured in electrons per second. Water pressure is measured in pounds per square inch, while electricity pressure is measured in volts. Table 4.1 gives some equivalents that may make is easier to understand the various electrical terms. Of course, the analogy can only be taken so far. In the case of alternating current, it is difficult to imagine a pipe in which the pressure changes from positive to negative many times per second.

TABLE 4.1 Equivalents for Electricity

Term	Description	Units
energy	joules or Btus	1 Btu = 1055 joules = 0.293 Wh
coulomb	number of electrons	1 coulomb = $6 \cdot 10^{23}$ electrons
current	flow rate of electricity	1 ampere (1 A) = 1 coulomb / second
volts	"pressure" of electricity	1 volt (1 V) = 1 joule / coulomb
resistance	obstruction of current	1 ohm (1 Ω) = 1 volt / 1 ampere
capacitance	"pressurized storage" of electricity	1 farad (1 F) = 1 coulomb / volt
power	rate of energy	1 watt (1 W) = 1 joule / second

There are many different types of electronic components. Most modern electronic HVAC controllers use a combination of *semiconductors* to process and generator electronic signals. The operation of these devices can be pretty complex, so we're not going to cover them in this book. However, there may be times when you need to identify certain components on a wiring diagram so that you can track down problems. Table 4.2 gives a description of different electrical components that you may encounter.

DC CIRCUITS

Direct current (DC) circuits are used in HVAC systems primarily for sensors, electronic controllers, and electronic actuators. The basic operation of DC circuits is governed by two concepts: (1) Ohm's law, which states that the voltage drop across a resistor found by multiplying the current i and the resistance R, $V = i \times R$, and (2) Kirchhoff's voltage law, which says that the sum of voltages around a closed loop must be equal to zero. Using the water analogy, these two laws say that there is a electrical pressure drop when the current passes through a valve, and that the total pressure rises around a circuit must add up to all the pressure drops. Resistors are usually color-coded (see Figure 4.1) to identify both the magnitude of the resistance as well as the accuracy of the rated resistance.

TABLE 4.2 Electronic Component Symbols

Component	Symbol	Function
Capacitor	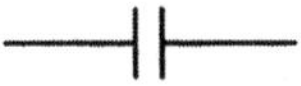	Maintains constant voltage when current varies
Circuit breaker		Breaks current flow when current is too high
Diode	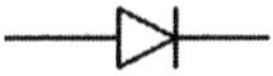	Limits current flow to one direction
Fuse		Breaks current flow when temperature gets too high
Ground		Reference for voltage in circuit
Inductor		Maintains constant current when voltage varies
Meter	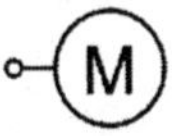 M	Measures voltage, current, or total power consumption
Operational amplifier	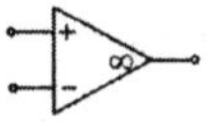	Amplifies an electronic signal
Relay	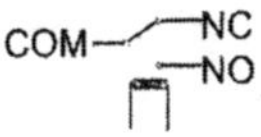 COM NC NO	Switches current between two or more different terminals
Resistor		Impedes the flow of electricity
Switch		Makes or breaks the continuity of a circuit
Transformer		Converts AC power at one voltage to another voltage
Transistor	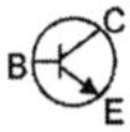 B C E	Electronic switch/amplifier
Voltage source	 + −	Provides voltage to a circuit

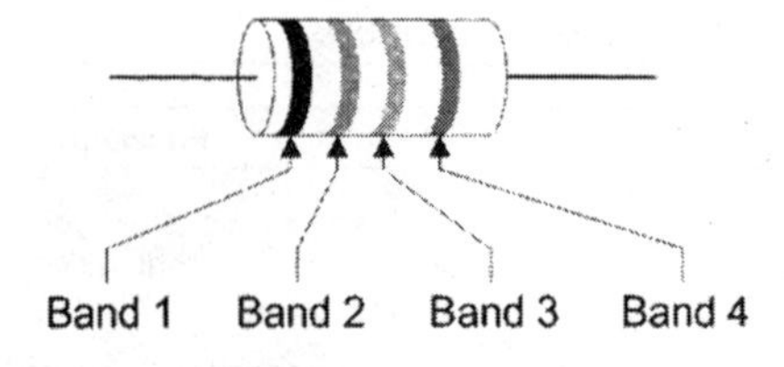

Color	First band	Second band	Third band	Fourth band
	First digit	Second digit	Multiplier	Tolerance
none				± 20%
Silver			0.01	± 10%
Gold			0.1	± 5%
Black	0	0	1	
Brown	1	1	10	± 1%
Red	2	2	100	± 2%
Orange	3	3	1,000	
Yellow	4	4	10,000	
Green	5	5	100,000	± 0.50%
Blue	6	6	1,000,000	± 0.25%
Violet	7	7	10,000,000	± 0.10%
Grey	8	8		± 0.05%
White	9	9		

FIGURE 4.1 Resistor showing location and value of color bands.

Some common resistor color combinations are:

- Yellow/Violet/Black = 47Ω
- Brown/Black/Brown = 100Ω
- Yellow/Violet/Brown = 470Ω
- Green/Black/Brown = 500Ω
- Brown/Black/Red = 1000Ω
- Green/Black/Red = 5000Ω
- Brown/Black/Orange = 10000Ω
- Green/Black/Orange = 50000Ω

Many temperature sensors used in HVAC measurements are resistors where the resistance changes according to the temperature. Remember that resistors act like a valve in a water stream—they decrease the voltage. Resistors behave differently when connected in series or in parallel (see Figure 4.2). The total resistance of series resistors is simply the sum of the individual resistances, $R_{TOTAL} = R_1 + R_2$, where the total resistance of the parallel resistors is taken from the sum of the reciprocals, $R_{TOTAL} = (1/R_1 + 1/R_2)^{-1}$

Voltage dividers are made up of two series that reduce the voltage provided to a sensor or controller. For DC circuits, a voltage divider works as illustrated in Figure 4.3. In some HVAC control loops you find slide wire resistors (Figure 4.4). These are resistors with three connections: one at each end and a movable tap in the middle. The resistance between the terminal at either end and the tap depends on the tap position. Standard end-to-end resistances of slide wire resistors are 100Ω, 135Ω, and 1000Ω.

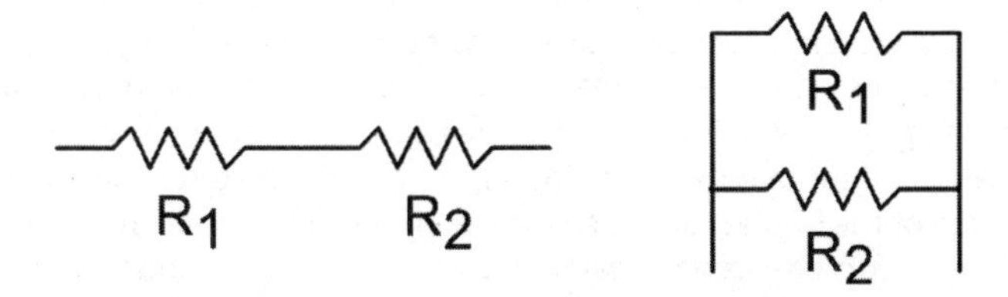

FIGURE 4.2 Series (left) and parallel (right) resistors.

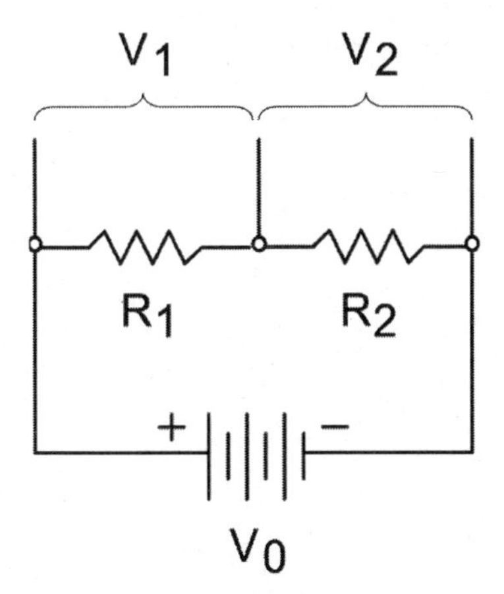

FIGURE 4.3 DC voltage divider.

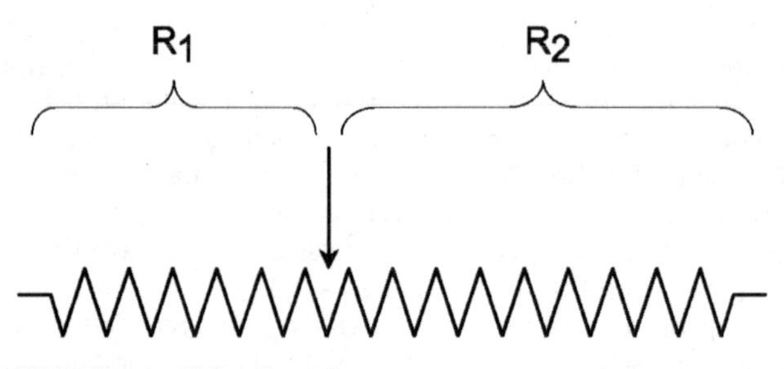

FIGURE 4.4 Slide wire resistor.

Suppose you had a controller that needed to know the difference between the outside air temperature and the supply air temperature in a building. You could use a Wheatstone bridge like that shown in Figure 4.5. These kinds of devices are often used in HVAC sensors that must respond to a small change in voltage or current. If all the resistances in this circuit are equal, then the output voltage V_o equals zero. Now suppose R3 is a variable resistor that measures the outside air temperature (this is called the compensation resistor, R_c) and R2 is a variable resistor that measures the supply air temperature. The output voltage of the bridge varies depending on the difference in resistance.

Capacitors are used to maintain a constant voltage under varying current conditions. They are an integral part of electronic filters that can are used, for example, to eliminate noise from sensor measurements using a filter like in Figure 4.6. This is called a low pass filter because low frequency signals are not affected while high frequency signals (the noise) get eliminated.

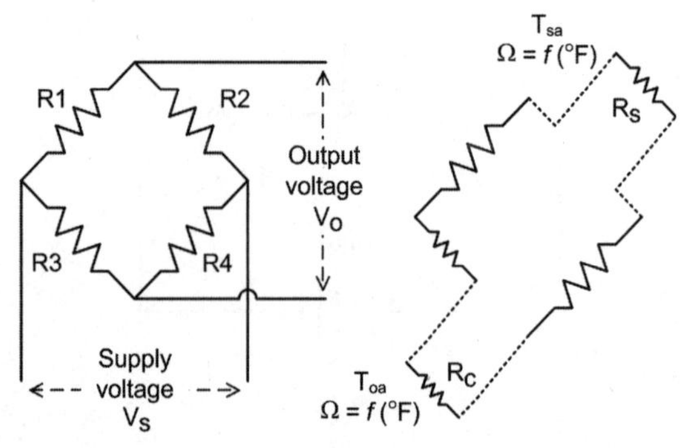

FIGURE 4.5 Schematic of a Wheatstone bridge circuit.

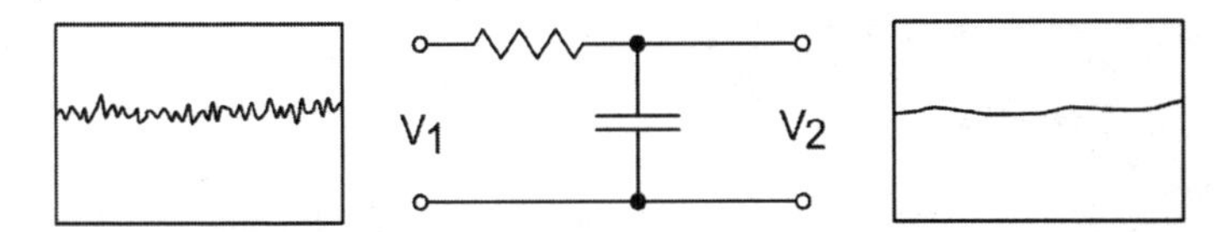

FIGURE 4.6 Low-pass filter.

A diode is like a check valve in that it only allows current to flow one way but not the other. You can use diodes to create a DC power supply from an AC line signal to use for powering sensors. A *rectifier* (Figure 4.7) is used to do this. The AC signal enters at the left and passes through the diodes that are used to create a full-rectified signal. This is then passed through a low-pass filter as described in the previous paragraph.

A transistor is like a valve with three connections: the collector, the emitter, and the base. When a current is applied to the base, the valve opens up and lets current flow from the collector to the emitter. In fact, even the current applied at the base goes out the emitter. In some DC circuits, power transistors are used to amplify sensor or control signals to useful levels. These components can produce a lot of heat that must be dissipated to prevent burning up the transistor. Often there is a heat sink (see Figure 4.8) attached to the side of a control box. Always insure that the heat sinks are clean and in a free air stream. Do not cover them with rags, papers, or anything else. If the heat sink is not hot, then it's doing its job. By covering it up, you are risking both a fire hazard and failure of the circuit.

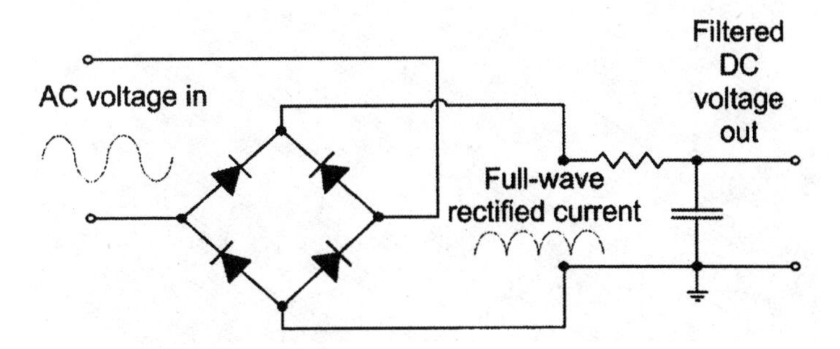

FIGURE 4.7 Rectifier with low-pass filter.

FIGURE 4.8 Heat sink.

fastfacts

Low-voltage DC wiring does not carry much current or power. However, if such signal or control wiring enters a high voltage box, local codes usually dictate that the temperature and voltage ratings on the wire must be the same as high voltage/high power wiring that enters that box.

AC CIRCUITS

The majority of large HVAC equipment is powered by alternating current (AC). The alternating part of this term is appropriate: the voltage and current in AC circuits flips back and forth 60 times per second. The primary service enters the building at 120/240 VAC if single phase, or 120/208 VAC, 277/480 VAC, and higher if three phase. The first number is the voltage between any single phase and ground and the second number is the voltage between any two phases. In commercial buildings, electricity enters a building though the primary service feeder (as shown in Figure 4.9) and then passes through a series of transformers, switchboards, and panels before arriving at the load. The HVAC equipment is usually powered directly from the switchboards. If a switchboard is devoted to just HVAC equipment, it is called a *motor control center.*

Most residential buildings are 120/240 VAC single-phase, three-wire systems. In this arrangement, a single-phase transformer is center tapped with this tap referenced to ground. The two hot wires from the transformer ends have 240 VAC, and each is 120 VAC with respect to ground. So plug loads can use the 120 VAC while larger loads such as the range, furnace fan, electric water heater, and central air heat pump can take advantage of the 240 VAC.

Most commercial buildings are three phase at either 120/208 VAC, 277/480 VAC, on up to 2400/4160 VAC. This latter is found only in the largest buildings, although some older commercial buildings still utilize high voltage HVAC equipment for the chillers and boilers.

fastfacts

Sometimes 120 V equipment is listed at 115 V, and 480 V equipment at 460 V. These differences are due to a standard 4% feeder voltage drop that is experienced almost everywhere.

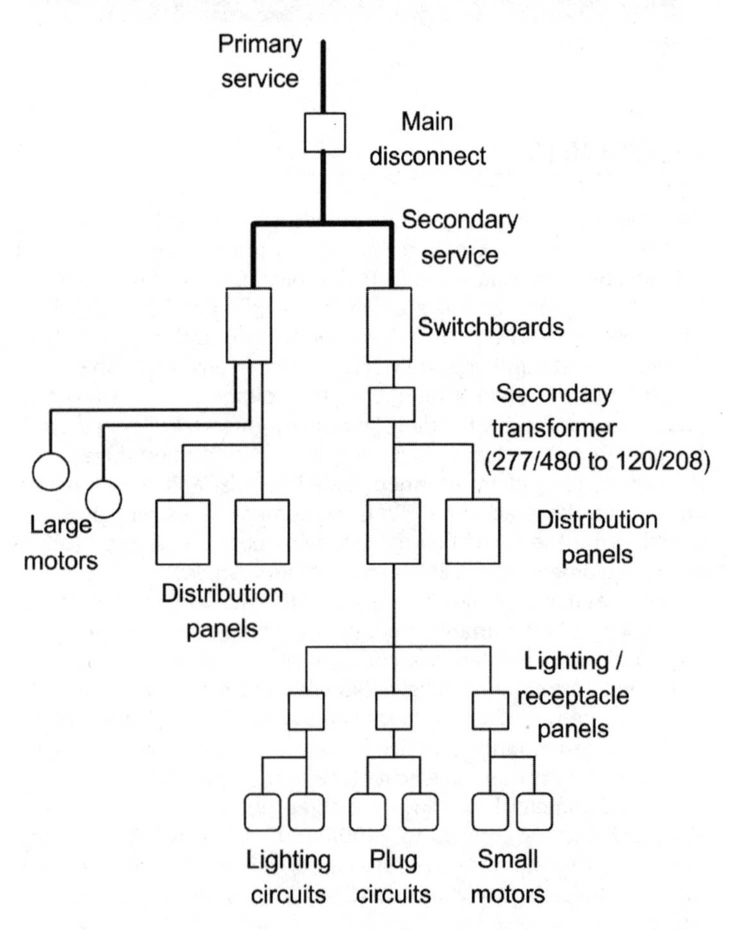

FIGURE 4.9 Typical commercial building electrical distribution system. *(Adapted from Stein and Reynolds)*

Ideally, all the varying current and voltage in an AC circuit would be in phase. That is, they would both reach their highest values at the same time. Since total power is voltage multiplied by current, this would give the highest power as well. However, as the electricity enters motors and other equipment, the voltage and current can get out of phase due to the presence of inductors and capacitors. Figure 4.10 shows what the voltage and current would look like if they were out of phase—in this case the current leads the voltage by 25° out of the full 360° required for one crest-to-crest cycle. The *power factor* is used to describe how much the current and voltage are out of phase. In this case, the power factor is the cosine of 25°, or 90 percent. What this means is that only 90 percent of the possible power is going to run the device, while the remaining 10 percent does nothing. With large HVAC equipment, you want to keep your power factor as high as possible because sometimes the utility penalizes you if your building has a low power factor.

Wye and Delta Circuits

Large fan and pump motors use either a wye or delta wiring configuration, although delta configurations are much more common. The names of these circuits come from the connections of the motor windings as shown in Figure 4.11. The different configurations can be used to control the *inrush current;* that is, the amount of current

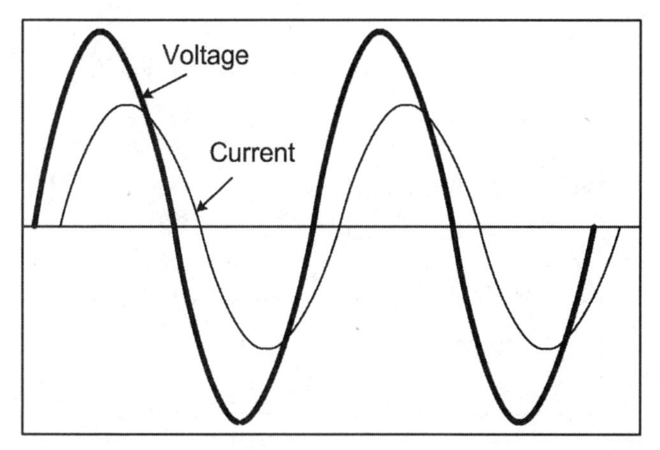

FIGURE 4.10 AC voltage and current 25° out of phase.

fastfacts

The electric grid in the United States uses 60 Hz power, but in other countries the grid operates at 50 Hz. If you use electric components manufactured overseas, make sure they are compatible with the electrical service you are using. Equipment specified for one frequency should not be used at another.

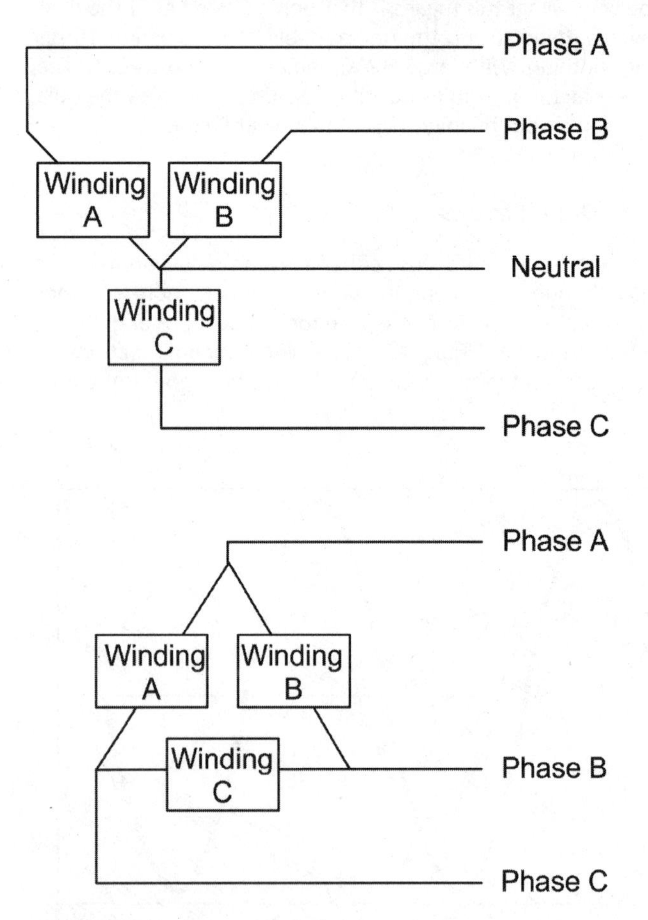

FIGURE 4.11 Wye and delta circuits.

that flows into the motor windings on startup. When starting a large motor, it is desirable to minimize this inrush current because too large a value can overheat the motor windings.

Variable Speed Drives

Motors spin at a speed related to the line cycle frequency. In the United States, the line cycle frequency is 60 Hz, so most motors operate at some multiple of this such as 1800 or 3600 RPM. However, often it is useful to have a motor operate at a different or variable speed. A clutch and gearing system can be used but is expensive and difficult to maintain. Since the late 1980s, *variable frequency drives* (VFDs or sometimes *variable speed drives*, VSDs) have become quite popular for applications that need changing motor speeds. Many variable air volume systems and large water distribution systems now use these drives, along with cooling towers and some kinds of chillers. VFDs change the actual speed of the motor by creating a new AC line voltage at any desired frequency. Figure 4.12 shows the principle behind the operation of a variable frequency drive. The

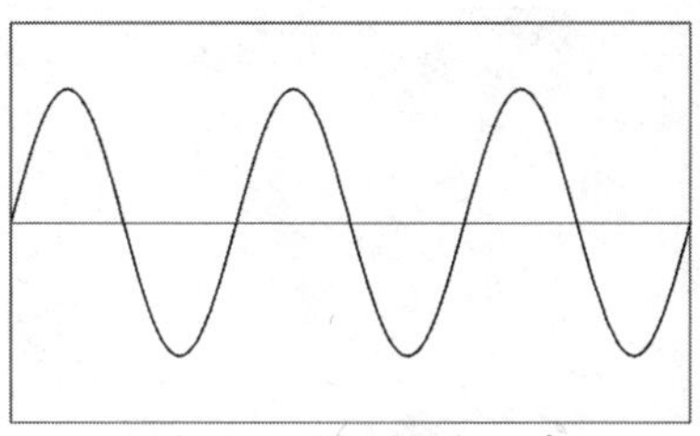

a. input 60 Hz signal

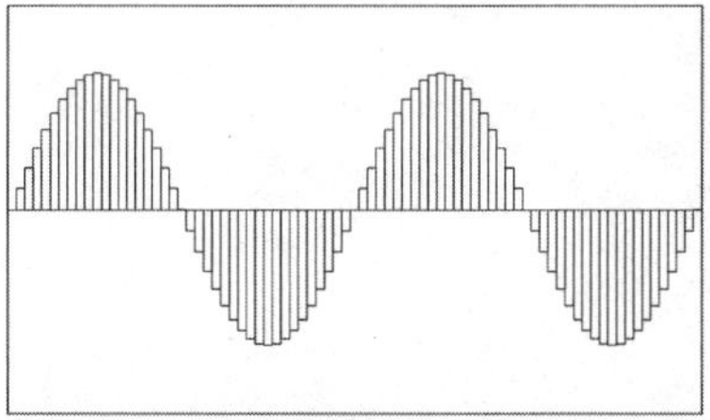

b. synthesized 40 Hz signal

FIGURE 4.12 Comparison of 60 Hz and 40 Hz signals.

standard line frequency (top) is decomposed and then used to generate a series of impulses that "blend together" to create a new periodic wave at a different frequency. Figure 4.12 shows how 60 Hz power is converted to a 40 Hz signal to slow a motor.

With VFDs it is important to remember that the motor may not be not operating at the design frequency, and that there may be certain frequencies that cause excessive vibrations or harmonics in the motor and attached fan or pump. Most VFDs have *critical speed step-over* circuits that are adjustable ranges that can be locked out to avoid any resonant speeds. Two step-over ranges of individually adjustable widths are sufficient for most applications. In addition, the VFDs can introduce noise back on to the *incoming* electrical lines that can affect the other motors in the building as well as sensitive electronic equipment. In many cases the VFDs are located near the motor control center (as in Figure 4.13) so that such problems can be easily traced. Identifying the culprit for such *harmonics* usually requires the use of an oscilloscope or other specialized equipment.

FIGURE 4.13 Motor control center (cabinet on left) and variable frequency drives (two boxes on right).

Transformers

Transformers are used to change the voltage in alternating current circuits. A *step-up* transformer increases the voltage, a *step-down* transformer decreases the voltage. Electromagnetic induction is used to transmit the energy from the input (primary) side of the transformer to the output (secondary) side of the transformer (see Figure 4.14). Because the two windings of a transformer are not directly connected, the transformer provides isolation, that is, it protects the load from noise on the line and protects the line from noise produced by the load. An isolation transformer may have the same input and output voltage. Transformers may also have multiple windings to allow a single input to produce multiple output voltages.

Transformers come in a wide variety of sizes, from the very small (as used in control systems) to very large (to drop kilovolt line voltages down to the 480 or 120 VAC used in the building). Transformers are rated according to their maximum power-handling capacity. This value is given in volt-amperes (VA). A typical low voltage transformer used for an actuator might have a rating in the 100 VA range (Figure 4.15) while a large, whole-building transformer is in the 1000 kVA range or larger (Figure 4.16).

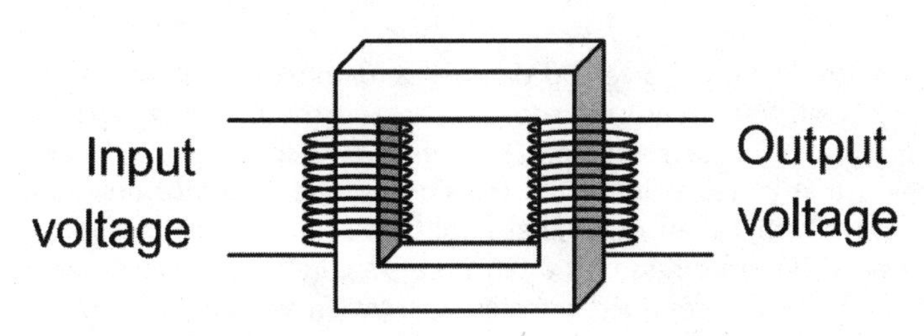

FIGURE 4.14 Transformer.

fastfacts

Transformers only work with AC voltages. Use a voltage divider on a DC circuit if you want to provide a stepped-down voltage to a sensor.

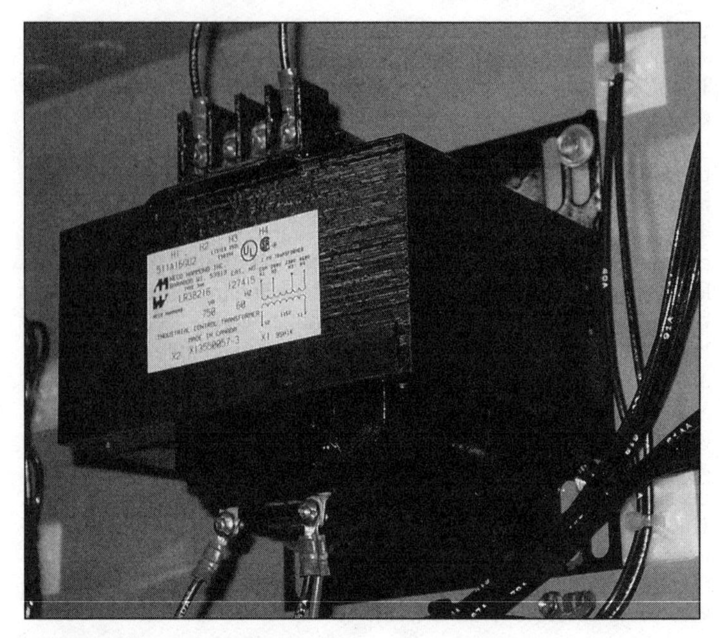

FIGURE 4.15 750 VA Transformer.

Most electronic valve and damper actuators use a dedicated 24 VAC or 48 VAC transformer for power. You should use a separate transformer for each actuator. There are two reasons for this, the first being that of reliability: if the transformer fails, you lose only one actuator rather than many that might be connected to that transformer. The other reason is a bit more obscure and has to do with floating versus grounded transformers. When working with transformers, remember that there is no physical connection between

fastfacts

If a modulating electronic actuator moves on its own or does not respond properly to the control signals, check to make sure that it has a floating transformer.

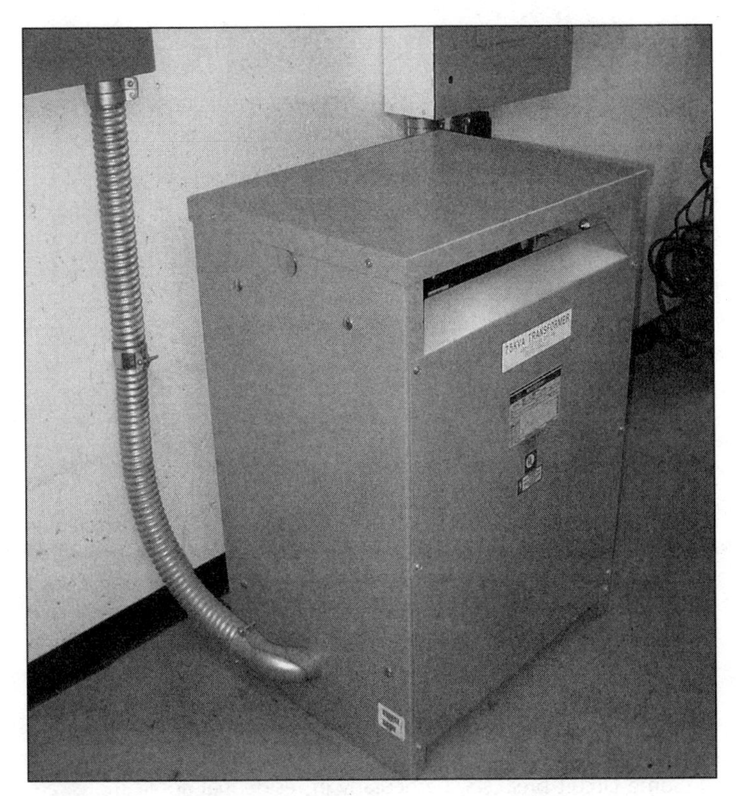

FIGURE 4.16 750 kVA Transformer.

the primary and secondary side. You can therefore reference the secondary to any arbitrary ground, or not attach it to ground at all. In the latter case, this is called a *floating* transformer. Figure 4.17 shows the difference between a grounded and floating circuit. Many actuators expect to see a floating circuit. If you power more than one actuator with a single transformer, each actuator will be referenced to the others and the circuit will not strictly be floating.

Note, however, that most large motor and fan secondary circuits are grounded, primarily for human and equipment safety reasons. In these cases, the ground should never be defeated by switches, circuit breakers, or fuses.

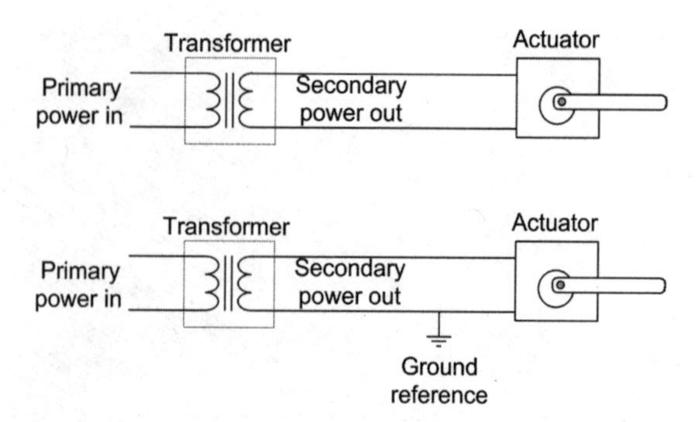

FIGURE 4.17 Floating (top) versus grounded (bottom) transformer circuits.

Circuit Breakers and Fuses

Practically all electronic equipment, whether DC or AC, is protected by circuit breakers, fuses, or some combination of the two. Circuit breakers are usually found in a breaker panel or a motor control center and can trip out when the current goes above a certain level or the breaker temperature gets too high. In both cases, the purpose of the breaker is to cut power to a device to prevent damage from too high a current flow. Breakers are generally resettable, meaning that you can close the circuit again once the breaker has tripped.

Some circuit breakers, particularly in residential applications, are *ground fault interrupts* (GFIs). These kinds of breakers measure not only the current heading to the appliance on the hot lead, but also the return current on the neutral. Any difference between the two implies that there is a path for the current other than the desired electrical circuit. If this occurs, the breaker will trip off. These kinds

fastfacts

Some thermal breakers are more susceptible to line noise. For example, variable frequency drives and unbalanced motors can cause repeated trip-outs of thermal breakers even though the actual current is not above the rated limit.

fastfacts

Fuses provide protection not only to equipment but to personnel as well. Never defeat fuse protection on a circuit. If a fuse continues to fail, identify the cause before returning the equipment to service.

of breakers are used especially in residential bathroom circuits where there are water, electricity, and occupants all together. However, you may also find GFI breakers on residential furnace fans and heat pumps that are meant to act as protection against any unintended short circuits.

Fuses are similar to breakers in that they halt the flow of current if it gets too high. The difference is that fuses must be replaced once they have broken the circuit. Fuses are located in both motor control centers as well as at the line input for most HVAC equipment. Figure 4.18 shows a fuse bank for an electric resistance heater in an air-handling unit. These fuses are rated at about 30 amperes. These fuses are classified as *slow-blow* fuses, meaning that there is some

FIGURE 4.18 Fuse bank.

delay between the occurrence of high current and the fuse failing. The reason for this is so that the fuse can tolerate brief periods of high inrush current to motors.

Electric Switches

Electric switches can be used for turning devices on and off, and for opening and closing valves and dampers. While the functions and sizes of the switches can differ, the principle is based on the energizing of a small electromagnet that is then used to close a set of contacts or to move a two-position actuator. The switches can be broken down into three different categories

A *relay* is used to switch AC and DC circuits up to about 120 V. The basic principle of a relay is shown in Figure 4.19, where the top portion of the figure shows the relay in a de-energized state and the bottom portion shows the energized version. The relay electromagnet (coil) is powered by a 24 or 120 VAC signal. While the figure shows a single contact between the terminals and common, in practice relays can have multiple contacts, or *poles*. This figure shows a *double-throw* relay in which connections are made in both the de-energized and energized states. A *single-throw* relay would not have two sets of contacts but instead would just have a NO or NC contact. The type of relay is often abbreviated; for example, a DPDT refers to a double-pole, double-throw relay that could connect two separate circuits, each with a NO and NC contact. Figure 4.20 shows a picture of a standard three-pole, double-throw relay. On these kinds of devices the switching mechanism can detach from the terminal base so that replacements can be made without having to remove the wires.

A balance relay (also known as a mouse trap relay) is a three-position switch used to run small actuator motors. One set of contacts drives the motor in one direction, the other drives the motor in the opposite direction, and the third is used to stop the motor. This kind of tri-state actuator is used on damper and valve motors. A *contactor* is basically a three-pole, single-throw relay that is used to switch high voltage multiphase equipment on and off. Contactors are used to activate pretty much all motors, fans, pumps, and compressors in HVAC applications. Figure 4.21 shows a picture of a three-phase contactor rated for 480 V service at 30 amps. One problem that can happen in contactors occurs just as the coil is energized. The contacts can hit each other and then bounce off before hitting again. Multiple bounces of the contacts is referred to as chattering. A spark can jump across the small gap between the contacts

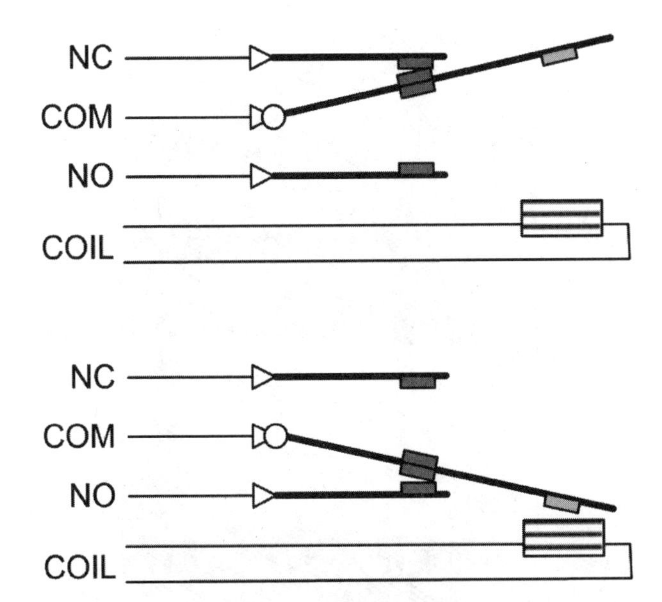

FIGURE 4.19 Schematic of electronic relay.

FIGURE 4.20 Electronic relay.

FIGURE 4.21 Three-phase contactor.

and too much chattering and can lead to eventual pitting of the contacts and poor electrical connections. This can, in turn, lead to an overheating of the contact and a failure of the contactor to either close that particular phase, or, in a worse case, to fuse the contacts so that the equipment remains powered even when then contactor is de-energized. For this reason, the contactor usually has internal springs that help make and keep the connection once the solenoid is energized.

Solenoids are two-position electromagnetic actuators that are used to open small valves on refrigerant lines, steam lines, and gas lines. They work in much the same fashion as contactors: a spring-return shaft is forced open by the energizing of a small coil. However, instead of closing electrical contacts, the electromagnet moves a rod of metal that can be used to close and open valves.

WIRES

Wire ratings are used to determine the size and current-carrying capacity of the wire. If you use wire that is too small for the amount of current, the wire will get too hot and will be a fire hazard. There are usually markings on the side of the wire jacket (see Figure 4.22) that list information about the wire. In the United States, wire sizes are listed in American Wire Gauge with smaller numbers being larger wires as shown in Table 4.3. For large commercial HVAC systems you find wire sizes in 14, 12, 8, 6, 4, 3, 1, 0, 00, 000, and 0000 gauges. 14 gauge wire can carry between 25 and 35 amperes, depending on the insulation, while 0000 gauge wire can carry between 300 and 400 amperes.

The dampness ratings of cable are DRY, DAMP, and WET. In the first case, the cable is suitable for indoor applications above floor level where there is no chance the cable will get wet. DAMP rated wire should be used in locations such as interior locations near the floor, while WET rated cable should be used in all outdoor or buried wire runs.

Wire is also listed by a series of letters that describe the insulation, ampacity, and temperature ratings. Some of these ratings are listed in Tables 4.5 and 4.6. This allows you to identify some of the more common wiring types such as T, TW, THW, RHW, and RHH.

Residential HVAC systems are often wired with standard residential cable such as BX or ROMEX. When using these types of wiring, make sure to follow local residential codes. For example, the National Electric Code has a minimum of 12 gauge wire for residential use.

Certain wire color conventions are used in residential and commercial electrical systems. Follow these conventions when wiring single-phase HVAC systems to avoid confusion and to facilitate future troubleshooting. Typically the black wire is hot, the white wire is neutral, and the green wire is ground.

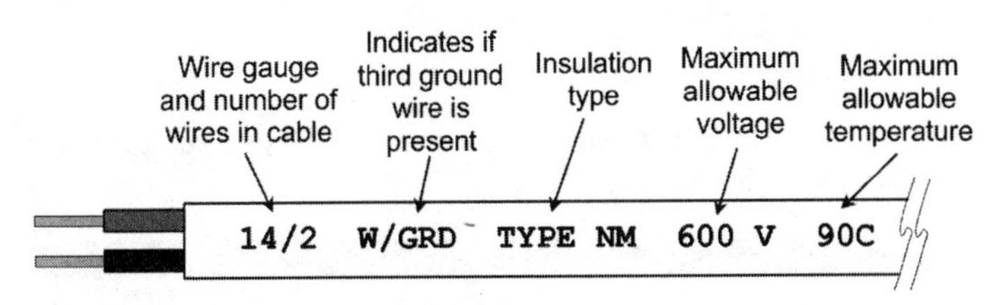

FIGURE 4.22 Example electrical cable and associated markings.

TABLE 4.3 Properties of Standard Copper Wire

Gauge, AWG	Ohms per 1000 feet	Diameter, inches	Weight, lbs/1000 feet
0000	0.05	0.460	641
000	0.07	0.410	508
00	0.09	0.365	403
0	0.11	0.325	319
1	0.14	0.289	253
2	0.17	0.258	201
3	0.22	0.229	159
4	0.27	0.204	126
5	0.34	0.182	100
6	0.43	0.162	79.5
7	0.55	0.144	63.0
8	0.69	0.128	50.0
9	0.87	0.114	39.6
10	1.1	0.102	31.4
11	1.4	0.091	24.9
12	1.7	0.081	19.8
13	2.2	0.072	15.7
14	2.8	0.064	12.4
15	3.5	0.057	9.86
16	4.4	0.051	7.82
17	5.6	0.045	6.20
18	7.0	0.040	4.92
19	8.8	0.036	3.90
20	11	0.032	3.09
21	14	0.0285	2.45
22	18	0.0253	1.94
23	22	0.0226	1.54
24	28	0.0201	1.22
25	36	0.0179	0.97

fastfacts

Never put BX or ROMEX wire inside a conduit. The wire can overheat and be a fire hazard.

TABLE 4.4 Prefixes Used in Wire Ratings

Prefix	Meaning
T	Dry only, maximum temperature of 140°F
CF	Fixture wire with cotton insulation and maximum temperature of 190°F
AF	Fixture wire with asbestos insulation and maximum temperature of 300°F
SF	Fixture wire with silicon insulation and maximum temperature of 390°F
R	Rubber covered
S	Extra hard service appliance cord
SJ	Lighter service appliance cord
SV	Light service appliance cord
SP	Lamp cord with rubber insulation
SPT	Lamp cord with plastic insulation
X	Moisture resistant synthetic polymer insulation
FEP	Fluorinated Ethylene Propylene insulation for dry service over 190°F

TABLE 4.5 Suffixes Used in Wire Ratings

Suffix	Meaning
B	Indicates outer braid of glass fiber or other material
H	Loaded current temperature rating up to 165°F
HH	Loaded current temperature rating up to 190°F
L	Lead jacket
N	Nylon or thermoplastic polyester jacket
O	Neoprene jacketed wire
W	Wet use type

In commercial HVAC applications, large current draws come from motors and compressors. Typical full load current from such devices can vary as shown in Table 4.6. These currents are given for loads at 460 VAC. If the device is running at 230 VAC then you need to double the values in the table. As the amount of current running through a wire increases, so does the temperature of the wire. For this reason the National Electric Code puts limits on how many wires can be run through a single conduit. These values are summarized in Tables 4.7 through 4.8 for standard conduit sizes.

TABLE 4.6 Full Load Current Draw From Motors óf Different Sizes

Motor HP	Full load amps at 460 V
1	1.7
2	3.0
3	4.5
5	7.5
10	14
20	26
30	39
40	51
50	63
75	90
100	123
150	180
200	240
300	348

TABLE 4.7 Maximum Allowable Number of Conductors in Conduit

AWG	Conduit size in inches								
	½	¾	1	1¼	1½	2	2½	3	3½
6	1	2	4	7	10	16	NR	NR	NR
4	1	1	3	5	7	12	17	NR	NR
2	1	1	2	4	5	9	13	NR	NR
1	NR	1	1	3	4	6	9	NR	NR
0	NR	1	1	2	3	5	8	12	NR
00	NR	1	1	1	3	5	7	10	NR
000	NR	1	1	1	2	4	6	9	12
0000	NR	NR	1	1	1	3	5	7	10

TABLE 4.8 Maximum Allowable Number of Conductors in Conduit (THHN Wire Type)

AWG	Conduit size in inches							
	½	¾	1	1¼	1½	2	2½	3
6	1	4	6	11	15	26	NR	NR
4	1	2	4	7	9	16	NR	NR
2	1	1	3	5	7	11	16	25
1	NR	1	1	3	5	8	12	18

fastfacts

Always check local codes for any further restrictions on types of wire and allowable conduit wire density that can be used in your application.

TABLE 4.9 Maximum Allowable Number of Conductors in Conduit (XHHN Wire Type)

AWG	Conduit size in inches							
	¾	1	1¼	1½	2	2½	3	3½
0	1	1	3	4	7	10	15	NR
00	1	1	2	3	6	8	13	NR
000	1	1	1	3	5	7	11	14
0000	1	1	1	2	4	6	9	12

POWER MEASUREMENT AND ELECTRIC RATES

Electrical meters are used to measure both the instantaneous power consumption of equipment as well as the total energy use over time. Instantaneous power is measured in watts (W) or kilowatts (kW) and is analogous to water flow rate: a watt meter measures how many electrons flow into a device every second, while a turbine meter measures how many gallons of water flow into a device every minute. Total energy is the total number of electrons that have entered a device over a period of time. This is measured in watt-hours (Wh) or kilowatt-hours (kWh). To use the water comparison again, an electric meter shows how many kWh a device has used over the course of a month, while a water meter shows the total number of gallons that a device has used.

Knowing your electricity use is important for two reasons. The first, of course, is that the utility charges you for the amount of electricity you use (and also for the peak rate at which you use it). The second reason to measure electricity use is to see how well your equipment is running. For example, if you measure the electricity consumption of a chiller and compare that to the actual cooling produced by the chiller, you can get a good idea as to whether or not the chiller is operating correctly.

fastfacts

Instantaneous power in watts is a rate. *Total energy in watt-hours is an* amount.

There are many ways to measure the power consumption of chillers, fans, and boilers, ranging from the very simple up to the very complex and expensive. One of the easiest ways to measure power consumption is to use current transducers (see Chapter 5) to determine the ampere draw of a device. Assuming that the voltage will not vary all that much, and also assuming that the legs of a multi-phase motor are pretty well balanced, you can use Tables 4.10 through 4.12 to find the power as a function of the current. This also works if you take spot measurements of the current using a clip-on ammeter. Note that to use the data in this table, it is necessary to know the power factor of the device you are working with. Unfortunately, it is not always easy to determine the power factor and you may have to purchase or rent a special meter for this.

The local electric utility charges for electricity use. The rate schedules used to calculate the bills can be complex. It is the goal of the building manager to choose rates that best fit the building energy use so that the operating costs are minimized. It is therefore important to understand the various components of utility rates since they directly affect the total monthly energy bill.

Unfortunately, there are thousands of electric and gas rates in the United States, each with its own structure and quirks. While many rates appear to be similar, the details of each can determine whether or not it is appropriate for a given building. For example, some utility

fastfacts

In the same way you can tell if there is a problem with your car by keeping track of the miles per gallon of gas, you can identify problems with HVAC equipment by tracking the electrical efficiency.

TABLE 4.10 Calculation of Power in Single Phase 120/240 Circuits

i = current and PF = power factor

Circuit Type	Power
1 phase, 2 wire, with neutral	i x 120 x PF
1 phase, 2 wire, no neutral	i x 240 x PF
1 phase, 3 wire, with neutral	I x 2 x 120 x PF

TABLE 4.11 Calculation of Power in Three Phase 120/208 Circuits

i = current and PF = power factor

Circuit Type	Power
1 phase, 2 wire, w/ neutral	i x 120 x PF
1 phase, 2 wire, no neutral	i x 208 x PF
1 phase, 3 wire, w/ neutral	I x 2 x 120 x PF
3 phase, 3 wire, Delta	i x 1.732 x 208 x PF
3 phase, 4 wire Wye	i x 1.732 x 208 x PF
3 phase, 4 wire Delta	i x 1.732 x 240 x PF

TABLE 4.12 Calculation of Power in Three Phase 277/480 Circuits

i = current and PF = power factor

Circuit Type	Power
1 phase, 2 wire, w/ neutral	i x 277 x PF
1 phase, 2 wire, no neutral	i x 480 x PF
1 phase, 3 wire, w/ neutral	i x 2 x 277 x PF
3 phase, 3 wire Delta	i x 1.732 x 480 x PF
3 phase, 4 wire Wye	i x 1.732 x 480 x PF

rates charge more for energy in the middle of the day than at night. While the *average* rate is low, the coincidence of the high energy prices with high daytime use can be very expensive. It may be better to find a rate that has a higher average price but does not penalize for the on-peak periods.

The overall use of electricity (in kWh) or gas (in Btu or therms) over a given interval of time is called the total *consumption*. The consumption charge can change depending on the total consumption up to the current time within a given billing period, or it can change depending on the time of day. The highest instantaneous use of electricity (in kW) at a given interval of time is called the *demand*. Like the consumption, the cost of demand can depend on the time of day.

Utility rates are made up of consumption, demand, surcharge, adjustment, meter charges, and tax components. In addition, there are riders and provisions that may affect the rate or even if the building or campus can use a particular rate. In addition, the various consumption and demand costs can change from season to season or based on whether or not it is a weekday, weekend or holiday.

Consumption

Fixed fee consumption charges are typical for most residential bills. These kinds of charges are very simple: the charge is the same throughout the billing period and throughout the day, regardless of the time of day or how much energy has already been used. For electricity rates this corresponds to a fixed fee per kWh consumed.

Block rates vary the cost of electricity depending on how much electricity is used. The first block has a given size and cost. When that block is "filled" you move on to the next block, and so on. Figure 4.23 shows an example of a declining block rate, where the more electricity you use, the lower the unit cost becomes. Many commercial electricity rates are block rates.

Time-of-use or *time-of-day* rates vary the cost of electricity at each hour of the day as illustrated in Figure 4.24. The highest rates are typically from noon through 3:00 PM, corresponding with high electricity use for cooling. Many commercial electricity rates incorporate some degree of time-of-use billing. The number of time-of-use periods can vary from two (on-peak and off-peak) up to five.

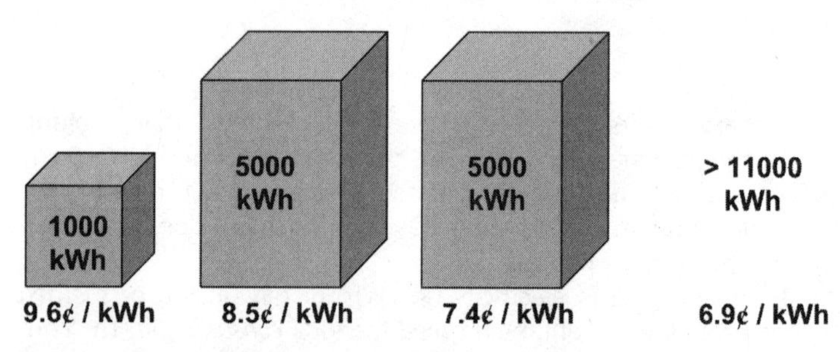

FIGURE 4.23 Example of declining block rate.

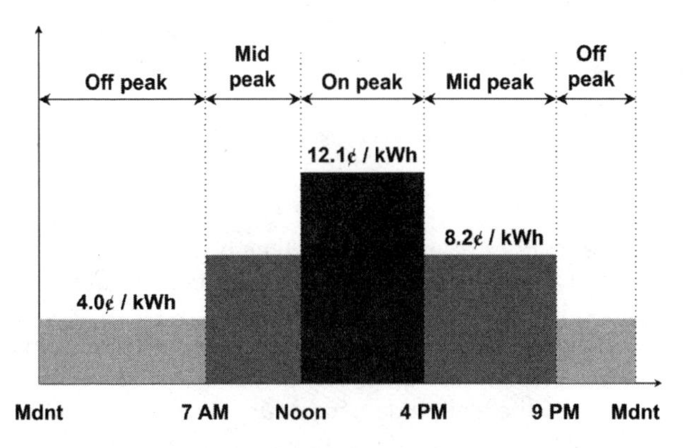

FIGURE 4.24 Example of time-of-use rate.

A large number of commercial and industrial electricity rates are combined block and time-of-use rates. For example, the on-peak period may have a block component so that there is block pricing structure that applies between certain hours of the day. Table 4.13 shows an example of a combined rate structure. Other adjustments on basic consumption charges include day type adjustments (where weekdays and holidays are metered at a different rate than weekdays) and seasonal adjustments where there are summer and winter differences, or four seasonal changes, or even cases where the rate structure changes each month.

Real-time pricing is a relatively new method available to large customers for electricity. In real-time pricing, the electricity rate changes every hour of the day depending on the utility's cost for producing the electricity. While the cost is generally very inexpensive most of the time, there are about 40 to 80 hours per year when

TABLE 4.13 Example Combined Time-of-Use and Block Electricity

Period	Cost	Comments
On-peak	9.362¢ for first 2500 kWh 12.48¢ for next 5000 kWh 15.66¢ for anything above 7500 kWh	On peak is 9:30 AM - 4:30 PM Monday through Friday except holidays.
Off-peak	5.763¢ for first 1700 kWh 6.311¢ for next 1300 kWh 8.210¢ for anything above 3000 kWh	All other periods are off-peak.

the cost can increase by a factor of ten or twenty. In these kinds of price structures, there is typically only a consumption charge with no demand charge.

Demand

The demand charges have block and time-of-use rates very similar to those used in consumption charges. The demand over a given billing period is usually set by the highest rate of energy consumption during any 15 minute period, although sometimes 30 minute or 60 minute intervals are used. The demand charges for a billing period are set by the highest demand *at any point* during that billing period. For example, Figure 4.25 shows the hourly demand values for a large office building where the monthly demand is set on the first Monday. Even though the rest of the month has a lower demand value, the monthly energy bill uses that peak demand for the calculation of the demand charge. Some energy rates have both on-peak and off-peak demand charges.

In some cases, the demand charges are *ratcheted* from month to month. This means that the peak demand charge can carry over from one month to the next, and sometimes even over the next year. There are many variations on this, such as having the demand charge for a given month be the greater of either the measured demand for that month or some percentage of the peak demand for any of the previous 11 months.

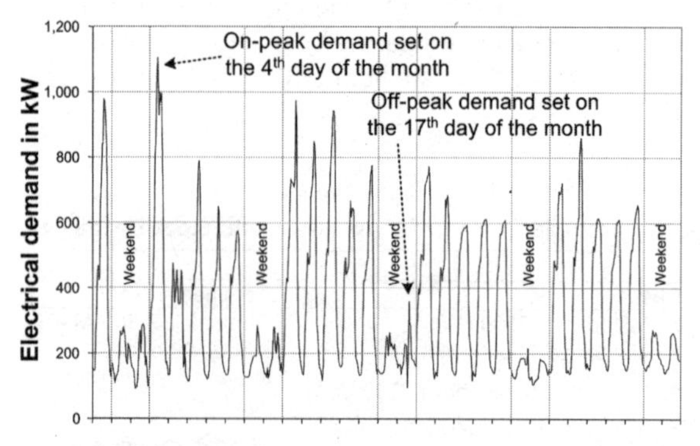

FIGURE 4.25 Example of monthly demand values and demand charge values.

For the HVAC engineer, an understanding of the demand rate used for a given building can play a big role in deciding what kind of energy conservation measures should be used. The cost of electricity demand can be almost equal to the cost of electricity consumption, and while it is difficult to reduce total consumption, it is somewhat easier to reduce the peak demand. For this reason, many energy saving techniques refer to *peak-shaving*, which tries to limit the peak demand for on-peak periods. Many larger chillers (see Chapter 8) have built-in controls that allow the user to set an upper limit on the kW use by the chiller compressor. Other pieces of HVAC equipment, such as thermal energy storage systems (see Chapter 10) are meant exclusively to perform peak shaving.

Cost Adjustments and Other Charges

On top of the consumption and demand charges, the utilities often apply cost adjustments that reflect their costs of producing or transporting the energy from the power plant. For electricity bills this is usually listed as the *energy cost adjustment* and is given in as additional cost per kWh or as a percentage of the total consumption and demand bill components. The cost adjustments can be positive or negative, depending on whether or not the cost to the utility for providing the service is more or less than expected.

Surcharges are additional fees applied to your bill that reflect charges applied for a variety of reasons, many of which are not directly related to your energy use. For example, many states impose an energy assistance surcharge that is used for a fund that helps subsidize low-income residential energy bills. Other surcharges come from exploration fees (often seen on your bill as "energy procurement") and energy transport fees ("interstate transition"). The surcharges can be a fixed fee per month, a function of total consumption, or a percent of the total bill.

Meter charges are seen in just about every energy bill, both gas and electric. The meter charges reflect the cost of installing and reading the meter every month. These charges can vary from several dollars per month for residential meters up to several hundred dollars per month for commercial and industrial customers. Generally, the more complex your rate, the more expensive the meter and the higher your monthly meter charge. For example, if you have a fixed consumption fee then the electrical meter can be relatively simple and cheap. But if you are on a time-of-use rate then the meter must be able to account for consumption at different times of the day and will be a more expensive meter.

Local, county, state, and federal *taxes* are all applied to various bill components. The total taxes can add up to anywhere from nothing to about ten percent of total bill. Finally, *riders* and *special provisions* are rules applied to each rate that dictate the use of the rate by a certain customer, or can provide incentives or added costs depending on the type of customer or the maximum load.

MEASUREMENT AND CONTROL

Proper HVAC control is essential to maintaining occupant comfort and ensuring the safety of the equipment and personnel. Too often, however, a control and monitoring system is installed, commissioned, and promptly forgotten. And yet the sensors provide the crucial information about how the system is behaving, so if the sensors are allowed to fall out of calibration or the control system is not periodically tuned, the results can require more work that it would normally take to keep the system in good order.

This chapter introduces the basic concepts of HVAC sensors and controllers. Keep in mind that every building has its own unique set of control needs and peculiarities. The information here to meant to help you design and debug control systems, but not how to address the specific control problems of each and every control loop.

TEMPERATURE SENSORS

The temperature of air and water streams is measured in any number of ways. Pneumatic sensors rely on the expansion of a refrigerant (toluene) within a capillary or bulb to drive a bellows that, in turn, creates a control action. Almost all electronic temperature sensors use thermocouples, thermistors, or RTDs. The output of these sensors is often amplified or otherwise modified to provide a more meaningful signal to the control system.

Pneumatic Temperature Sensors

Many buildings use bulb and capillary sensors for transmitting temperature signals to a control system. The bulb (see Figure 5.1) contains a refrigerant that expands or contracts depending on the temperature. Since the bulb and capillary are in a closed system, this results in a change of the pressure in the capillary. The capillary is connected to a diaphragm that either activates a snap-action switch within a thermostat (for on/off control) or moves the center tap of a slide-wire resistor (for modulating control). In on/off control, the thermostat can be set to close or open the switch at a given temperature. Usually there is some differential (*hysteresis*) built into the switch so that it does not open and close quickly when the bulb temperature is near the setpoint temperature. When used for modulating control, the circuit operates similar to a Wheatstone bridge with the motor balance relay trying to match the slider position at the thermostat.

These kinds of sensors have many advantages, including that they do not require an electric power source to operate. They are also quite reliable and are used in practically all commercial HVAC systems. They are particularly useful for freeze protection control on a coil, since the bulb can be of any size. Figure 5.2 shows an averaging bulb in the mixed air plenum of an air handling unit. Note that the bulb extends from one side of the coil to another as well as from the top to the bottom of the duct. If any point on the coil is below freezing, the bulb pressure reflects this even if there are warmer sections of the coil.

One of the main disadvantages of bulb and capillary systems is that they are difficult to fix if the bulb or capillary spring a leak. If this occurs you may have to throw away the entire thermostat and replace it with another. It is also easy to kink the capillary and either create a leak or prevent the bulb from working altogether.

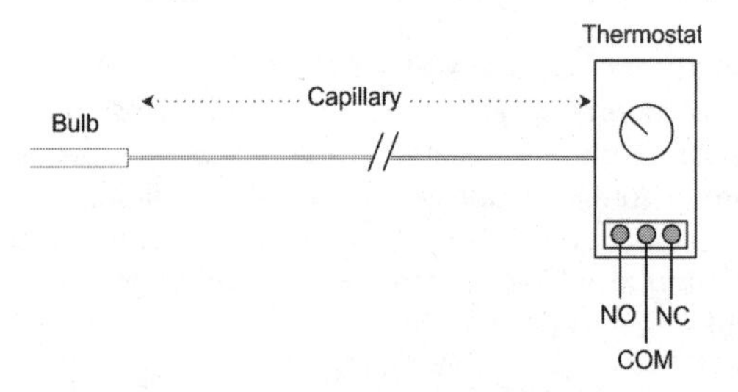

FIGURE 5.1 Bulb and capillary thermostat.

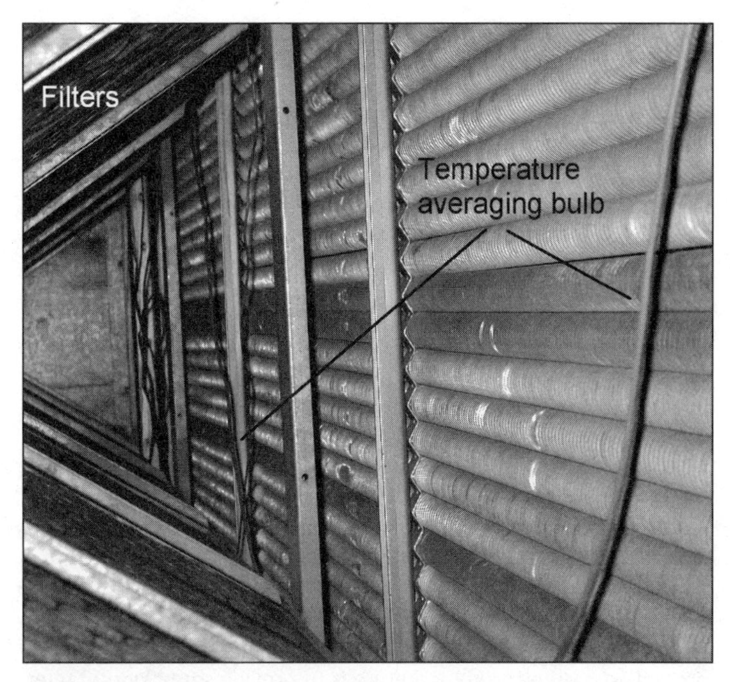

FIGURE 5.2 Mixed air averaging bulb.

Most zone thermostats use some form of coiled bimetallic strip to either swivel a mercury filled bulb to make contacts on an electric thermostat or to change the pressure control bellows on a pneumatic thermostat. In the electric version of the coiled bimetal strip sensors, the signal provided is a simple on/off as with the bulb and capillary. These kinds of thermostats are common for residential furnace control and baseboard heating control. The pneumatic version of these sensors can be used to provide a proportional signal for modulating control valves.

Electric Temperature Sensors

Two different metals in contact produce a small voltage corresponding to the temperature of the junction. The junction is called a *thermocouple*. By choosing the right metals it is possible to create a low-cost temperature sensor. There are many different types of commercially available thermocouple wire that can be used for specific temperature ranges and desired accuracy. Thermocouples are inexpensive, rugged, and pretty easy to use since they provide their own voltage. However, this voltage must be referenced to a fixed temperature source as well as amplified because it is quite small (in the microvolt range). In addition, a thermocouple signal may not be linear in the range you are measuring and may be accurate only to within several degrees—both disadvantages in situations where tight control is critical.

Table 5.1 shows the ranges and normal errors of several different types of thermocouples. The sensitivities are graphed in Figure 5.3. Note that all of the responses are referenced to the freezing point of water at 32°F.

Care must be used when measuring the resulting voltage from a thermocouple. For example, consider a type T thermocouple that produces a voltage proportional to the temperature at the junction of the two metals as shown in Figure 5.4a. If the voltmeter has copper terminals, it is not possible to measure the thermocouple voltage directly because the connection to the voltmeter will also be a thermocouple. In this case, the voltage measured by the voltmeter corresponds to the difference between the thermocouple junction temperature and the voltmeter terminal temperature. The way to properly measure this temperature is to provide some kind of reference temperature as shown in Figure 5.4b. Here an ice bath is used to establish a known temperature at the second thermocouple junction, thus allowing the desired temperature to be measured at the voltmeter. An ice bath was traditionally the standard reference point for thermocouples; that's why the response curves in Figure 5.3 all

TABLE 5.1 Characteristics of different thermocouples. *(From Omega Engineering, 2000)*

From Omega Engineering, 2000

Type	Seebeck Coefficient		Std. Error	Range
	μV / °F	ref. °F	°F	°F
B	3.3	1112	8 – 15	32 to 3300
E	32.5	32	3 – 8	-454 to 1800
J	27.9	32	2 – 5	-346 to 1400
K	21.9	32	2 – 5	-454 to 2500
R	6.4	1112	3 – 7	-58 to 3200
S	5.7	1112	3 – 7	-58 to 3200
T	21.1	32	1 – 5	-454 to 750

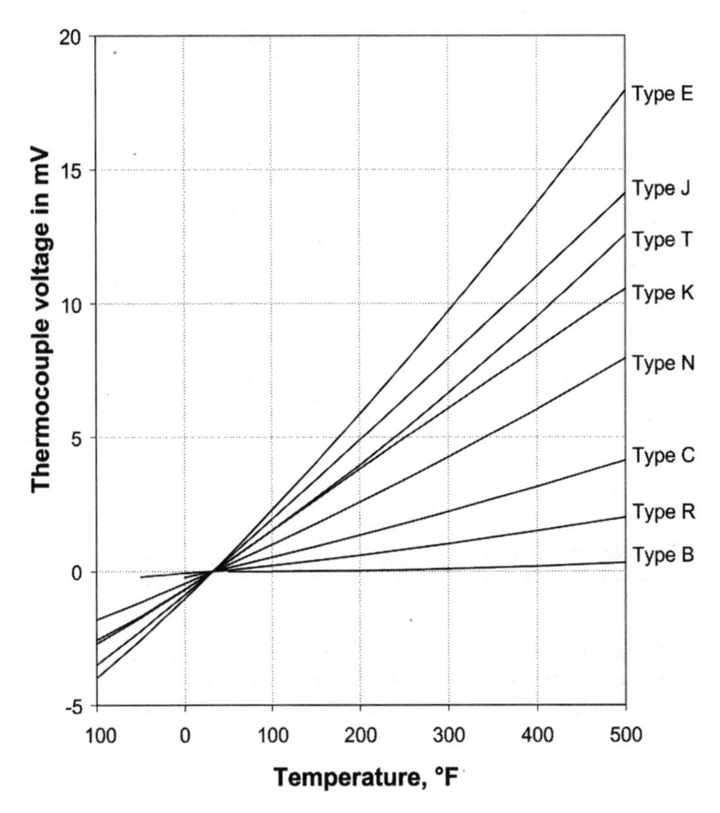

FIGURE 5.3 Response curves of different thermocouples.

go through 32°F. However, if the temperature of the voltmeter terminal is known then a reference temperature is not necessary and the voltmeter reading can be adjusted accordingly.

Thermocouples can be used to measure the temperature *difference* between two points. Any even number of thermocouples can be arranged to produce a *thermopile* that can be measured with a standard voltmeter. Figure 5.5a shows a basic type T thermopile. Since the same metal is in contact with the voltmeter terminals, no reference temperature is required. Multiple thermopiles can be used (Figure 5.5b) for averaging purposes or to amplify the signal of a point measurement.

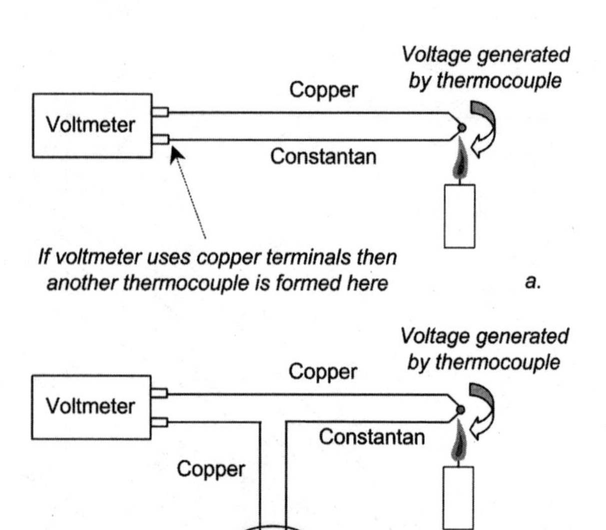

FIGURE 5.4 Measuring thermocouple voltage; (a) without reference temperature; (b) with reference temperature.

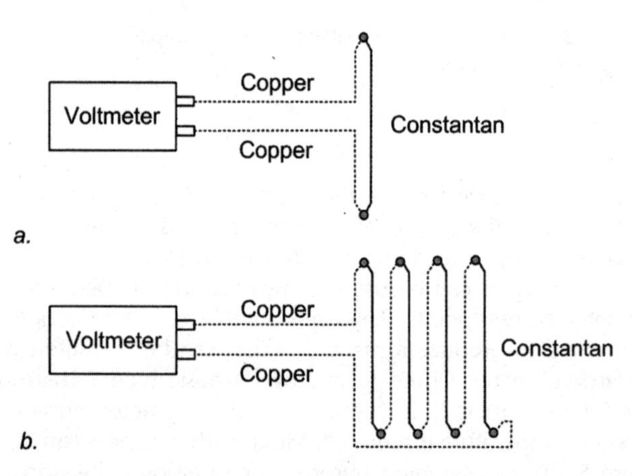

FIGURE 5.5 Thermopiles; (a) basic thermopile; (b) multiple point thermopile.

A *resistance temperature detector* (RTD) is a thin piece of metal with a precise and repeatable relationship between the temperature and resistance. Platinum and Balco (40% nickel, 60% iron) are two commonly used RTD metals. The temperature of an RTD is found from:

$$°C = (\Omega_{RTD} - \Omega_{ref}) / (100 \times \alpha)$$

where α is the average slope of the Ω_{RTD} versus temperature line between 0 and 100°C. Platinum RTDs have values of $\alpha = 0.00385$ or $\alpha = 0.00392$ for a Ω_{ref} of 100Ω at 0°C. Balco RTDs have values of $\alpha = 0.0041$ for a Ω_{ref} of 100Ω at 0°C. Figure 5.6 shows a comparison of resistance versus temperature curves for platinum RTDs at the two different sensitivities.

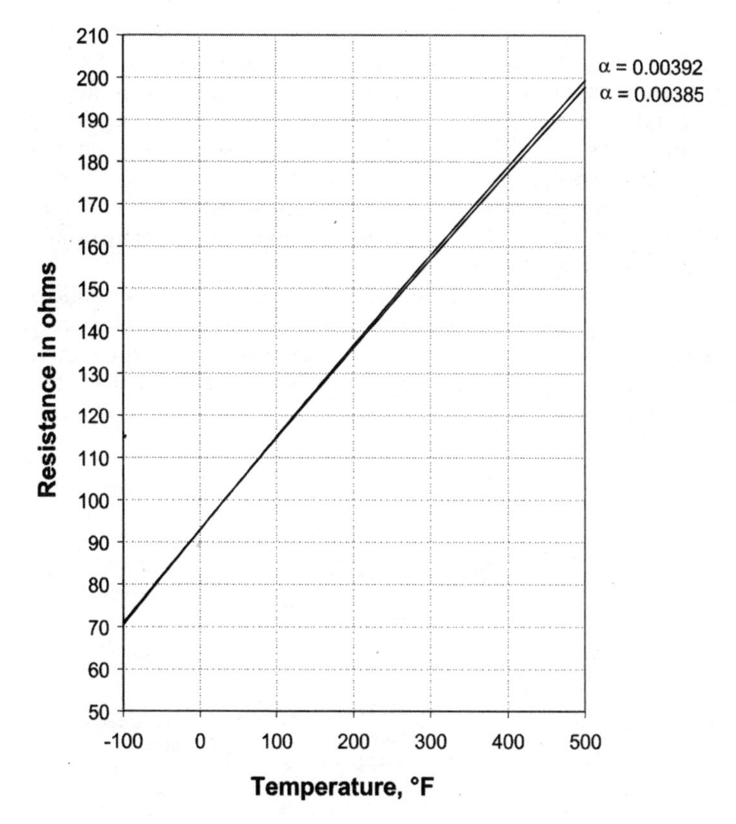

FIGURE 5.6 Response curves for platinum RTDs.

RTDs are quite accurate, with most manufacturers claiming error ranges of less than a quarter of a degree. They also seem to last a long time without requiring recalibration and are linear over a very wide range. However, they are relatively expensive, running about $50 per point just for the sensor. Also, since their resistance values are fairly low, care must be taken when making the measurements. For example, RTD sensors have nominal resistance values of 100 to 150 ohms under normal conditions experienced in HVAC systems. Standard 18 gauge wire has a resistance of about 0.67 ohms per 100 feet. It is therefore necessary to use 3 or 4-wire resistance measurements to avoid introducing too much error into the resulting signal when measuring RTD resistance.

In 3 wire or 4-wire measurements, a power source is used to send a small current (typically around 1 mA) through the resistor and the resulting voltage is measured as shown in Figure 5.7. The current flows only along the paths marked by the dotted line. If the current is accurately controlled, then the resistance across the sensor can be found through Ohm's law (from Chapter 4). It is important not to allow current to flow through the sensor continuously as this leads to heating of the resistor and can cause erroneous readings. If you do not have the equipment to make these kinds of measurements, adapters are available that convert a 2, 3, or 4-wire RTD reading into a 4–20 mA signal that can be read directly by any data acquisition system.

The RTD sensor can be a single point or a long wire. A standard RTD comes protected in a metal sheath that can be used for point measurements, such as to take water temperatures using a thermowell in a pipe as shown in Figure 5.8. Note that in this image the surrounding insulation has been removed to show the RTD body.

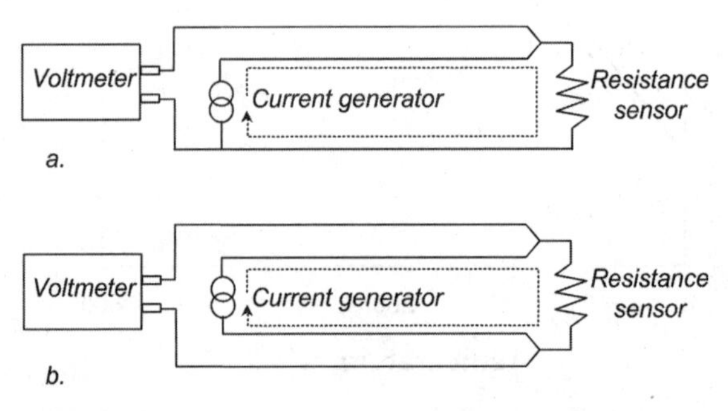

FIGURE 5.7 Three and four-wire resistance measurements.

FIGURE 5.8 RTD in thermowell.

Always put the thermowell in a piping tee or reducing elbow so that at least 90% of the bulb or sensor is in contact with the fluid (see, for example, Figure 5.9). Also make sure to use a thermal conducting grease in the well so that the sensor is in good thermal contact with the well.

It is also possible to use long RTDs to take average temperature readings across the face of a duct or coil. Figure 5.10 shows a 24 foot long RTD that has been coiled to provide a good indication of what the average duct temperature is. Compare this picture to the bulb and capillary freeze protection sensor in Figure 5.2. The main difference between these two is that the RTD provides an average signal while the bulb and capillary responds to the lowest temperature at the coil. They are two different sensors with two very different functions.

Thermistors are small semiconductors whose resistance changes as a function of temperature. As the temperature increases, the resistance decreases as shown in Figure 5.11. The main advantages of thermistors is that they are small, inexpensive, and respond very quickly to changes in temperatures. They also have a relative high resistance so the lead wire resistance introduces very little error as

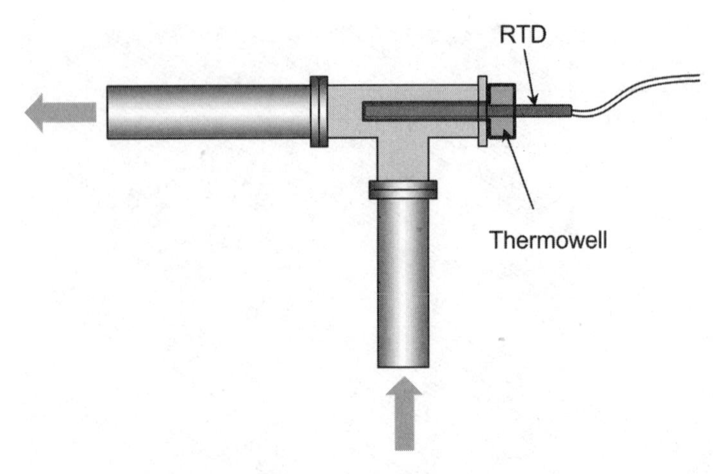

FIGURE 5.9 Placement of RTD in piping tee.

FIGURE 5.10 Duct averaging RTD sensor.

compared to an RTD. However, these kinds of sensors are very nonlinear and have a limited range. The temperature of a thermistor is found from the relationship:

$$1/T = A + B \times \ln(\Omega_{thermistor}) + C[\ln(\Omega_{thermistor})]^3$$

where T is the temperature in Kelvins, $\Omega_{thermistor}$ is the resistance of the thermistor, and A, B, and C are curve fitting constants. Thermistors come in a variety of nominal resistances at 77°F: 2.252 kΩ, 3 kΩ, 5 kΩ, 10 kΩ and 30 kΩ.

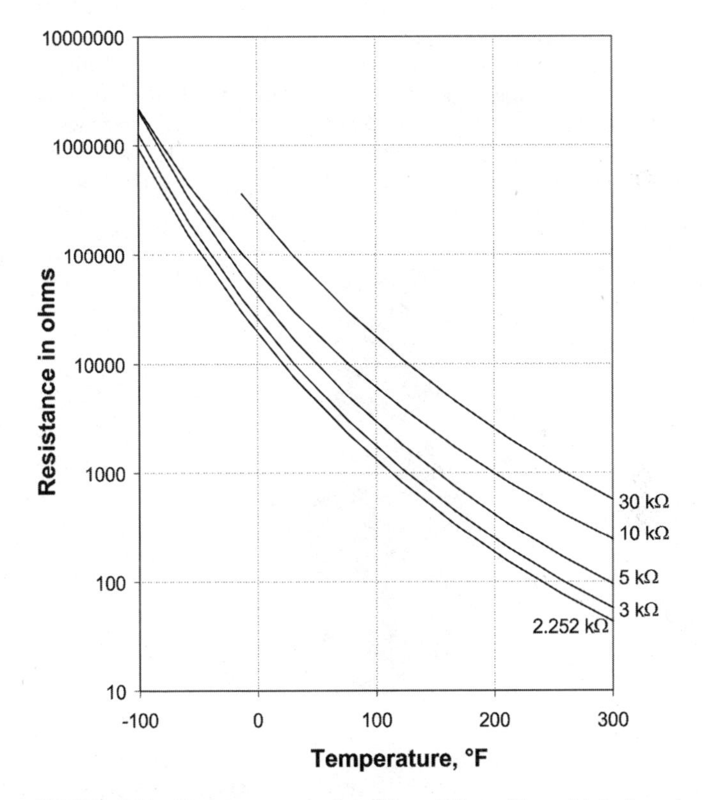

FIGURE 5.11 Response curves for different thermistors. (Note logarithmic scale on y-axis)

HUMIDITY SENSORS

Some traditional ways of measuring moist air properties have included using materials such as wood or natural fibers that change shape or length depending on the ambient humidity. While some of these kinds of sensors are still in use today, most humidity measurements are made using an electronic device such as a thin-film resistor (see Figure 5.12) in which the resistance changes as a function of the humidity. Such devices use alternating current to measure the electrical properties of the thin film. A variation of this kind of sensor is the thin-film capacitor where the capacitance changes as a function of the humidity. In both cases, the response of the sensor is relative slow because it takes time for the water to migrate into and out of the sensor body. Also, care must be taken not to saturate the sensor with very humid air for too long since this can introduce permanent offsets into the sensor calibration.

These kinds of sensors are pretty expensive—around $100 per point—and have an accuracy of about ±5 percent relative humidity. In theory, they should be replaced every year or so, but in practice you will find humidity sensors that have been in place for years without maintenance. Fortunately, most HVAC applications have only a

FIGURE 5.12 Thin-film resistor used for measuring humidity. (From Kreider et al., with permission)

fastfacts

Always use the proper excitation voltage type for a given humidity sensor according to the manufacturer's instructions. Many thin film humidity sensors want to see AC current. Use of direct current can permanently damage the sensor.

few points where humidity measurements are necessary so the maintenance costs can be kept relatively low.

Chilled mirror sensors use reflecting light to determine the dew point temperature of an airstream. A beam of light is bounced off an electrically-cooled mirror as shown in Figure 5.13 while an airstream passes over the mirror. The mirror is then cooled just until the point where condensation begins to form. The resulting water on the mirror scatters the beam of light so that the sensor does not pick it up. In theory, these kinds of sensors are extremely accurate provided that they are kept clean and within a specified temperature range. In practice, it seems like chilled mirrors are only appropriate for a few HVAC applications such as clean rooms and hospitals. The mirrors get dirty very easily and can be a big maintenance headache. In addition, chilled mirrors that experience outside air can rapidly become useless due to dust, pollen, and any other material that could get caught in the condensed water on the mirror. Given the cost of chilled mirrors (over $1,000 per point), they should only be used when absolutely necessary.

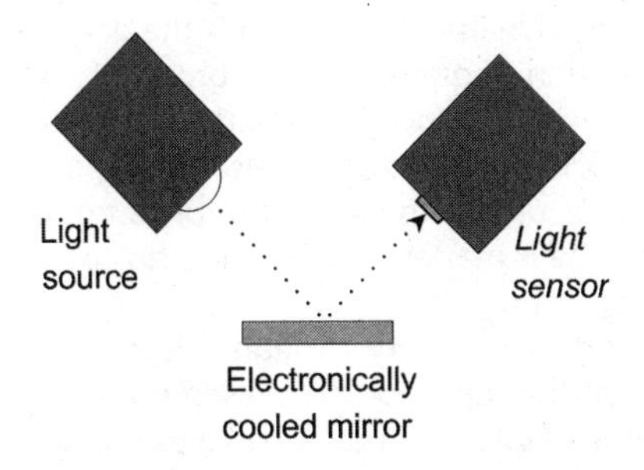

FIGURE 5.13 Chilled mirror.

fastfacts

A chilled mirror sensor was installed in the mixed air plenum of a supermarket where humidity control was very important. After just a few months the sensors developed very large errors. Unfortunately, the outdoor air inlet louvers were located near the loading dock and diesel particles from idling trucks had condensed on the mirror surface, ruining the sensors.

AIR FLOW SENSORS

Knowing the air flow rate in ductwork is important for ensuring that all zones are receiving enough air and that the fans are operating correctly. Flow rates are measured both in main supply ducts as well as at the entrance to zone terminal boxes.

One of the simplest ways to measure airflow is with a pitot tube. This is basically a tube within a tube (see Figure 5.14) that has a tap at the front to measure the total pressure of the moving airstream and taps on the side that measure just the duct static pressure. The difference between the two is taken to get just the pressure from the moving airstream. This is called the *dynamic* pressure and is related to the air velocity by

$$v = \sqrt{\frac{2 \cdot P_d}{\rho} g_c}$$

where P_d is the dynamic pressure, ρ is the density of the air, and g_c is the acceleration of gravity (32.17 lbm·ft/lbf·s^2). For hand measurements of the duct velocity, a liquid manometer like that illustrated in Figure 5.14 is easy to use. These are sloped tubes filled with water, mercury, or oil that measure the pressure differential across the ports. The tube is marked to give the pressure difference in any number of units.

The profile of the air velocity through a duct is usually not uniform across the face of the duct. For this reason it is better to take multiple readings of the airflow in a duct and average the result. Split the duct up into equal area sections as shown in Figure 5.15 and take the

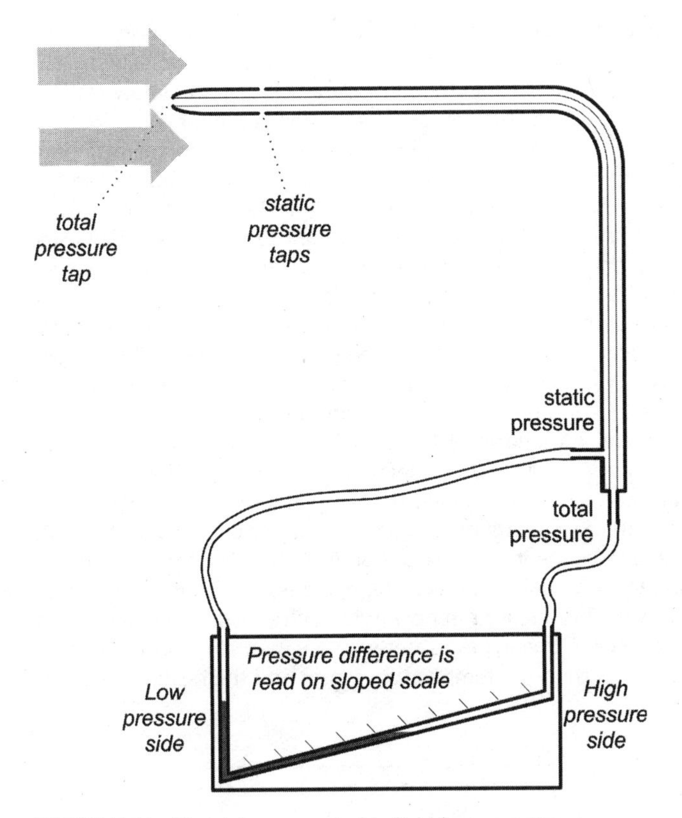

FIGURE 5.14 Pitot tube connected to liquid manometer.

fastfacts

When using a liquid manometer make sure that you connect BOTH pressure taps before inserting the pitot tube into the duct. The total pressure (high side) usually has enough pressure to raise the liquid level past the overflow of the low pressure side. If this is a water or oil manometer you will have a mess; if it is a mercury manometer you will have a hazardous waste cleanup.

average of the readings found at each point. If it is necessary to have a constant measurement of the air flow, consider using an airflow station such as that diagrammed in Figure 5.16. In an airflow station, multiple measurements are taken with modified pitot tubes that are basically just sections of copper pipe with holes in them (Figure 5.17). Since the signal will be noisy if there is any turbulence in the air stream, flow straighteners are usually included just upstream of the pitot arrays. In addition, it is important to situate an airflow station in a long, straight region of duct, several duct diameters downstream of any kind of obstruction or elbow in the duct.

Another method used for measuring air flow rates is a *hot wire anemometer*, sometimes called a thermal anemometer. These types of sensors can have several different working configurations. One is illustrated in Figure 5.18 where the current passing through a heating element is varied in order to maintain a constant temperature (usually around 200°F) at a downstream thermistor. The amount of heating necessary to maintain the temperature is a function of the airflow rate. By measuring the amount of current passing through the heating element, the velocity of air passing by the heating element can be calculated. Since the response of the thermistor also depends on the actual air temperature, it is necessary to measure this value as well. Other types of hot-wire anemometers have the thermistor act as both the temperature sensor and the heating element.

a.

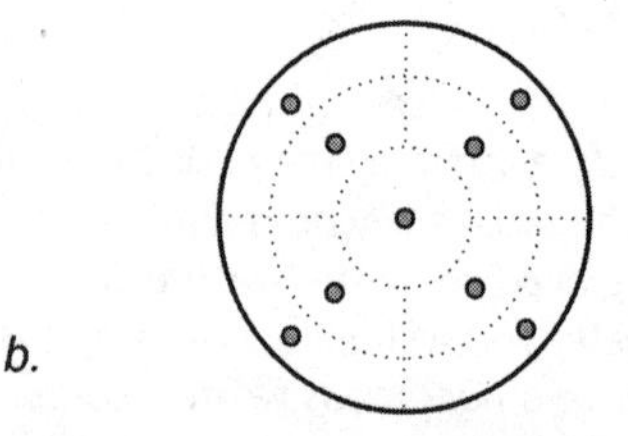

b.

FIGURE 5.15 Equal area regions for rectangular (a) and round ducts (b).

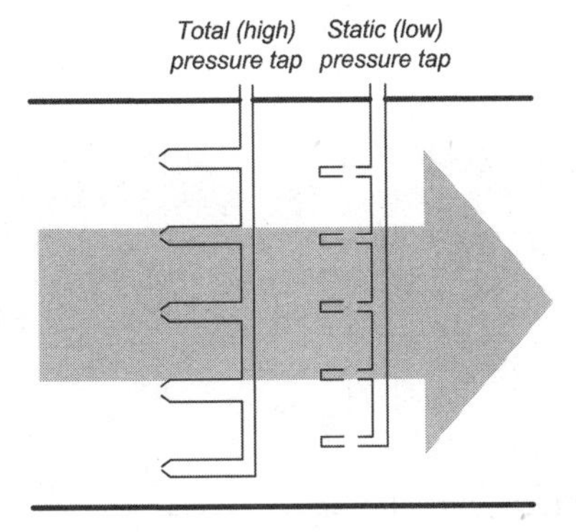

FIGURE 5.16 Cross section of an airflow station.

FIGURE 5.17 View of airflow station from downstream side.

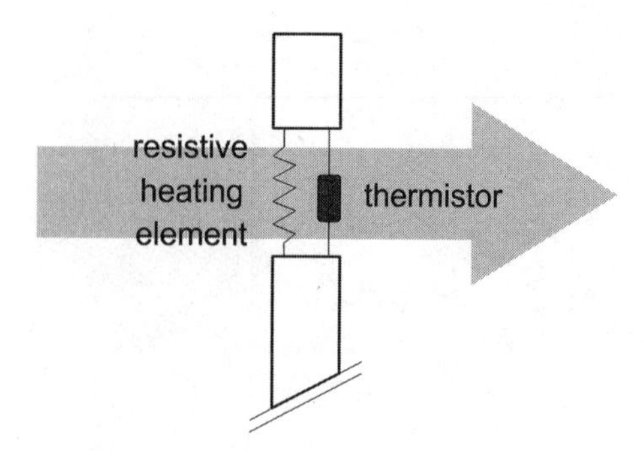

FIGURE 5.18 Hot-wire anemometer.

Airflow stations similar to those using pitot tubes are made from hot-wire anemometer arrays, but these tend to be quite expensive and should be used only when necessary.

WATER FLOW SENSORS

Measuring water flow is important for calculating equipment efficiencies and for the control of chiller plants. As with air flow measurements, there are many different ways to measure water flow.

Flow switches are used to confirm flow once a pump has started and are often in series with chiller and condenser start-up circuits. A flow switch is a simple paddle (Figure 5.19) installed into a pipe that makes or breaks a circuit depending on whether or not there is flow. Pneumatic flow switches are also available. These sensors are spring loaded and adjustable so that you can preset the amount of water flow rate that causes the switch to close.

fastfacts

Single element hot-wire anemometers are insensitive to the direction of the air stream. In some cases these sensors are used to measure supply air flow rates into a zone and cannot determine if negative duct pressure is actually drawing air out of the zone.

FIGURE 5.19 Flow switch.

Orifice plates (Figure 5.20) and *venturi meters* (Figure 5.21) are constrictions in the pipe that create pressure drops in the fluid stream. The pressure drops are correlated to the flow rate given the fluid's density and kinematic viscosity. Orifice plates create higher system pressure drops than venturi meters, but they are much less expensive. These devices are permanently installed in the piping system and are connected to pneumatic or electric transmitters (Figure 5.22) that send the flow signal back to the control system.

Pressure-driven flow meters like venturis and orifice plates should not be used as flow switches to start chillers or condensers. Low flow rates, dirt in the line, or failure of the electronics can cause the sensor to indicate flow when it does not exist. Always use a hard-wired mechanical flow switch for start-up or safety purposes.

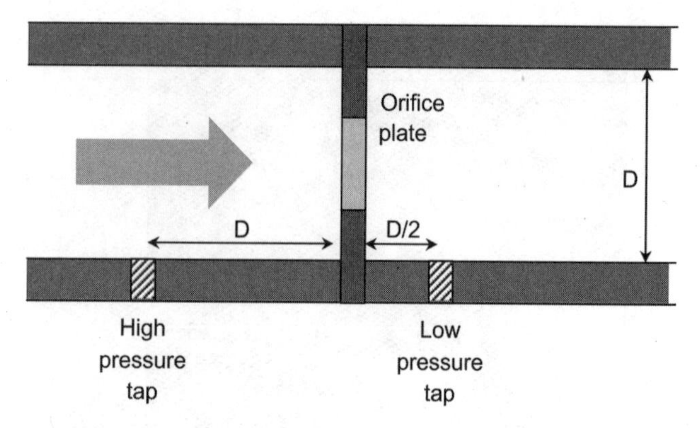

FIGURE 5.20 Cross section of orifice plate flow meter.

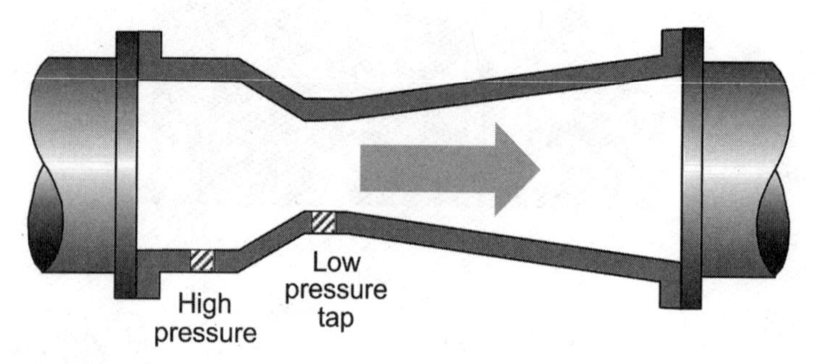

FIGURE 5.21 Cross section of venturi flow meter.

A *turbine meter* (Figure 5.23) is a volumetric flow meter that consists of a propeller with a magnet attached to the shaft. The propeller makes a certain number of turns per volume of fluid that passes through the blades and a magnetic pick-up on the outside of the pipe counts the number of times the propeller turns over a given time interval. The turns/volume ratio is called the k-factor. The k-factor should remain constant, but can tend to drop off at low flow values due to friction in the turbine bearings. Turbine meters are very expensive and require annual maintenance to replace the bearings, so they should only be used if very accurate flow measurements are needed.

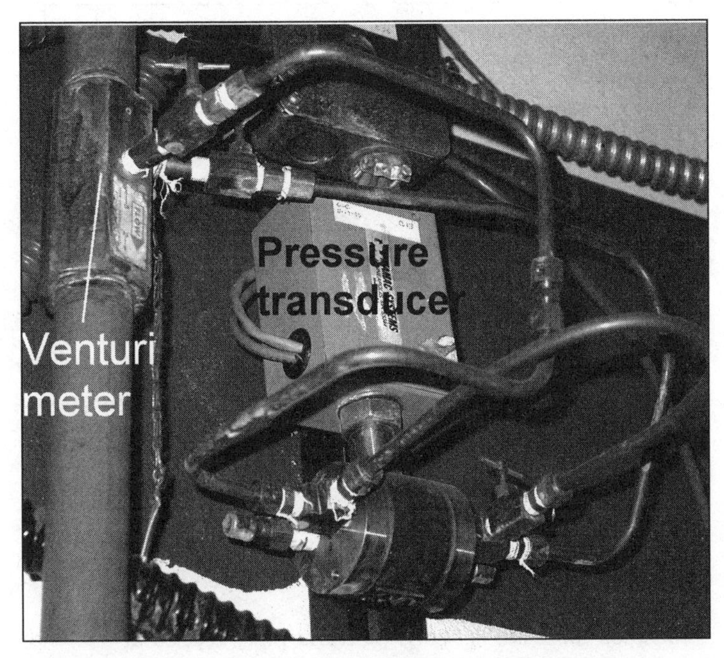

FIGURE 5.22 Venturi meter with attached pressure transducer.

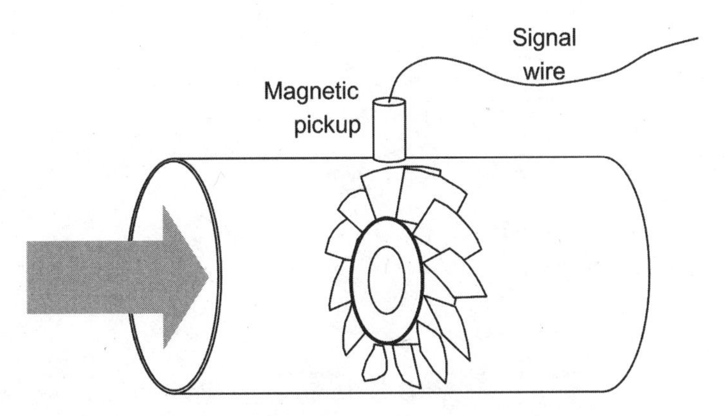

FIGURE 5.23 Cut-away view of turbine flow meter.

fastfacts

For both orifice plate and venturi meters, the low pressure tap is downstream of the high pressure tap.

PRESSURE SENSORS

Pressure is cited in either *gauge* (sometimes *gage*) or *absolute* pressure. Gauge pressure refers to the pressure above the ambient atmospheric pressure, while absolute pressure uses a complete vacuum as the zero reference.

A *bellows* sensor (Figure 5.24) uses a flexible coupling to amplify changes in pressure and translate the pressure change into a physical motion. The opposite end of the bellows is attached to some kind of armature that performs an action depending on the displacement of the bellows.

Bourdon tubes are flattened pieces of pipe, capped at one end, that flex slightly when a pressure is applied to the open end (see Figure 5.25). This motion is then translated into an actuator motion or dial adjustment through an armature connected to the pipe. Bourdon tubes are used in a large number of dials and gauges and can provide a signal to pneumatic actuators.

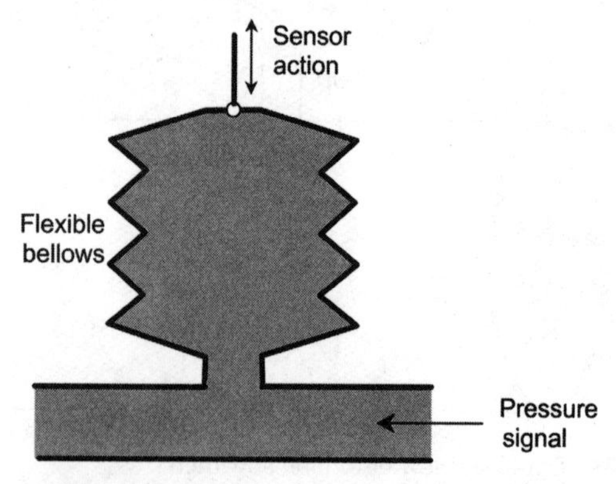

FIGURE 5.24 Bellows pressure sensor.

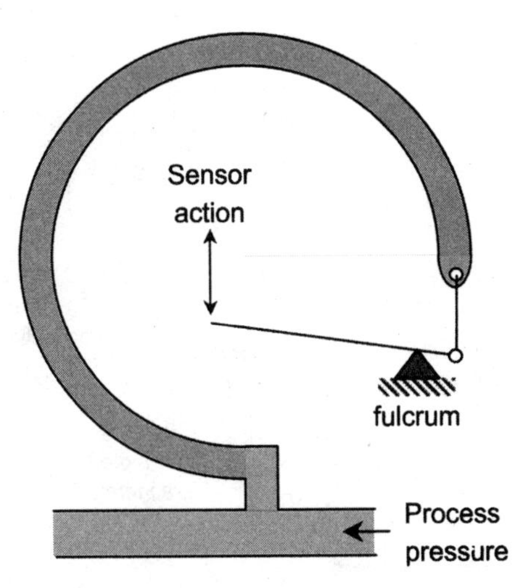

FIGURE 5.25 Bourdon tube pressure sensor.

There are also electronic pressure sensors that use several different methods for determining the pressure applied to the sensor. One method uses resistive meshes consisting of grids of thin wires as shown in Figure 5.26. As the mesh is flexed by a pressure difference, the resistance across the wires changes. This resistance can be measured by passing a small, constant current through the mesh and then measuring the resulting voltage using a Wheatstone bridge as discussed in Chapter 4. A similar kind of sensor uses *piezoelectrics* to measure pressure. The piezoelectric effect is seen in some crystals that produce a small current when stressed. This current can be measured and used as the basis for determining the pressure on the crystal. In pressure transducers, the crystal is a thin wafer that spreads out over a fairly large area. As the crystal bends, it produces a current that is then amplified and made available at the sensor outlet terminals.

POWER SENSORS

Electrical current is measured with current transducers (sometimes called "current donuts"). This kind of transducer is simply a long, continuous winding of wire that uses the induced magnetic field from current flow in a power line to generate a proportional measurement signal.

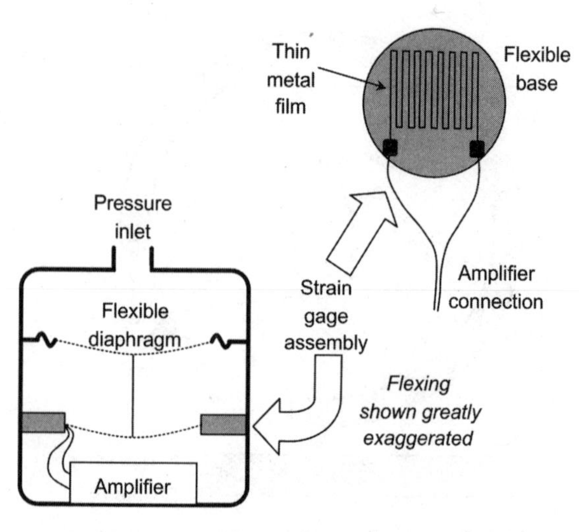

FIGURE 5.26 Schematic of resistance-type pressure sensor.

The current transducers can be solid-core (for permanent installations) or split-core (for temporary installations). Solid-core transducers (Figure 5.28) must be installed when the wiring is installed; otherwise the wiring must be disconnected in order to slide the transducer over the end. Split-core transducers (Figure 5.29) can be added or removed as desired.

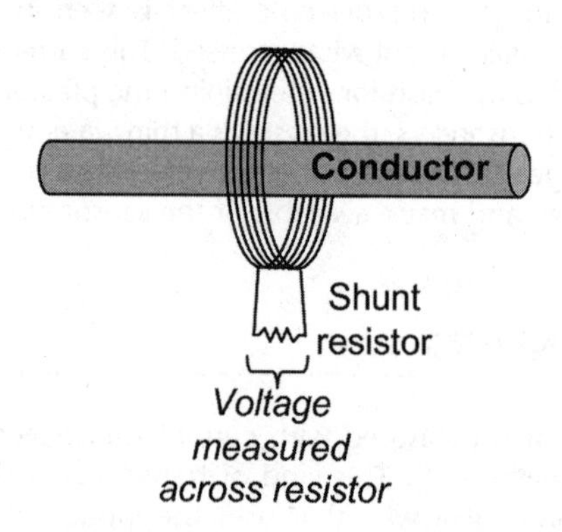

FIGURE 5.27 Current "doughnut" around conductor.

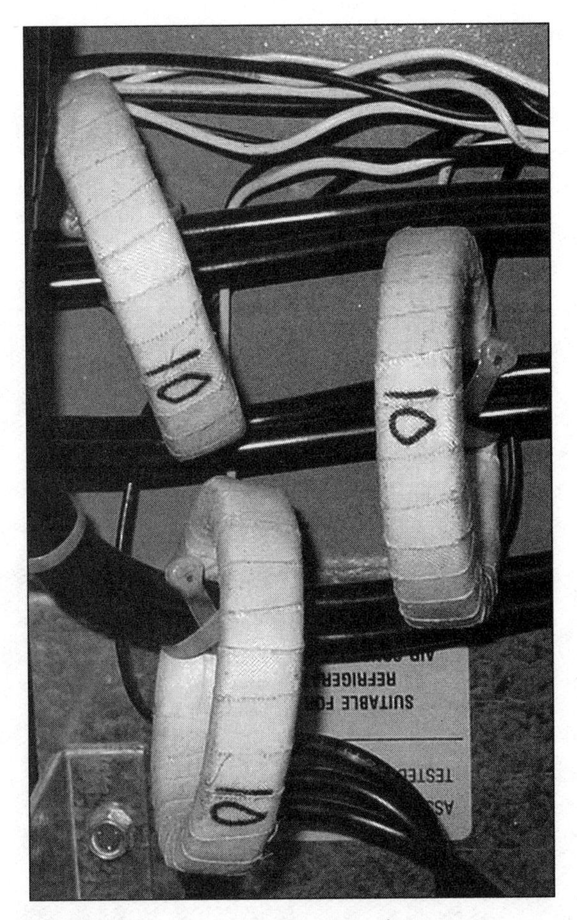

FIGURE 5.28 Solid-core current transducers.

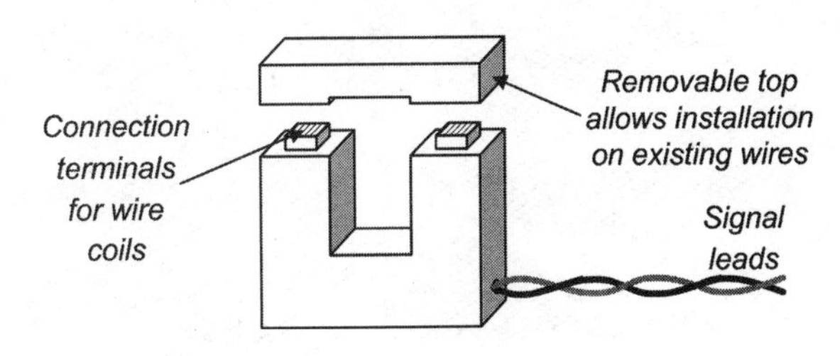

FIGURE 5.29 Split-core current transducers.

Watt transducers (Figure 5.30) are used to measure the instantaneous power of equipment. These devices use current transducers in tandem with transformers (that drop the line voltage down to 24 or 48 VAC) and calculate the power factor internally to produce a signal proportional to the true power consumption. These are not cheap devices—several hundred dollars just for the watt transducer—but their accuracy may be worth it in power management strategies in large systems.

FIGURE 5.30 Watt transducers.

PNEUMATIC ACTUATORS AND CONTROLLERS

Typical air pressures used in pneumatic control systems vary from 3 PSI to 20 PSI. Pneumatic actuators take advantage of the air pressure by spreading it out over diaphragms that, in turn, produce great force to move valves and dampers. The force required to move the damper or valve determines both the working pressure and the size of the diaphragm used in the actuator.

Because they respond to changes in pressure, pneumatic actuators operate very fast and can develop huge torques. Suppose the surface in Figure 5.31 is subjected to a standard HVAC control pressure of 15 psi. If this surface is 1 inch in diameter then the total force exerted on the surface is almost 12 pounds. Since the area of the surface varies with the square of the diameter, if the surface is 2 inches across the total force is almost four times as much—about 47 pounds. This is the principle behind the use of pneumatic controls and actuators. Even a slight change in the pressure coming from a pneumatic sensor can lead to tremendous changes in the force exerted by the actuator. The right side of Figure 5.32 shows a 2 inch diaphragm used on a ½ inch steam valve, while the left side of the figure shows a 12 inch diaphragm used on a 6 inch steam valve. A 1psi change in control pressure on these diaphragms would change the actuator force by 3 pounds and 113 pounds, respectively.

Figure 5.33 shows a cross section of a spring return pneumatic actuator body. The control signal enters at the bottom and applies pressure to the diaphragm that then forces the stem to the left, shown here about half way closed. As the control pressure is reduced, a spring forces the stem back to the full open position. In some cases the spring is encased within the actuator housing (see the left side of Figure 5.34) while in other cases the spring is separate and requires only a one-direction actuator (right side of figure). In both cases, the spring tensions can be chosen to change the characteristics of the actuator.

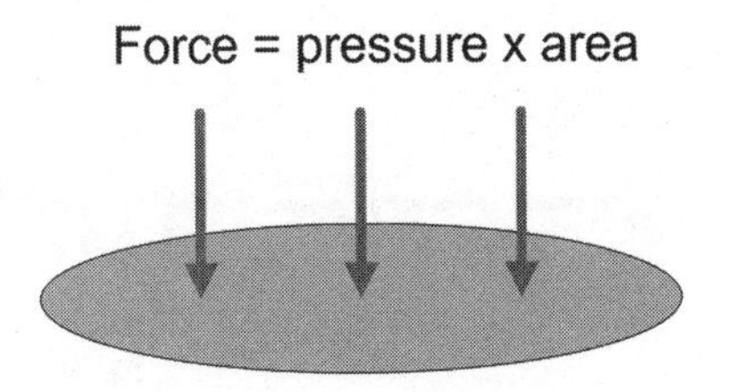

FIGURE 5.31 Force of pneumatic diaphragm.

FIGURE 5.32 Large and small valve diaphragms.

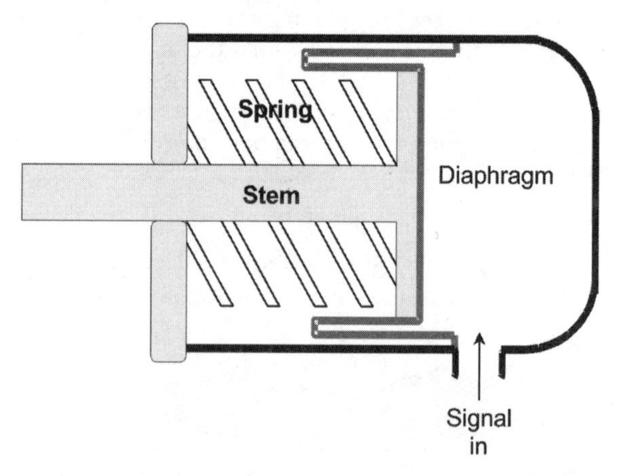

FIGURE 5.33 Cross section of pneumatic actuator.

FIGURE 5.34 Pneumatic actuator connected to valve (left); spring return valve (right).

Figure 5.35 shows a pneumatic actuator connected to a damper control linkage. This damper requires considerable torque to open and yet is accomplished by a 3 inch pneumatic actuator. Note also that the linkage can be adjusted to set the starting and stopping position of the damper.

Most actuators, valves, and dampers are subject to friction and sticking. This can be annoying if you are trying to obtain exact control of a process. To overcome this problem, many valves use a *positioner* (diagrammed in Figure 5.36) to convert the pneumatic signal from the controller into the proper pressure sent to the actuator. Positioners are also used to sequence several valve or damper actuators connected to the same controller.

Pneumatic controllers are broken down into two different types: *bleed* and *non-bleed*. In bleed type controllers (for example, Figure 5.37), the air is constantly leaking out and must be replenished by the compressed air system. In a non-bleed controller (Figure 5.38) the sensor action is relayed to a closed volume that does not allow

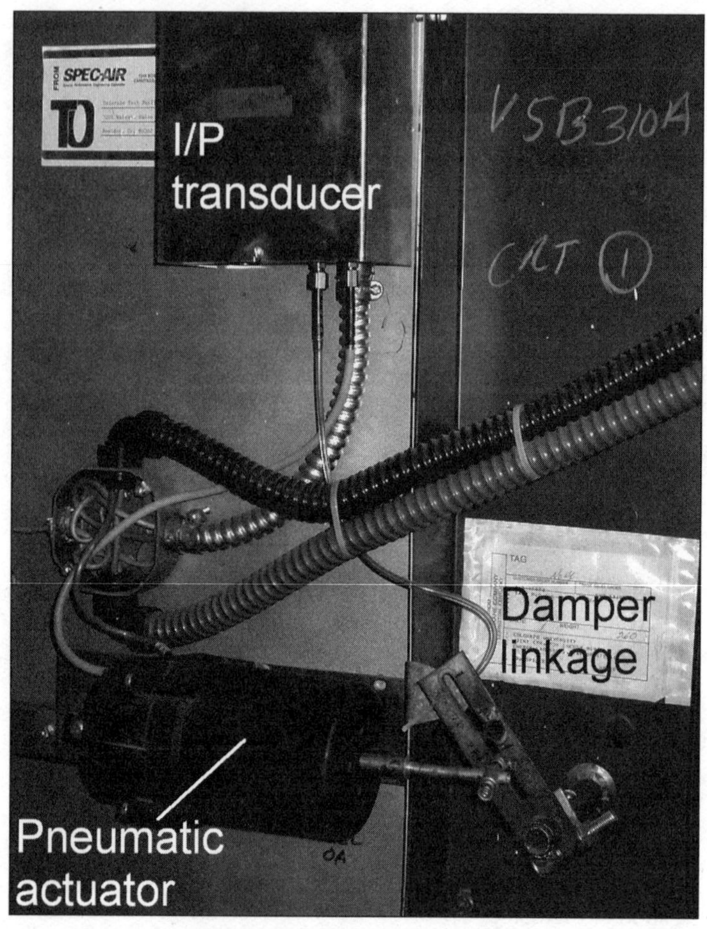

FIGURE 5.35 Pneumatic damper actuator and linkage.

for air leakage. The main advantage of pneumatic controllers is that they are self-powered, easy to debug, and have a fast reaction time. The main disadvantage is that it can be difficult to implement complex control algorithms.

Often you will find the pneumatic controls arranged in a panel box such as that in Figure 5.39. Here there are timers, switches, and other pneumatic relays that run the building HVAC. As more and more buildings move to computerized building management systems, there is

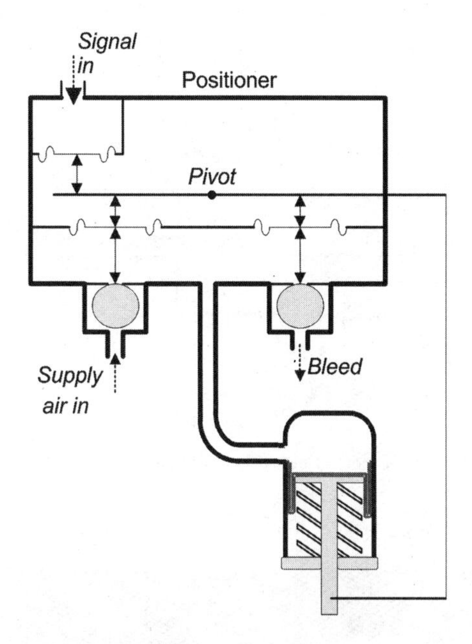

FIGURE 5.36 Pneumatic positioner.

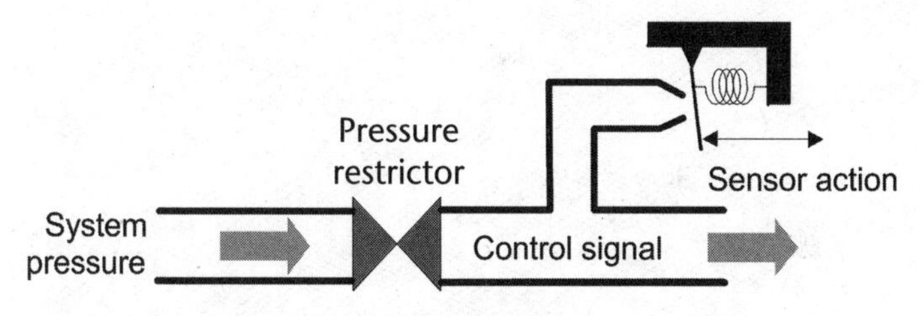

FIGURE 5.37 Bleed-type pneumatic temperature sensor.

the problem of interfacing pneumatic sensor and control signals to computer-based control systems. The solution is to use pneumatic/electric transducers (like that shown in Figure 5.40) that convert the pneumatic signal into a standard 4-20 mA or 0-10 VDC signal. Note that the conversions can run both ways so that electric/pneumatic control signals can be sent to the pneumatic relays.

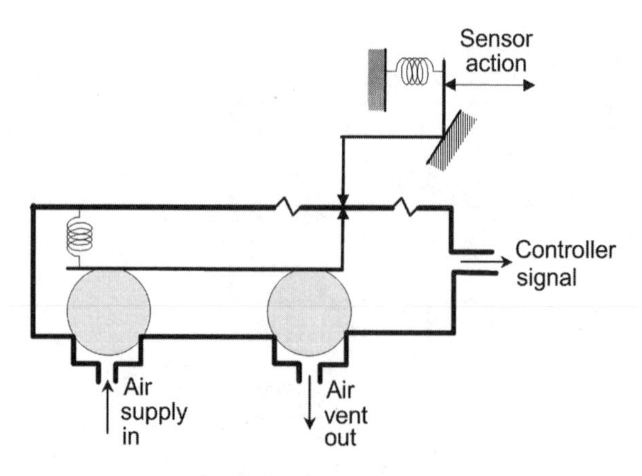

FIGURE 5.38 Non-bleed pneumatic controller.

FIGURE 5.39 Traditional pneumatic control panel showing relays, timers, and switches.

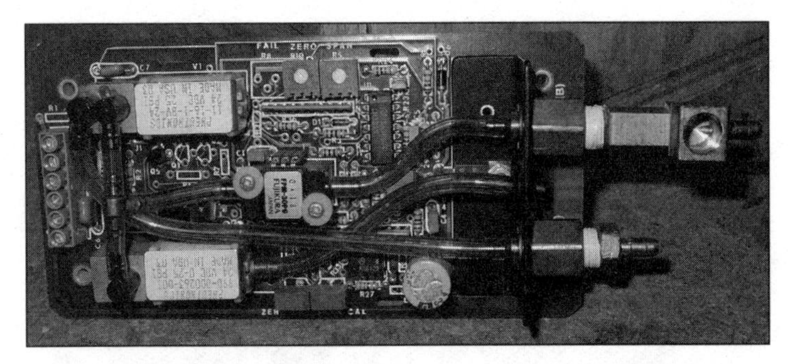

FIGURE 5.40 E/P transducer.

fastfacts

Electronic actuators generally do not move very fast yet they can develop tremendous torque. Always exercise caution around these devices because they can slowly "grab" you.

ELECTRONIC ACTUATORS AND CONTROLLERS

Electronic actuators use a series of motors and reduction gears to move valves and dampers. They accept control signals up 20 VDC or 20 mA and translate the signal into an actuator position. Because the motors are often geared down significantly to achieve the desired torque at the valve or damper shaft, the travel time of electronic actuators can be tens of seconds.

Figure 5.41 shows an electronic damper actuator with linkage to two separate dampers. This kind of sensor requires 24 VAC power to run the motor and accepts a 0 to 5 VDC control signal. Figure 5.42 shows an electronic valve actuator connected to the valve housing. Note that both actuators have position indicator scales on the front.

FIGURE 5.41 Electronic damper actuator and linkage.

Electronic controllers are like small computers that record sensor values, accept user-defined setpoints, and perform the necessary calculations to create control signals. These kinds of controllers can vary greatly in size and capacity and are specified in terms of the number and types of input and output signals they can accommodate.

- *Analog inputs* are used to measure modulating current or voltage signals. These are usually rated for 4 to 20 mA or 0 to 10 VDC signals. Special analog inputs are used for measuring thermocouple signals.
- *Analog outputs* are output channels used to send a modulating control signal to an actuator.

FIGURE 5.42 Electronic valve actuator and valve housing.

- *Digital inputs* measure on/off signals such as switch closures.
- *Digital outputs* send on/off controls. These can be used to open or close two-position devices such as solenoid valves. Special high voltage digital outputs can be used to activate large pieces of equipment.
- *Frequency inputs* (sometimes called *counters*) are used to measure the rate at which something occurs. For example, an unprocessed turbine flow meter produces a sequence of pulses that must be counted in order to determine the flow rate. These are also used for measuring electric, gas, and water meter signals.

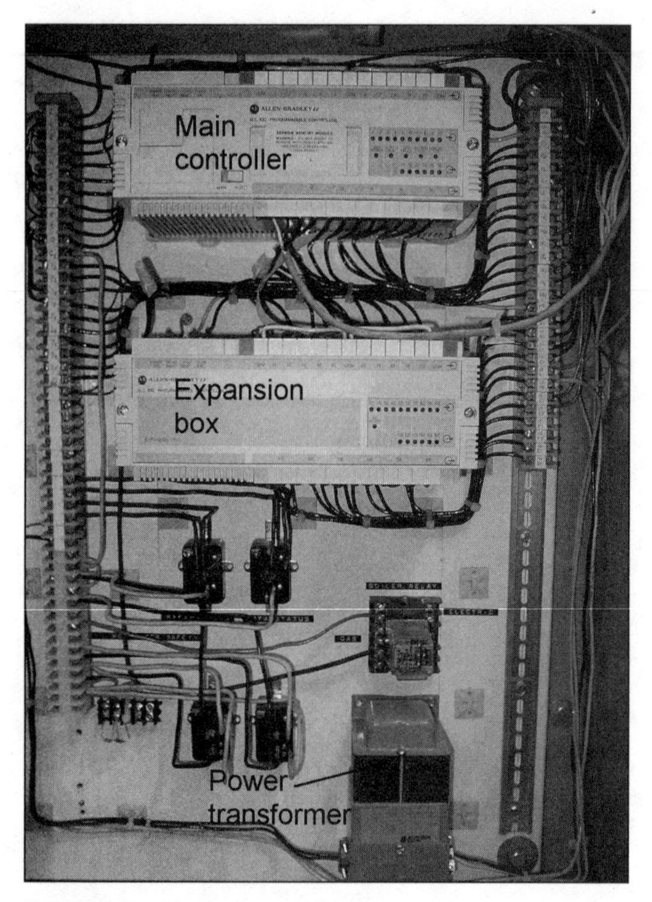

FIGURE 5.43 Electronic controller.

Sensor Noise Rejection and Shielding

Most sensors produce a voltage or current signal in response to a change of measured value. Typical voltage ranges for sensors are 0 to 1 VDC, 0 to 5 VDC, 0 to 10 VDC, 6 to 18 VDC, and 0 to 24 VDC. Typical current ranges are 0 to 16 mA, 0 to 20 mA, and 4 to 20 mA with the 4 to 20 mA being by far the most common. When measuring voltages or currents it is important to prevent electronic "noise" from being introduced to the useful signal. Unfortunately,

fastfacts

Never run signal cable or wire through the same conduit as AC wiring.

since these low-voltage and low-current sensors are often located near large motors and power cables, the potential for induced noise is high. The simplest method for preventing noise in signals is careful placement of the signal wires with respect to the high power equipment. Signal cables should never be placed parallel to high voltage wires since such an arrangement greatly increases the susceptibility of the signal to magnetic fields and the corresponding noise. If possible, signal cables should be placed several feet away from any high-power lines and, if required, should cross high power lines at right angles.

Most sensor cable (smaller than about 16 gauge) can be obtained with shielding (see Figure 5.44). The shielding consists of a wrapping of metal foil around the sensor cables and is, in turn, wrapped with a shielding wire. The shielding wire should be connected to a reliable earth ground *at one end of the wire only* to prevent any current flow through the shield itself.

CONTROL LOOPS

Feedback loop controllers measure sensor values, compare them with setpoints, and calculate the appropriate control signal. The components of a typical feedback loop are shown in Figure 5.45. For

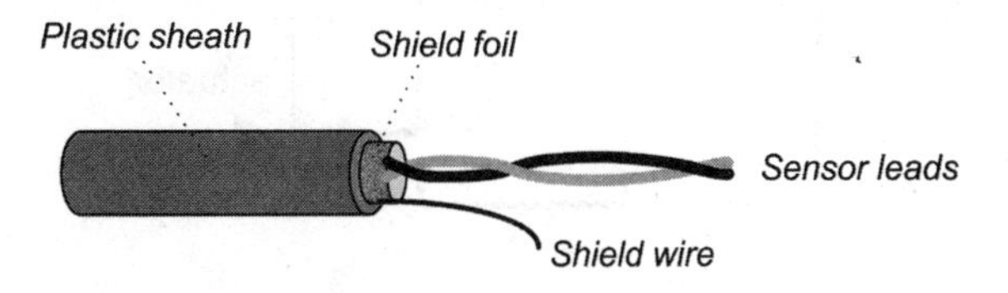

FIGURE 5.44 Detail of shield wire and foil.

example, in the process shown in Figure 5.46, a mechanical feedback loop is used to maintain the water level in the tank even if the flow rate of water entering the tank is always changing. In this system, the setpoint is the level in the tank, the controller is the combination of the sensor and armature, the error is the difference between the setpoint and the tank level, and the external disturbance is the varying input flow rate.

In most HVAC applications, a proportional-integral-derivative (PID) equation is used to generate a control signal for a given error. The PID equation is:

$$Output = K_p \cdot \left[E + K_i \int E\,dt = K_d \frac{dE}{dt} \right]$$

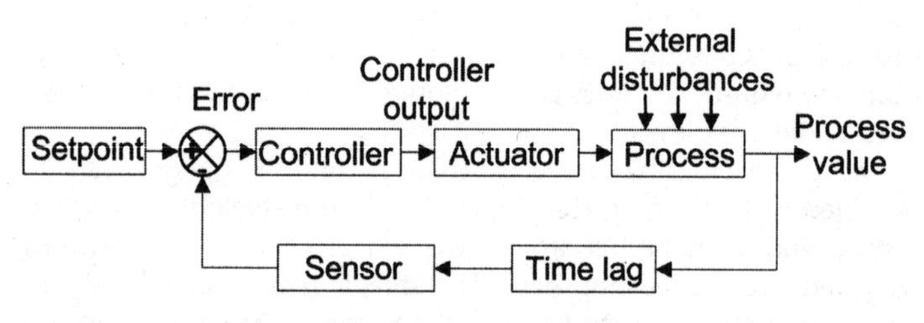

FIGURE 5.45 Components of a feedback loop.

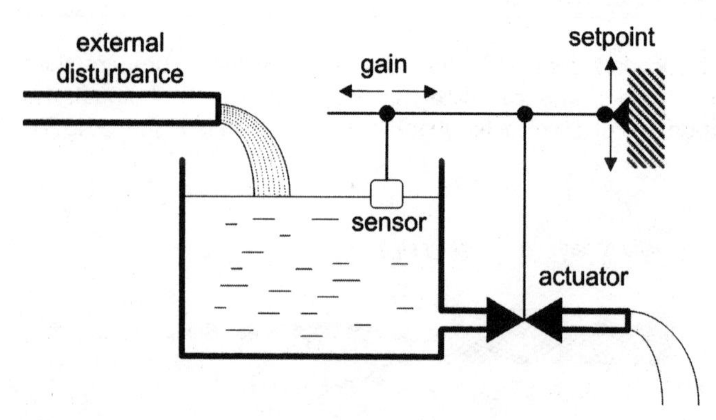

FIGURE 5.46 Typical feedback loop.

where

E = error (setpoint–process value)

K_p = proportional gain constant

K_I = integral time constant

K_D = derivative time constant, and

dE/dt = change in error from previous measurement.

- The *proportional term*, $K_P \cdot E$, has the greatest effect when the process value is far from the desired setpoint. Very large values of K_P tend to force the system into oscillatory response. The proportional gain effect of the controller goes to zero as the process approaches set point.
- The *integral term* K_I is the reciprocal of the reset time, T_R, of the system. Integral control is used to cancel any steady-state offsets that would occur using purely proportional control.
- Derivative control is used in cases where there is a large time lag between the controlled device and the sensor used for the feedback. The derivative term has the overall effect of preventing the actuator signal from going too far in one direction or another, and can be used to limit excessive overshoot.

You can spend a lot of time trying to tune a controller so that it gives nice, steady control over the entire year. The problem is that when the external disturbances start changing, so do the correct control constants. For example, a hot water coil used to heat the outside air will perform very differently if it is cold outside versus when it is hot outside. The same set of control constants may not work at both times. Some people use *adaptive control* that attempts to change the control constants based on the external disturbances.

The *reaction curve technique* (see Letherman, 1981) was developed to help find PID constants. To use this method, you must disable the feedback loop and manually change the controller output to see how this affects the process output. Imagine that you disconnect the control signal to a hot water coil and then force the control valve to go from full closed to full open. If you plot the temperature of the leaving air temperature versus time you might end up with a curve as shown in Figure 5.47. In this figure, Δc is the change of process output (that is, the total change of the coil outlet air temperature) and Δu is the change of controller. You can also draw some lines on this chart and get L (the time between the change and intersection) and T (the time between the lower and upper intersections). If you are using a PI controller, then "good" control constants can be found from

$$KP = 0.9 \times (\Delta u \div \Delta c) \times (T \div L)$$
$$K_I = 0.3 \div L$$

If you are using a full PID control, then

$$KP = 1.2 \times (\Delta u \div \Delta c) \times (T \div L)$$
$$K_I = 0.5 \div L$$
$$K_D = 0.5 \times L$$

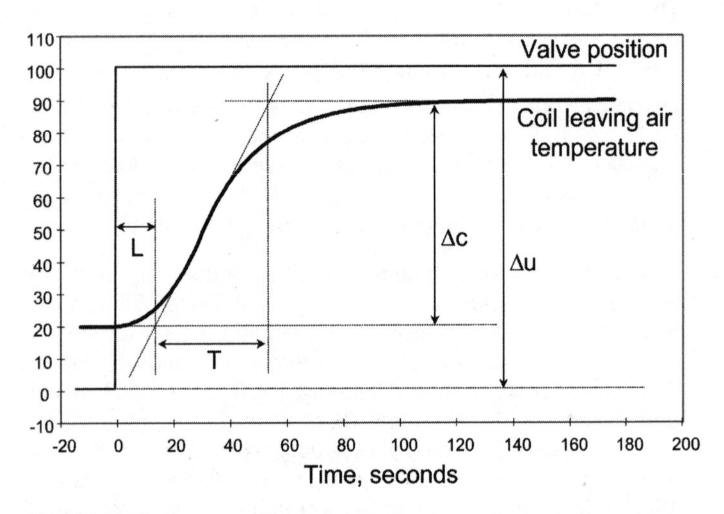

FIGURE 5.47 Response of process to step change in controller output.

chapter 6

PUMPS AND VALVES

Water is used to transfer cooling and heating energy around a building. This section describes the fundamentals of pumps and valves used in HVAC systems. For more information on water and steam distribution systems, see Chapters 7 and 12.

THEORY OF OPERATION

Pumps increase the velocity and pressure of fluids so that they can circulate throughout buildings. There are many different ways to pump fluids, but centrifugal pumps are used in practically all HVAC applications.

Centrifugal Pumps

Centrifugal pumps use an *impeller* encased in a pump housing to increase the velocity of the fluid that enters the pump. The fluid enters at the center of the impeller that is, in turn, spinning at a high velocity. Centrifugal forces act on the fluid and it moves to the outside of the impeller, increasing in both speed and pressure. The components of a centrifugal pump are shown in Figure 6.1. Figure

6.2 shows an impeller removed from the pump housing. Water enters at the middle of the spinning impeller and is then centrifugally forced to the outer points.

Pumps and motors that share the same shaft are said to be *close-coupled* (see Figure 6.3). This is typical for smaller pumps at low horsepower. If the motor shaft is separate from the pump shaft, a *coupler* is used to connect the two as in Figure 6.4. Note that the pump in Figure 6.4 must be mounted in a specific position to keep the oil from running out the lubrication hole.

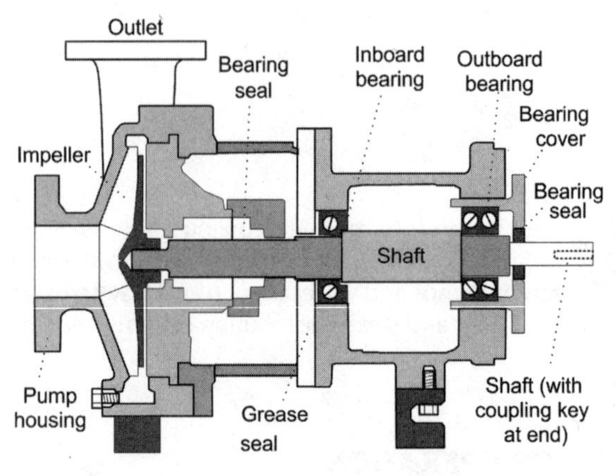

FIGURE 6.1 Centrifugal pump components.

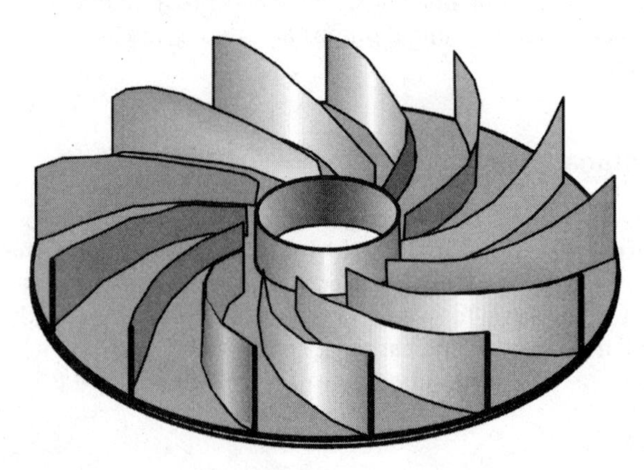

FIGURE 6.2 Centrifugal pump impeller.

FIGURE 6.3 Close-coupled, sealed bearing pump.

FIGURE 6.4 Centrifugal pump separate from motor.

The performance of a pump depends on the power of the motor and the shape and size of the impeller. A fluid passing through a pump increases in pressure; this is called the pump head (as in *head pressure*). The power added to the fluid is found by multiplying the flow rate and the total head:

$$\text{water HP} = \frac{\text{GPM} \cdot \text{feet head} \cdot \text{SG}}{3960}$$

where SG is the *specific gravity* of the fluid being pumped. The specific gravity is the density of the fluid related to the density of water. So the specific gravity for water is 1.0 while the specific gravity for lighter fluids like glycols is less than one.

The *pump affinity laws* (Avallone and Baumeister, 1996) are used to describe the behavior of pumps under various conditions. If the impeller size of the pump is a constant, then the flow rate through the pump is directly proportional to the pump speed,

$$\frac{\text{GPM}_2}{\text{GPM}_1} = \frac{\text{RPM}_2}{\text{RPM}_1}$$

The pressure rise across the pump varies as the square of the pump rotation speed,

$$\frac{\Delta P_2}{\Delta P_1} = \left(\frac{\text{RPM}_2}{\text{RPM}_1}\right)^2$$

and the pump power consumption varies as the cube of the rotation speed,

$$\frac{\text{kW}_2}{\text{kW}_1} = \left(\frac{\text{RPM}_2}{\text{RPM}_1}\right)^3$$

This last relationship shows how use of a variable speed drive on a pump can achieve significant energy savings.

If the pump speed is a constant, the flow rate is directly proportional to the impeller diameter,

$$\frac{GPM_2}{GPM_1} = \frac{D_{i,2}}{D_{i,1}}$$

the pressure rise varies as the square of the impeller diameter,

$$\frac{\Delta P_2}{\Delta P_1} = \left(\frac{D_{i,2}}{D_{i,1}}\right)^2$$

and the power consumption varies as the cube of the impeller diameter,

$$\frac{kW_2}{kW_1} = \left(\frac{D_{i,2}}{D_{i,1}}\right)^3$$

Pump and System Curves

The *pump curve* is a graph showing the pressure rise across the pump as a function of the flow rate. The *system curve* is a graph of the pressure drop through a system as a function of the flow rate. Figure 6.5 shows what these curves typically look like. As with the pump affinity laws, there are laws that describe the pressure drop across the supply and return lines of a water distribution system. For example, the pressure drop in a piping system varies as a function of the square of the volumetric flow rate of fluid passing through the pipe,

$$\frac{\Delta P_2}{\Delta P_1} = \left(\frac{GPM_2}{GPM_1}\right)^2$$

It is often convenient to superimpose the system and pump curves, as shown in Figure 6.5, so that the operating point of the system can be identified. The operating point is where the pump and system curves intersect and shows the actual flow through the system.

As mentioned above, the total power output of a pump is found from multiplying the flow rate by the pressure rise. If the pump outlet is valved off then you can spin the pump motor without producing any water flow. Obviously, this isn't such an efficient way of doing things. Similarly, if a lot of water is flowing through the pump

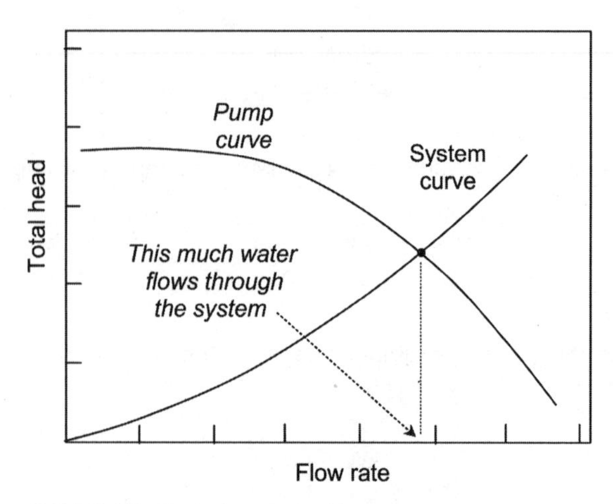

FIGURE 6.5 Typical pump and system curves.

but there is no pressure rise then you won't be able to force the water through the piping system. These two extremes are the lowest efficiency points for the pump, and you certainly don't want to operate like that.

The shape of the pump curve implies that there is some maximum efficiency point where the total pump output (that is, the flow rate multiplied by the pressure rise) is at its highest. Most pump curves show the lines of constant efficiency so that you can determine how well your pump is operating. Figure 6.6 shows a typical pump curve with the efficiency points superimposed. While it is desirable from an energy standpoint to have the pump operating at its highest efficiency, it is not always possible since the system curve may force the flow to another condition. Overpumping is worse than underpumping since it can cause too much system pressure and associated system leaks, so take care when sizing the pump.

Positive Displacement Pumps

While most of the pumps used in HVAC systems are centrifugal pumps, there are several situations where you may encounter positive displacement pumps. Rather than using centrifugal force to move fluids, positive displacement pumps do not use impellers but

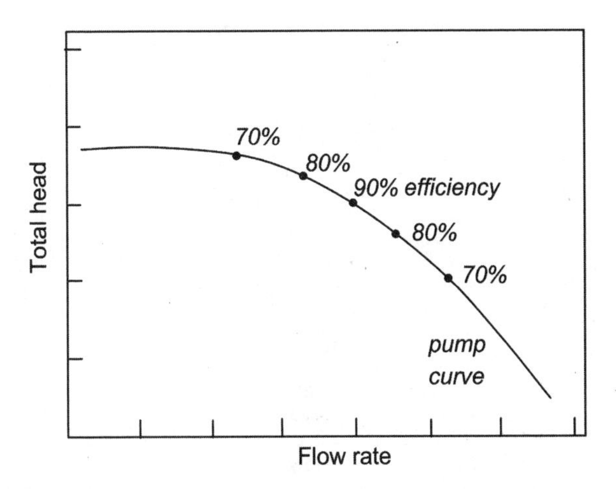

FIGURE 6.6 Typical pump curve showing pump efficiency points.

instead rely on filling up a volume with liquid and then mechanically forcing the liquid into a pipe or reservoir. There are two kinds of positive displacement pumps seen in HVAC applications.

Gear pumps, vane pumps, lobe pumps, and screw pumps are classified as *rotary pumps*. Rotary pumps are often used for corrosive and high viscosity fluids like brines and oils. While the overall capacity of rotary pumps is much less than that of centrifugal pumps, they can handle a much wider range of fluids and do so at higher efficiencies. These pumps are relatively small and are usually self-priming. By their design, rotary pumps can achieve a continuous flow regardless of the upstream and downstream pressures (up to a point).

Another type of positive displacement pump is the *reciprocal pump*. This kind of pump operates very similarly to that of an automobile engine, with the pistons and cylinders used to force fluid through a series of cams. This kind of pump is useful for moving slurries and sludge, but is most often found in HVAC applications as reciprocal compressors for refrigerants in vapor compression cycles.

There are many different kinds of positive displacement pumps with different pump curves for each. In general, however, the pump curves are approximated as shown in Figure 6.7. A crucial difference between positive displacement and centrifugal pumps is that the positive displacement pumps have no best efficiency point. If you need more flow, you simply apply additional power to the pump shaft.

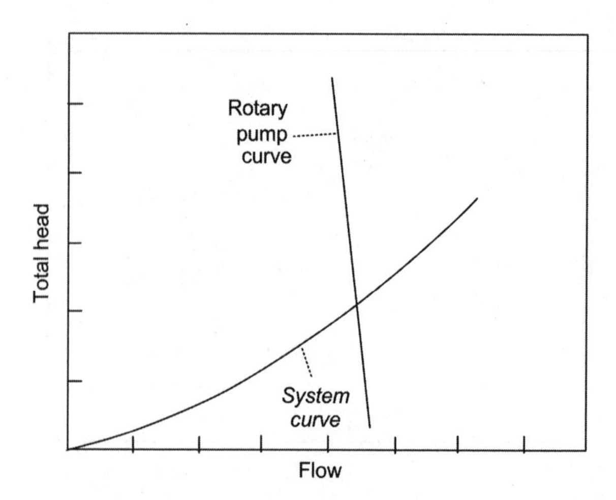

FIGURE 6.7 Pump curve for rotary positive displacement pump.

Valves

Water and steam flows in HVAC systems are controlled through valves. Valves are chosen based on the pipe size and the desired flow characteristics. For water valves, the most commonly used flow parameter is the valve *flow coefficient*, C_V. This is the number of gallons per minute of 60°F water that flows through the valve with a pressure drop of 1 psi when the valve is fully open. The flow under any other condition can be found from GPM = $C_v \cdot \Delta P^{0.5}$ where ΔP is the pressure drop across the valve.

Different types of valves can be chosen that have different flow rates as you open the valve. Of particular interest is the flow that occurs when the valve is initially opened. For example, when opened just as little as possible, many valves have a flow of about 3 to 5% of the maximum possible flow. This value is known as the *turndown ratio.* A typical commercial valve might have a turndown ratio of 5% (20:1) while industrial process control valves may have turndown ratios of 50:1 to 100:1. This kind of precision is generally not needed in HVAC applications unless the piping system is very oversized.

When multiple valves are installed in a system, some cause greater pressure drops than others. The *authority* of a valve is the ratio of the valve pressure drop to the total system pressure drop.

Figure 6.8 is a side view of a typical valve showing the individual valve components. The *valve stem* is the part that moves up and down and drives the valve open or closed. At the end of the stem is the *plug* that is, in turn, made up of the *disc* and the *seat*. As the valve closes, the disc comes into contact with the valve seat. A good seal between the disc and the seat is necessary to prevent leaky valves. Seats are usually made of brass, although some are made of stainless steel.

The valve stem moves up and down through a given distance between the full open and full closed position. This distance is called the *valve stem travel*. To prevent water or steam from leaking out around the stem, *packing* is used to create a tight seal between the stem and the valve bonnet. The packing is usually made up of some mixture of Teflon, rubber, or elastomeric compound. Older valves use rope packing while newer valves use Teflon or elastomeric cup packing.

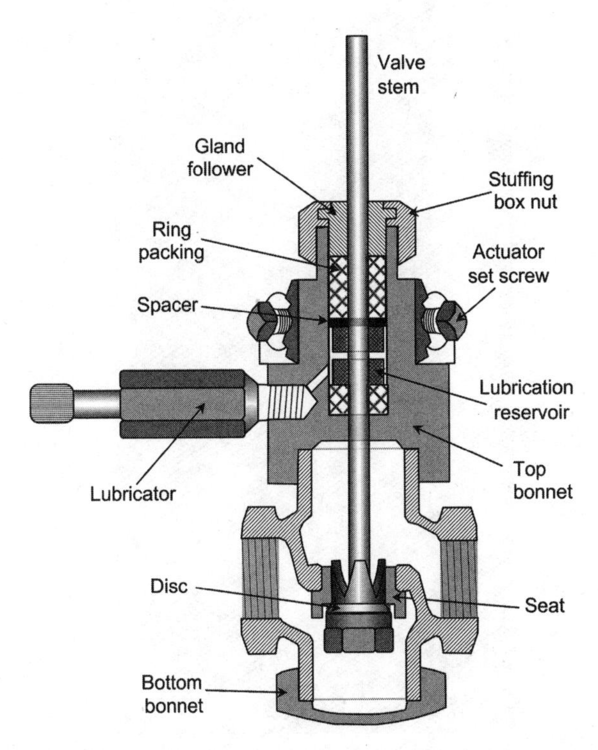

FIGURE 6.8 Break-away view of typical valve showing parts.

The size and shape of the plug varies from valve to valve. Figure 6.9 shows what a typical valve stem and plug combination look like for a larger valve, in this case a 6 inch steam valve. Note also that the valve stem in this picture is wrapped with tape. This is to protect the highly machined and polished surface of the stem. You want the valve stem to remain very smooth so that the packing can make good contact to prevent leaking.

Figure 6.10 shows various sizes and types of valve disks. The disk on the left is for a ½ inch valve, the three in the middle are for 3 inch valves, and the large disk on the right is for a 6 inch valve. The three disks of different sizes along the top of this figure are standard composition disks and are suitable for general use. The white 3 inch disk is Teflon and is used for high temperature applications, while the bottom 3 inch disk is of bakelite. Care needs to be taken with the bakelite disks since they can pit or shatter if dropped.

FIGURE 6.9 Plug for 6 inch valve.

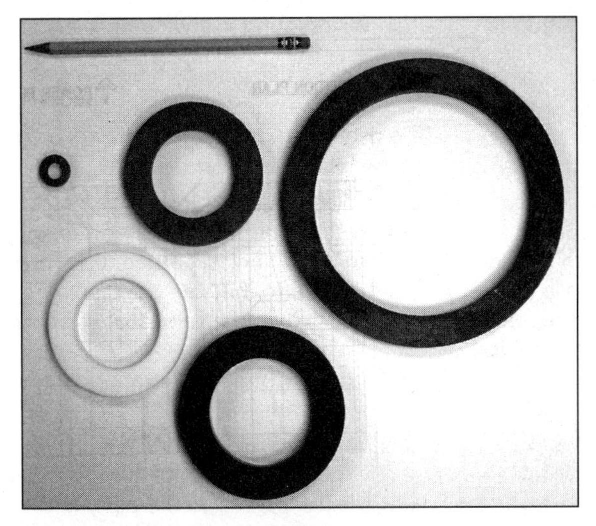

FIGURE 6.10 Various valve disks.

Valve Types

Valves have many different arrangements depending on the needs and function of the system. *Two-position* valves (Figure 6.11) are used when rapid flow is needed with minimal movement of the valve. Strictly speaking, these valves can be adjusted to any position but they are usually reserved for simple applications where accurate flow modulation is not required. Usually two-position valves are used for shut-off service.

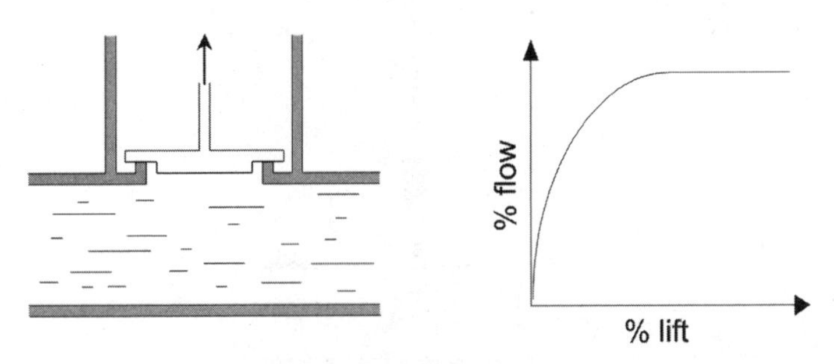

FIGURE 6.11 Two-position (quick opening) valve schematic and flow characteristic.

Common two-position valves found in HVAC applications are:

- *Gate valves* are slidable baffles that open or close to allow water flow (see Figure 6.12). The actuator is often an electric solenoid that rapidly slides the gate open or closed. These are often used on refrigerant systems and smaller water pipes.
- *Butterfly valves* use a thin plate within the pipe that can be turned perpendicular to the flow direction to impede the flow or parallel to the flow direction to allow flow (see Figure 6.13). They are usually used as on/off flow control but can occasionally be used for proportional flow control when the valve shaft is attached to a mechanical actuator. When operated as flow control, they behave as equal percentage valves. Butterfly valves come in sizes from half-inch to several feet in diameter.
- *Ball valves* are also used for on/off or proportional control. These valves use a spherical valve housing that contains a ball with a hole in it. When the hole is rotated parallel to the pipe, there is flow through the valve as shown in Figure 6.14. Ball valves are often brass or PVC and come in sizes from about half an inch up to about 4 inches. There are also ball valves that use a stainless steel contoured ball to provide total shut-off, such as the ball valve in Figure 6.15 that controls the steam flow to an absorption chiller.

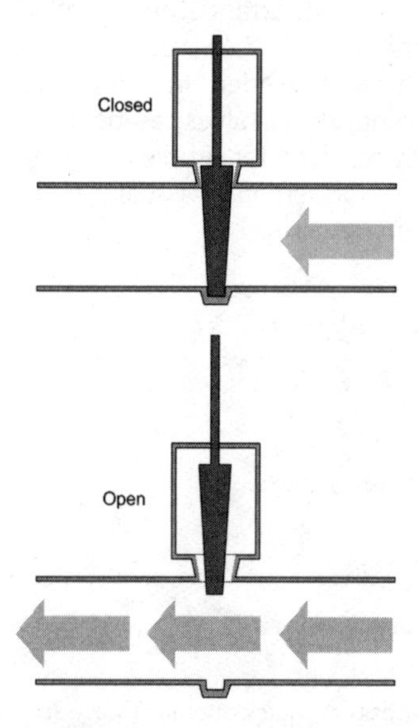

FIGURE 6.12 Gate valve.

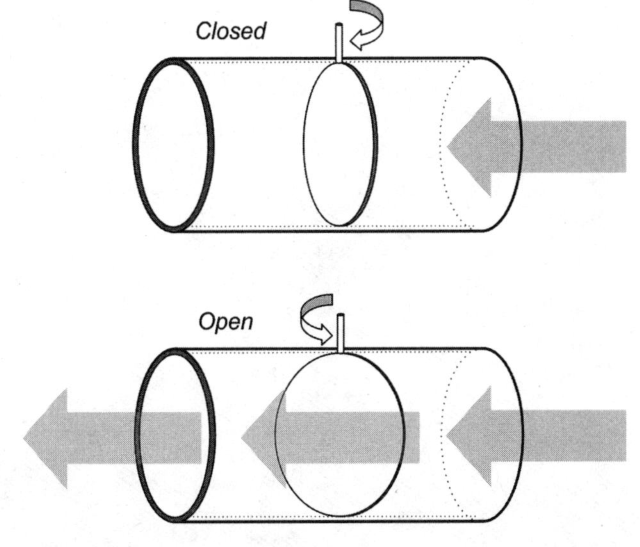

FIGURE 6.13 Butterfly valve.

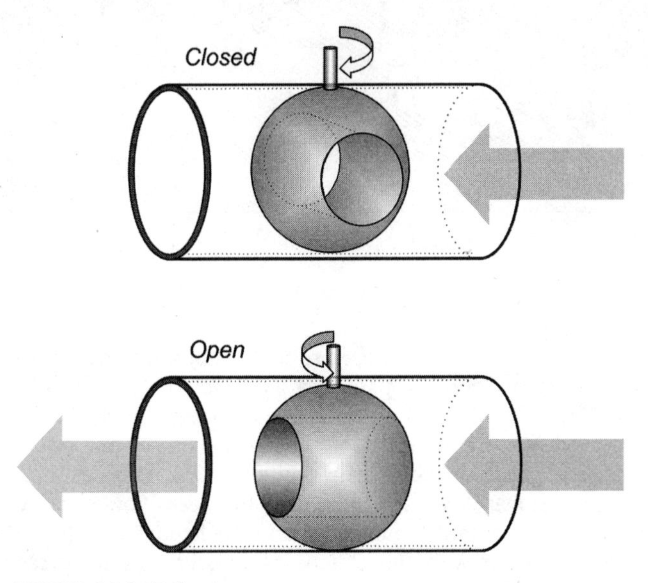

FIGURE 6.14 Ball valve.

FIGURE 6.15 Five inch ball valve used in steam service.

Other types of valves are used to provide proportional control to water or steam flow. The plug of a *linear valve* is shown in Figure 6.16. When the valve opens just a little bit, only a small flow area is made available. As the valve continues to open, more and more area is available, giving the flow versus position characteristic as shown in the figure. Linear valves are often used for steam flow because the steam is at a constant temperature and the latent heat of condensation is uniform with the change of pressure. *Equal percentage* valves have a plug shape as shown in Figure 6.17. Equal percentage

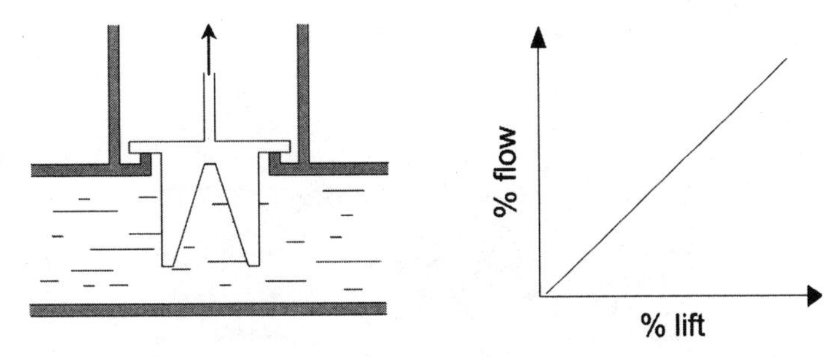

FIGURE 6.16 Linear valve schematic and flow characteristic.

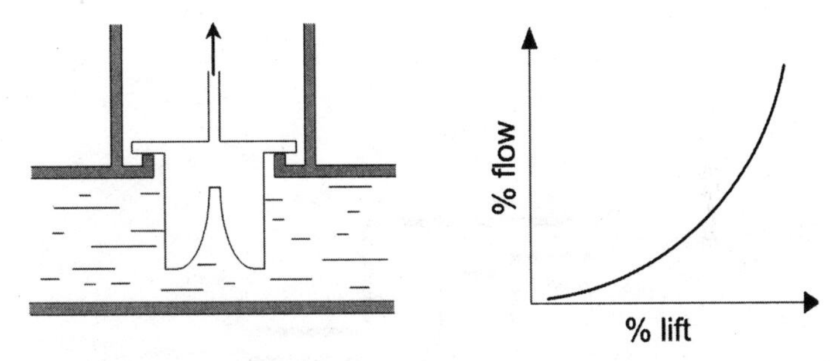

FIGURE 6.17 Equal percentage valve schematic and flow characteristic.

valves are good for cooling and heating coils because the combination of the typical coil with an equal percentage valve provides effectively linear control.

Check valves are used to prevent backflow through piping. There are many different kinds of check valves. Figure 6.18 shows a swing-type check valve that allows flow from left to right, but the valve will close and seal if reverse pressure is experienced. Check valves are used to prevent backflow on refrigerant pipes, city water make-up lines, and steam condensate lines.

Often a valve is used not just to shut off the flow but to redirect flow from one pipe to another. This is accomplished through the use of *three-way mixing* and *diverting valves* (see Figure 6.19). These two types of valves are constructed differently so they should not be used interchangeably. Most three-way valve bodies have clear markings indicating the proper flow paths (Figure 6.20) to help during

FIGURE 6.18 Swing type check valve.

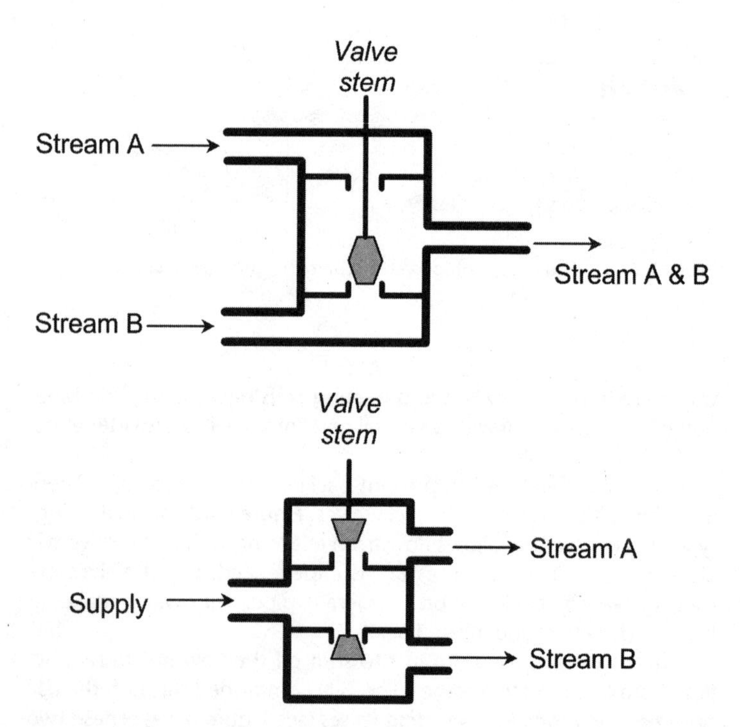

FIGURE 6.19 Schematic of mixing valve (top) and diverting valve (bottom).

FIGURE 6.20 Three-way mixing valve showing water flow.

installation. Never try to force water the wrong way through a three-way valve; you will have bad flow characteristics, leaky valves, and can damage the valve components.

When three-way valves are used in a piping system, there is usually a *balancing valve* in the bypass branch. These valves are manually adjusted to provide a desired pressure drop in the bypass so that the pressure drop across the bypass is the same as in the associated coil. Balancing valves use a dial like that in Figure 6.21 to indicate the relative position of the valve. The balancing valve ensures that there is always the same flow through the valve no matter what the valve position.

fastfacts

Modern balancing valves have spring-driven pressure regulators and must be installed with the fluid flow passing in a specific direction through the valve.

FIGURE 6.21 Adjustable balancing valve.

CONTROLS

Pumps and valves are controlled to maintain desired pressure and flow through water distribution systems, refrigerant lines, coils, etc. Pump control can be relatively simple: you turn the pump on or off. In the past decade, however, many pumps on water distribution systems have been fitted with variable speed drives. This allows for precise control of the differential pressure across the supply and return sides.

When working with variable speed drives, remember that there are certain operating frequencies that seem to create unacceptable levels of vibration in the pump and attached pipes. Most VFDs can be programmed to skip over these frequencies. Keep in mind, however, that if you lock out certain frequencies that the controller is requesting, it can cause some havoc as the pump and controller fight it out.

> # **fast**facts
>
> *Water is an incompressible fluid, so be very careful when using a variable speed drive on the motor to control the water pressure. A sudden increase in pump speed can cause dramatic and dangerous increases in the system water pressure.*

Large pumps should have an interlock with a flow switch so that the pump will stop if there is no flow confirmed once the pump has started. This is usually done with some kind of timing relay in line with the pump control circuit. Figure 6.22 shows a typical pump start-up control line diagram. Once the spring-loaded start button is pressed, the motor relay 1M and the timer relay 2M are activated and the pump starts. If the flow switch goes open for a period of time then the pump stops.

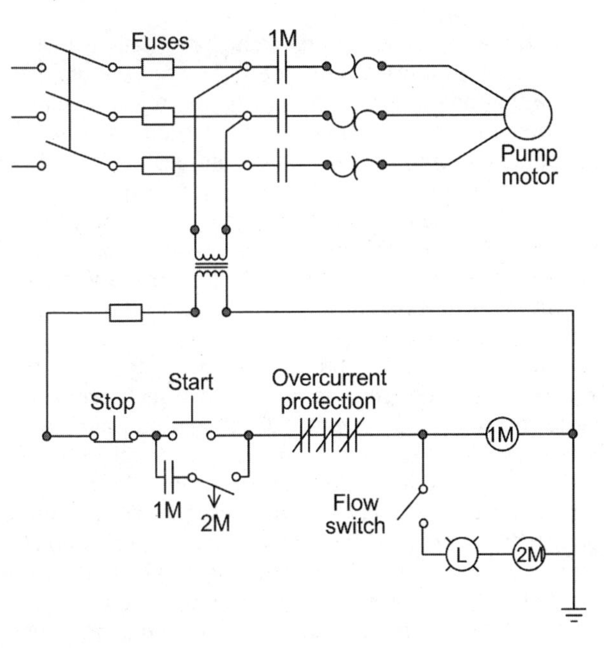

FIGURE 6.22 Pump start-up line diagram.

fastfacts

Selecting either pneumatic or electronic controls for your valves affects your maintenance costs. Electric actuators on baseboard plug and seat valves usually last about five years while a pneumatic actuator with a silicone diaphragm may last twice as long.

Valves are operated by pneumatic or electric actuators. Modulating pneumatic actuators are controlled by a 3 to 18 psi compressed air signal that comes from a pneumatic relay or thermostat. Modulating electric actuators use a 24 V or 120 V power supply to run the motor and a 4 to 20 mA or 0 to 10 V signal to set the valve position. Modulating electric valves can be fairly expensive and require electronic thermostats or controllers. In many cases, such as baseboard heaters, the electric actuator is two-position, non-modulating.

Pneumatic valves come with different spring ranges that can be used to synchronize a valve with another piece of equipment. For example, a thermostat in a room may use a single pressure line to control both the amount of cool air coming into a zone and the amount of heating supplied by a baseboard heater. To avoid simultaneously heating and cooling the space, you want the damper to respond to one range of the control signal and the heater valve to respond to another range. For example, the valve might respond to a controller pressure signal between 3 and 6 psi and a VAV box damper responds to 8 to 13 psi. Valve rebuild kits (as shown in Figure 6.23) offer a variety of spring ranges, typically 3-6, 4-8, 8-13, and 9-13 psi. You can swap springs as necessary to ensure the best control. For fine-tuning, you can do a "bench set" spring adjust to accommodate about 1 psi in either direction from the ratings.

The feedback loop on modulating valves is usually used to control the temperature or pressure in a fluid stream. Higher gain on the feedback loop can lead to faster control response but can also cause the valve stem to operate up and down more frequently and lead to wear on the valve stem seal. If the gain is set too high, the valve can oscillate between full open and full closed—you may not even notice a problem because the mass of the system or the size of the room tends to dampen out the resulting sensor signal.

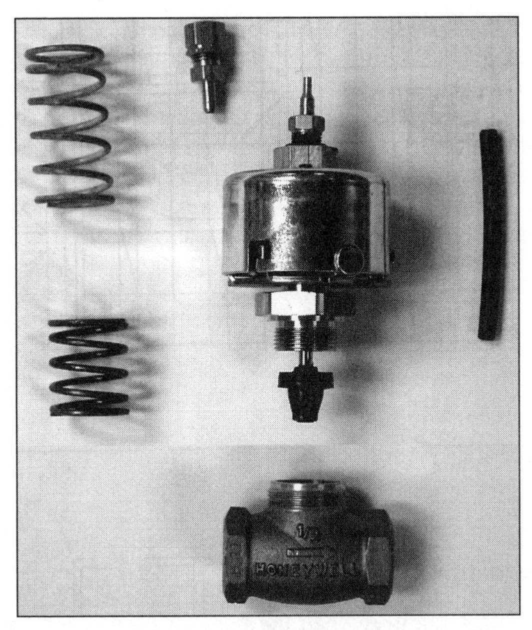

FIGURE 6.23 Valve rebuild kit showing different springs available.

The more the valve operates, the more frequent the need to repack the valve or replace the actuator. Imagine a valve with a one-inch travel that's connected to a thermostat with a high gain. Suppose this valve switches between full open and full closed every minute for 12 hours each day. This means that the valve stem moves past the valve seal an equivalent of over eight miles a year! Even with lubrication that's a lot of wear on the valve stem seal. Add in dirt,

fastfacts

Make sure you have a properly tuned controller and that the controller gain is not too high. You need good stable control not only for the process but also to reduce wear on the valve stem and seals.

sand, and various debris and you've got a pretty bad scraping problem. Leaking valve stem seals will damage the packing, valve stems, and even the actuators.

Problems can also occur between controllers for electric and pneumatic valves, leading to travel-time mismatch. Some buildings have pneumatic valve actuators while the building HVAC management system is computer based and configured for electronic valve actuators. Electric-to-pneumatic transducers are used to integrate the two systems, but there can still be problems if the control system assumes the actuator travel times to be slower than they really are. This can lead to excessive use and damage of the pneumatic actuators because they cycle open and shut much more than they should. The bottom line is that valve stems last much longer with appropriate control signals.

SAFETY

Deadheading

Deadheading a pump is when the pump is turning but there is no flow because the piping has been isolated for some reason. This can lead to cavitation and subsequent damage of the pump components. In some cases, someone valves the pump off for maintenance and then someone else turns the pump on and walks away. If the pump is

fastfacts

A building was retrofitted with 300 electric/pneumatic transducers for controlling pneumatic reheat valves. Unfortunately, during this retrofit the building management system was programmed with electric actuators in mind. The mismatch in assumed response time lead to a constant maintenance cost of replacing pneumatic valve stems. The technicians requested that the computers be reprogrammed, but the building owner refused to spend the $10,000 required to do this. Over the course of several years, the small, incremental costs of replacing valve stems ran to much more than the cost of reprogramming the computers.

fastfacts

Lock out and tag out procedures are essential and critical! If you remove a piece of equipment from service, make sure that others know they are not to operate that device. Use local power shut-offs, padlocks, and clearly written tags to identify what equipment is out of service and who is responsible for bringing it back on line.

working against a no-flow condition, the pump can get so hot that it burns off the paint, melts the insulation, and can flash the water in the pipes. If the situation is allowed to continue, the pump motor itself can catch on fire. This is an extremely dangerous situation.

Cavitation

Cavitation occurs when the pressure upstream of a pump is not high enough to force fluid into the pump intake fast enough. Bubbles or cavities form in the fluid and then "collapse" onto the impeller. When cavitation is occurring, the pump makes metallic pinging sounds and vibrates more than normal. Cavitation damages the impeller through pitting and erosion and can reduce the pumping capacity. It also can cause the pump shaft to bend, and the resulting shaft deflection can lead to seal and bearing failure.

The net positive suction head (NPSH) for a pump is the minimum system static pressure at the pump inlet required to prevent cavitation. System designers should always ensure that the available pump suction pressure is greater than the NPSH. The NPSH is usually listed on individual pump curves supplied by the manufacturer.

Personnel Safety

As with all electrical motors, pumps involve a rotating shaft and high voltages. Precautions should be taken to ensure that loose clothing, jewelry, and long hair do not get caught in any exposed sections of the motor or the pump shaft.

Valves with actuators can move without notice. Under no circumstances should fingers or hands be placed in areas where they could be caught between the valve stem and the actuator body. In particular, pneumatic valves can move quite rapidly and with tremendous force.

On a pump motor where the motor and pump are separate and connected using a coupler, you need to pay close attention to the coupler, particularly after maintenance on the pump or when the pump starts up. Couplers are usually made of rubber and have a tendency to fail once in a while. If it fails, the coupler can come loose with a tremendous force and velocity. If you hear the coupler making a rattling noise during normal operation, the best thing to do is shut the pump off, lock out and tag out the pump power, find out what the problem is and replace the coupler if necessary. Most couplers have a safety shield (see Figure 6.24) but it's best not to take chances.

FIGURE 6.24 Separate motor and pump showing coupler shield.

fastfacts

Do not stand over a coupler when a pump is started. If it comes loose it can cause severe injury to anyone standing nearby.

fastfacts

A technician was called in to investigate a 50 HP motor on a pump that was making an odd noise. By chance, the technician turned around and faced away from the motor about 10 feet away when the motor suddenly exploded in a big mushroom cloud. Grease from excessive lubrication had gotten onto the windings and the motor heated up to the point where it failed in a spectacular fashion.

If a motor is making a funny noise, shut it off and investigate! It is much safer to identify the problem and fix it than to ignore it and hope it goes away. Motor cooling ducts can get filled with dirt, grease, lint, and whatever else is floating around in the mechanical room. Make sure the cooling ducts are open and that the motor is not running too hot. In some cases a motor can get hot enough to catch fire. This is, of course, a serious hazard in a mechanical room where there may be many different chemicals and flammable materials. Usually if a motor starts burning the fire shoots out the bottom through the air vent, but in some cases it can be much more catastrophic.

Keep in mind that some locations don't allow for good motor maintenance. This is particularly true in tight spots where the motor is near the walls or ceiling. Unfortunately, in these circumstances there is a tendency to go for the quick fix. For example, if the motor bearings are making a noise, sometime the technician just gives it a shot of grease to make the noise go away. This can quickly lead to over-lubrication of the motor and after a couple of years the motor blows up because it is full of grease. The bottom line is that it is cheaper and safer to fix the problems as they occur rather than to wait for the pump to stop working altogether.

Often in large systems, the pump and motor are mounted separately on a base pad and are connected with a coupler. Usually in such systems there are shims used to adjust the respective heights of the pump and motor to ensure that the shafts line up correctly. If service is performed on the pump or motor and either one is moved, make sure that the shims are replaced as they were originally. If this is not done, the pump and motor can end up being misaligned. This can

fastfacts

Don't overgrease or over-oil a pump. In the case of a sleeve bearing, like a small circulator pump, over-oiling can lead to drooping of the rubbing mounting pads. This can tear the seal or coupler out of the pump. This can be dangerous, particularly if the coupler comes flying out of the pump.

twist the housing and put too much strain on the shafts, leading to motor overheating, pump leaks, strained shafts, and worn bearings.

Pipes may be used to support pumps up to a certain size. Larger pumps (> 10 hp) tend to vibrate and put torsional stress on the piping. If enough pressure is put on the pump by unsupported piping, the end of pump can break off, resulting in loss of pumping and serious system leaks.

Valve Disassembly and Repacking

Many valves can last for decades with no problem, but some must be repacked every few years to ensure they won't leak. When repacking valves, make sure the pneumatic air supply or electric power is shut off and that ALL isolation valves are shut off, not just those upstream of the valve.

Be aware that large pneumatic valves (and two-position electric valves) have very powerful springs in them. When disassembling the valves and removing the spring, make sure you follow manufacturer's recommendation exactly. Do not try to take shortcuts. If the valve stem is broken when you pull valve top off, the spring can come flying out with great force. Most eye protection will not stop a coiled spring, so you must be very careful. Also, some stems have a nut on the bottom that hold the plug on (see Figure 6.8). If this nut has come loose, you won't know until the whole stem and spring assembly has unexpectedly shot out of the valve body.

A similar problem can occur if the stem wears through. This is especially true with older valves that lay on their side. Gravity works on stem, disk, and seat as the valve modulates, and eventually the

fastfacts

If you are taking a valve apart and you feel tension in the valve housing, stop removing the actuator! The tension may mean that the spring is about to come flying out.

stem wears to the point where it breaks in two. When you pull off actuator, the spring can come flying out.

In pneumatic actuators there is a lot of force involved in actuator assembly from the control signal. Be sure to pull the air pressure tubing off before pulling the actuator off. If the actuator is loosened while it is still pressurized, if can smash your hands against the ceiling or anything else in the way. Once the containment is off the diaphragm, it can blow up and explode like a balloon. Electric actuators have gear assemblies and it is easy to get your fingers caught in the gears or any attached levers. These components can exert a surprising amount of force. Also note that these actuators are powered by 24 VAC (sometime 110 VAC), so be very careful working around water and grounded pipes.

Sometimes with electric actuators you can create short circuits inside the conduit when you pull the top off. For example, if the power wiring is not securely connected to the terminals, pulling the valve actuator can loosen the wire and bring it into contact with the conduit, which can, in turn, shock you. Sometimes metal flex conduit can cut into wiring insulation and now the whole assembly is hot.

When replacing packing on a hot water system, allow the system to cool down first before disassembling the valve. If there is a problem

fastfacts

Always shut the power off before removing an electric actuator, and always remove air pressure from pneumatic valves before removing a pneumatic actuator.

fastfacts

If possible, follow manufacturer's instructions for valve disassembly. Don't try to save time by taking shortcuts.

with the valve stem, sometimes the packing can blow out, followed by a stream of water. If the water is still hot, this can lead to scalding. Also note that sometimes there are springs behind the valve packing that shoot the packing out along with the water. Once the actuator is off and you start taking the packing out, take it out slowly and be ready to screw it back in if something goes wrong. If you're unsure, shut the whole system down, let the water cool down, and if necessary drain the system down.

Once the packing is out, determine if the stem is scratched up or gouged. If so, you will have to disassemble the whole valve and put in a new valve stem. If you don't, you can repack but it will be leaking again in a couple of weeks. When putting packing in, pay close attention to getting it in just right. For example, make sure it is not backwards and that it does not bind the stem. This is one of those jobs that is worth doing right the first time.

TROUBLESHOOTING

Failures of HVAC pumps or valves can have catastrophic consequences. For example, the pump on the primary loop of a chiller or boiler is responsible for guaranteeing constant heat transfer between the water and the equipment. If the pump should suddenly stop, the chiller evaporator tubes could freeze, or the water in the boiler piping could flash to steam. Both of these failures would be extremely expensive in equipment repair and potentially dangerous to facilities personnel.

Electric Pump

The heart of a water distribution system is the pump. In HVAC applications, the pump is most likely a centrifugal pump ranging anywhere from a couple of horsepower up to a couple hundred horsepower. If the pump is not operating correctly then the entire water distribution system will not provide the desired heating or cooling effects. Figure 6.25a shows a troubleshooting chart for electric pumps that do not start or have low flow. Figure 6.25b shows the procedures to follow if the pump is too noisy.

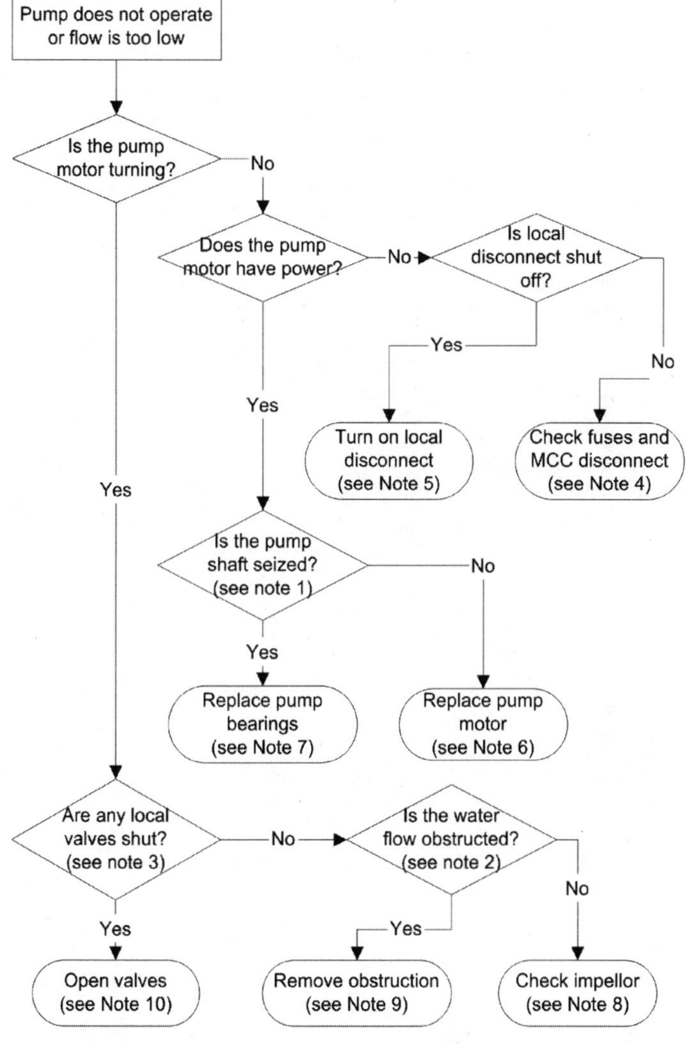

FIGURE 6.25a Troubleshooting chart for non-operating electric pumps.

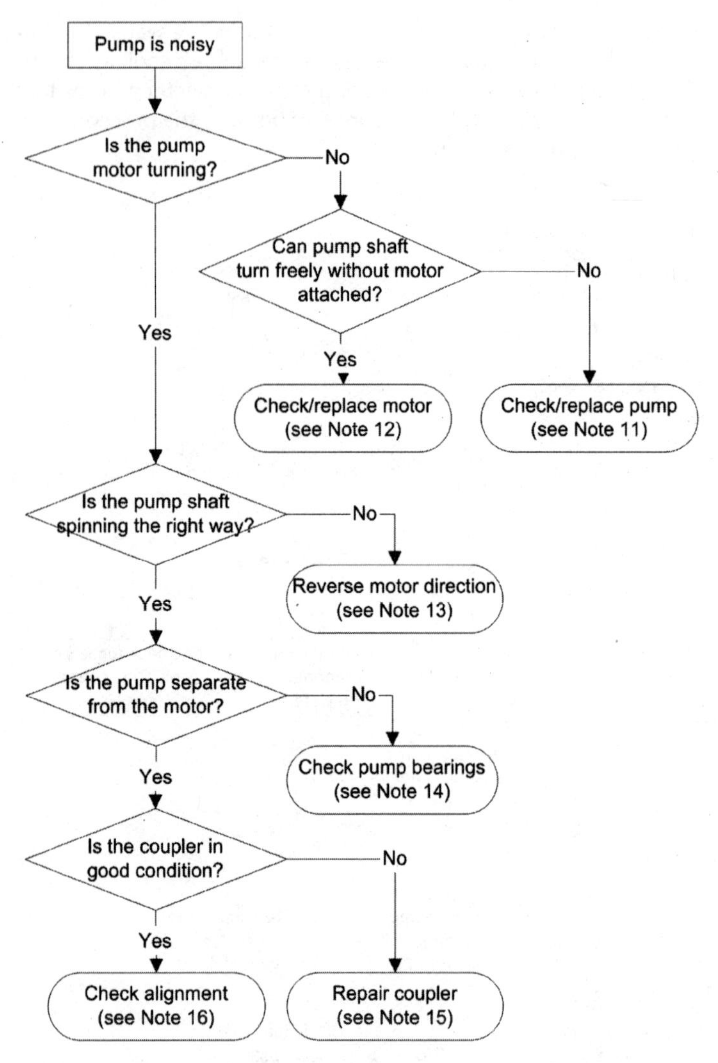

FIGURE 6.25b Troubleshooting chart for noisy electric pumps.

Note 1: If any of the bearing sets on centrifugal pumps (see Figure 6.1) fail then the shaft can seize and stop turning. To check for this condition, you have to remove the bearing cover and check both the inboard and outboard bearing races. Make sure that the local power disconnect is turned off before working on the pump. If the bearings appear to be intact then the problem may be a *locked impeller* (see Note 8).

Note 2: Some rather amazing articles can find their way into piping. Anything from small pieces of metal and wood to rubber gloves can show up even in small piping. Any of these items can get caught in filters, strainers, or in the pump impeller itself. Hopefully you will not have this problem, since pipe obstructions are notoriously difficult to remove. The first thing to do is check any strainers for obvious clogs and see if that cures the problem.

Note 3: Check to see if any upstream or downstream valves have been shut. Keep in mind that the valves may not be near the pump itself. While it is easy to identify if a butterfly valve is closed, double check to make sure that the handle isn't loose so that a closed valve actually appears to be open. Also check the valve stem positions on electric and pneumatic valves to see if they have shut off flow to or from the pump.

Note 4: If the electrical connections are all valid between the motor control center (MCC) and the pump motor, then there is a good chance the problem may be a blown fuse or that the power is shut off at the MCC. You can check for fuse continuity using a standard ohmmeter (see the appendices for details on how to do this). Make sure the MCC power is turned off before accessing the fuses. Do not attempt to override the safety latch on the MCC fuse access.

If fuses keep blowing, there may be a short somewhere in the line between the MCC and the pump motor. Check the connections in the local disconnect as well as the wire nut connections within the motor housing that connect the motor to the line voltage. If these connections are mechanically and electrically secure, then a constantly blowing fuse may indicate that the motor is nearing failure.

If thermal breakers in the MCC keep failing, check for the presence of variable speed drives anywhere on the system (not necessarily the pump motor). In some cases, VSDs can introduce harmonics into the main line voltage that can cause thermal breakers to trip repeatedly. This usually happens at certain VSD frequencies. If possibly, identify those frequencies and set the VSD controller to avoid them.

Note 5: Never activate a local disconnect until you have identified why it was shut off in the first place.

> **fast**facts
>
> *Never reapply power at the MCC if you do not know why the power was turned off. Normally the MCC would be tagged and locked for maintenance or other service reasons, but this step is sometimes overlooked. Never energize a circuit until you have verified that there is no threat to personnel or equipment.*

Note 6: If the pump shaft rotates freely without noticeable play then it is possible that the pump motor has burned out. You should replace the motor and return the pump to service. However, check the new motor frequently to ensure that it is not running too hot. Failure of a motor can be an indication of other problems such as worn bearings, water flow obstructions or pressure instability, etc. Rather than constantly replacing motors, try to identify the underlying root cause of the motor failure.

Note 7: Worn bearings allow extra play in the shaft motion, causing the shaft to wobble as it turns. Eventually, this can cause the pump to seize up. In addition, the pump motor can overheat and fail as well. It is much better to replace the bearings at the first sign of wear rather than waiting for more expensive failures. Periodically check pump shafts for any unusual side-to-side motion.

Note 8: There is a chance that the pump impeller is either locked or has come loose. You will probably have to disassemble the piping to identify if this is the case. If the impeller motion was obstructed while the pump shaft continued to turn, the shaft may be so damaged that you are not able to attach a new impeller to it. In this case you may have to replace the shaft or the entire pump body.

> **fast**facts
>
> *Motor burnout can results from a balancing valve set full open during maintenance and never returned to the correct position. This causes the pump to over-pump and quickly overloads the motor.*

fastfacts

Do not open valves that have been manually shut until you have determined why the valve was shut. There may in fact be a very good reason why the valve was closed, such as a leak or ongoing repairs. Ideally all manual valve shut-offs are tagged and explained, but this is not always the case.

Note 9: Check filters and strainers for any material that may have accumulated in the system. Also, check to see if there has been any maintenance recently performed on any part of the water system that may have introduced solder, PVC glue, or any other contaminants that could block systems. The idea is to identify locations where the obstruction could be without having to take apart significant lengths of the piping system. If this becomes necessary, first look in the obvious places such as at tees, elbows, and valves. Also, check any sensor locations such as thermowells, in-line flow meters, or flow switches that can act as catch points for any material in the water stream.

Note 10: If valves have been shut off during maintenance or seasonal switch-overs, you should be able to easily identify these and open as necessary. If the pump has operated for a long time under no-flow conditions, there is a good chance that the impeller has been damaged from pitting. Check for proper flow and excessive noise once the pump has been returned to service.

Note 11: If the motor is unable to turn because something has locked the pump shaft, you should consider yourself lucky that the motor has not burned out. You should disassemble the pump and fix the impeller or pump bearings or whatever else it is that has seized the pump shaft.

Note 12: The problem would seem to be a bad motor winding or a lost phase. Check the pump wiring and lubrication. If all appears correct, you probably have to replace the motor.

Note 13: Motor maintenance can lead to mis-wired motors. If this is a three-phase motor, reverse any two of the wires.

Note 14: Noisy bearings are an indication that you are about to seize the pump shaft. Replace the bearings as soon as you can.

fastfacts

Mistakes in wiring can occur anywhere in the system. If you have one pump operating backwards, look around to see if there are any more.

Note 15: If the pump is separate from the motor, check to make sure the coupler is in good condition. A broken coupler can rattle as it spins. This usually means it is getting ready to break apart—a pretty dangerous situation. Fix this as soon as you can.

Note 16: Good alignment between a separate motor and pump is crucial to avoid seized shafts, worn bearings, and overheated motors. Check to make sure the alignment is straight. Use shims to adjust the heights as necessary.

Valves

Sometimes valves or their associated actuators fail to operate properly leading to a valve that does not move or does not seal correctly. Figure 6.26a shows a troubleshooting flow chart for automatic control valves that do not operate; Figure 6.26b shows a procedure for dealing with control valves that leak too much.

Note 1: Most electronic valve actuators are powered by 24 or 120 VAC power supplies. If the power is not available to the valve motor, it may not move (some actuators are spring-loaded to return to a NC or NO position upon loss of power). If there is a transformer used to provide the 24 VAC power, make sure that the proper connections are intact on both the primary and secondary sides of the transformer. Also, many transformers have multiple winding configurations, allowing for a wide variety of output voltages. If the transformer is not providing the correct output voltage, you may need to re-wire the transformer depending on the input voltage.

Note 2: Many electronic actuators rely on isolated power supplies in order to work properly (see Chapter 4 for additional information on isolated power supplies). If the actuator is powered by a trans-

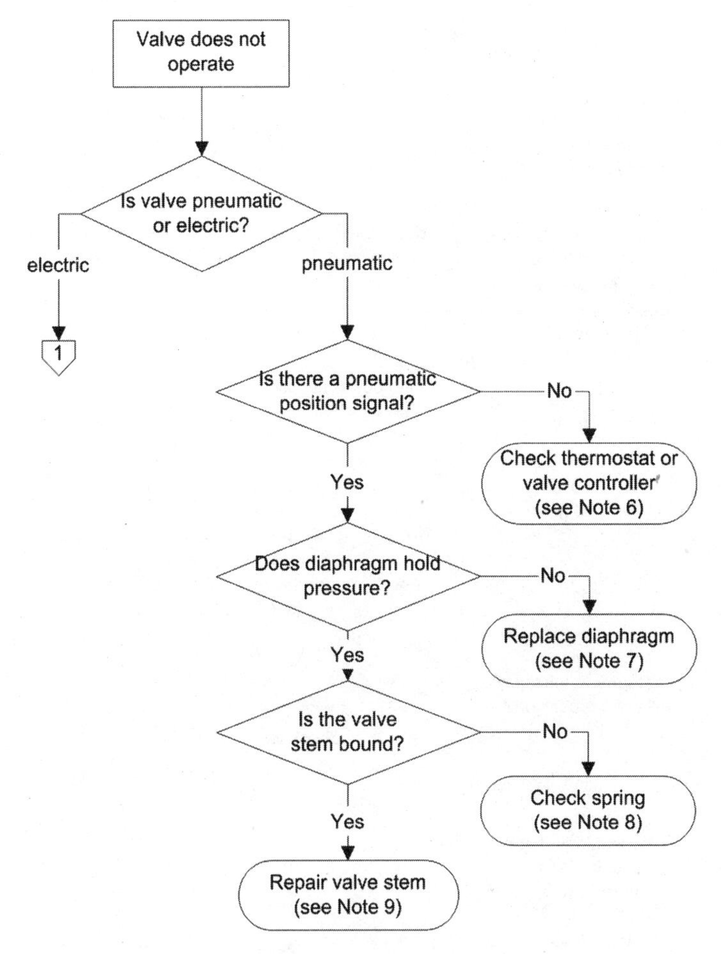

FIGURE 6.26a Troubleshooting flow chart for non-operating control valves.

former that also is connected to other actuators or is referenced to earth ground, then it is probably not isolated. A good rule of thumb is one transformer per actuator.

Note 3: The valve linkage can bind up if the linkage is rusty, bent, or at a minimum or maximum extreme. Examine the linkage for binding and repair or replace as necessary.

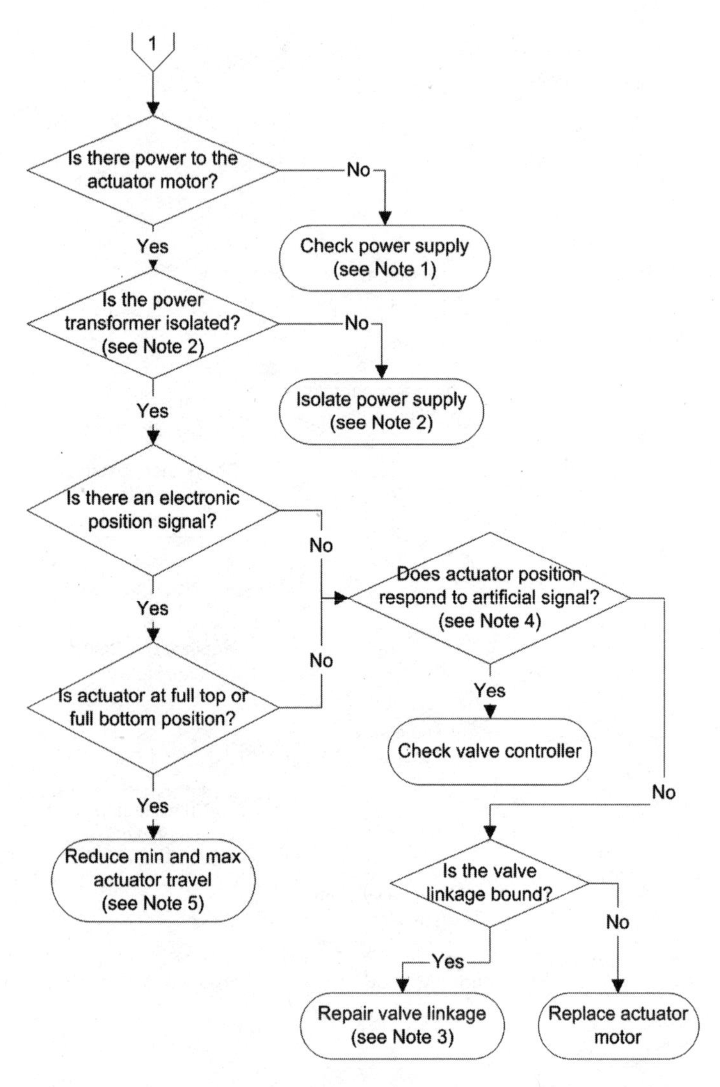

FIGURE 6.26a *(continued)* Troubleshooting flow chart for non-operating control valves.

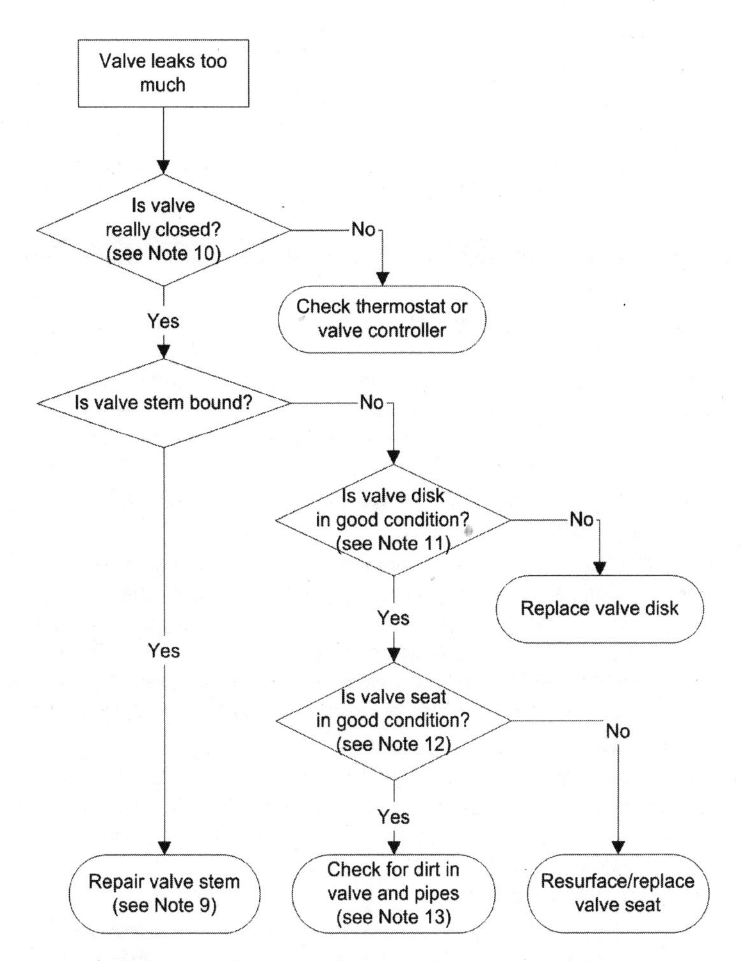

FIGURE 6.26b Troubleshooting flow chart for leaking control valves.

Note 4: If the actuator has a valid power source connected, you can check to see if the actuator responds to a control signal by connecting a 1.5 or 9 VDC battery to the electronic signal inputs. Disconnect the existing signal wiring first and make sure that you connect the positive side of the battery to the positive side of the input signal terminals. You should be able to drive the actuator from one end to the other if the signal is in milliamps or volts.

fastfacts

It is often easier to hear or feel the air coming from a hole in the diaphragm as opposed to looking for a hole.

Note 5: Under some circumstances, an electronic actuator motor can run the actuator shaft off the threads of the electronic motor. This can happen when the actuator shaft is at its minimum or maximum extreme. You can usually inspect this visually by taking off the actuator cover or by gently forcing the shaft up or down until the gears engage again. If this is the problem, you may need to reposition the motor on the actuator shaft or (if available) by setting the extreme limits.

Note 6: Pneumatic valves will not operate unless there is a control signal of sufficient pressure to drive the actuator. Check the signal pressure and make sure it is somewhere in the 3 to 20 psig range. If the valve is part of a baseboard or reheat system, check the thermostat to see if it has both inlet pressure and is providing outlet pressure. If not, the problem may be in the thermostat (see Chapter 16).

Note 7: You may be able to see if the actuator responds by forcing a signal using a squeeze bulb like the one in Figure 6.27 (you cannot generate the necessary pressure with your lungs). If the actuator does not respond to a signal, you may be able to open up the cover and check for holes in the diaphragm. Replace the diaphragm if you find any holes.

Note 8: If the diaphragm is able to hold pressure (as discussed in Note 7) then you should remove the actuator and see if you can move the valve stem up and down manually. Sometimes the plug gets stuck in the seat and you can force it out. If the stem moves easily, check to see if the valve spring is corroded. Sometimes you can get a pretty good idea of the condition of the spring by looking at the actuator (see Figure 6.28).

Note 9: The valve stem can bind up if it gets scored or worn. This happens if the valve seals or packing are dirty. You may need to take the valve out of service and disassemble it to remove the stem. Check the stem area shown in Figure 6.29 for any wear. The surface should be smooth and polished. If not, rebuild the valve and replace the stem before returning it to service.

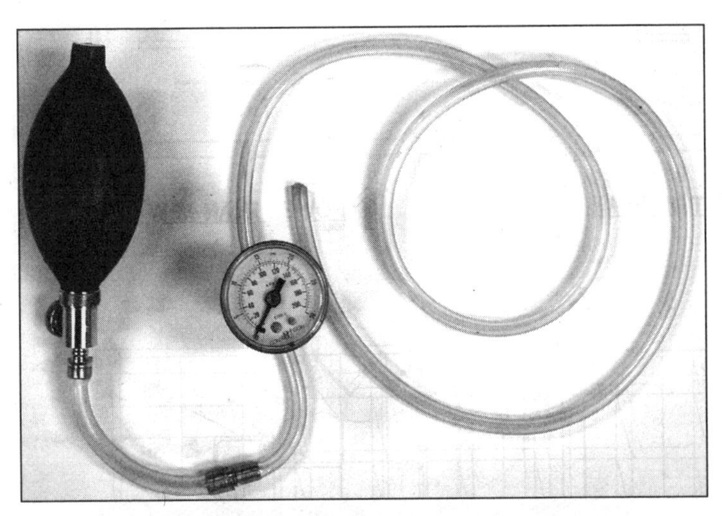

FIGURE 6.27 Squeeze bulb used for testing pneumatic actuators.

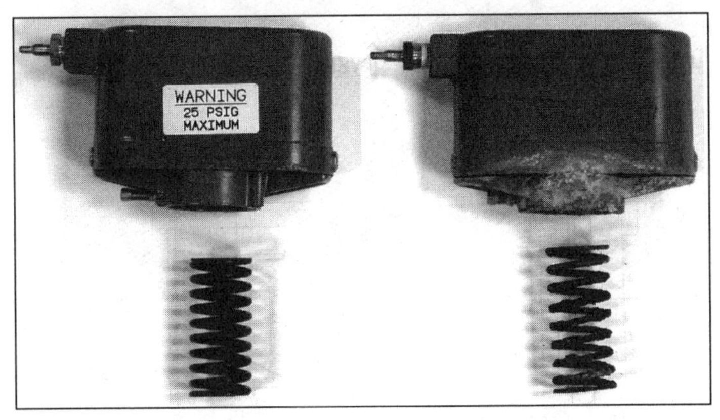

FIGURE 6.28 New (left) and corroded (right) valve actuator and spring.

Note 10: It is entirely possible that the valve is leaking through because it is never getting a full close signal from the thermostat or controller. If the actuator has a position indicator such as shown in Figure 6.30, check to make sure the valve is fully closed. If not, check the control signal to make sure it really is indicating a zero position. If the actuator is pneumatic, make sure that the spring range is properly sized to allow the valve to close.

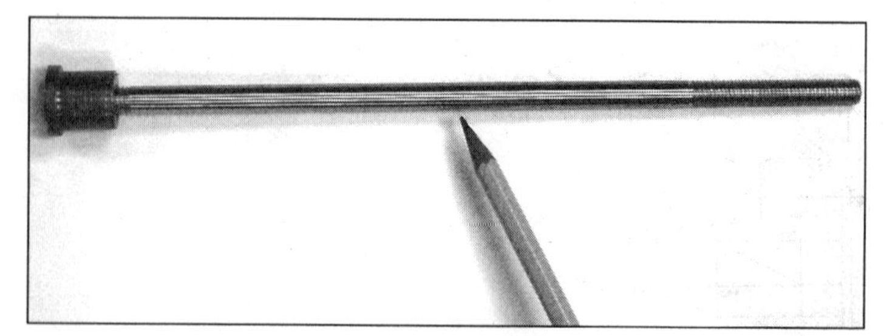

FIGURE 6.29 Valve stem showing area to check for wear.

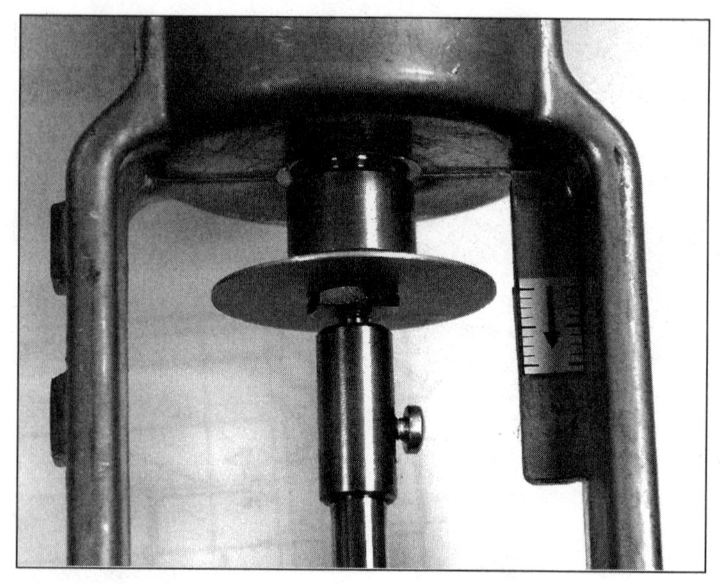

FIGURE 6.30 Picture of valve stem and position indicator.

fastfacts

If the valve is stuck closed, there is a good chance that the spring is rusted or broken.

Note 11: At this point you need to start disassembling the valve. Make sure all upstream and downstream isolation valves are closed and, if it is a hot water or steam valve, that the system has cooled off. Investigate the valve disk for wear or erosion at the location shown in Figure 6.31. Be sure to check the side that comes into contact with the seat and replace if necessary.

Note 12: If the valve disk is okay, check the valve seat at the location where the disk makes contact (see Figure 6.32). Make sure there are no uneven patches, worn spots, or wire draw. If possible, use a manufacturer's resurfacing kit to polish the seat. Larger valves (greater than 1 inch or so) are more prone to wire drawing and seating problems because of the greater pressures involved. If the valve body itself is leaking liquid or steam, it may be time to replace the seals and packing. Fortunately, most valve manufacturers sell rebuilding kits (as shown in Figure 6.33) that make it easy to replace the key components in specific valves. In fact, some valve manufacturers market equipment that allow you to disassemble the valve without removing it from the piping. Figure 6.34 shows a valve-rebuilding tool that is used to remove a valve plug without even having to turn off the water service to a valve. It allows for the replacement of the valve trim even when the system is under pressure. Of course, this is only practical on smaller valves, but it can still save you a tremendous amount of time.

FIGURE 6.31 Check valve disk for wear or erosion on side facing seat.

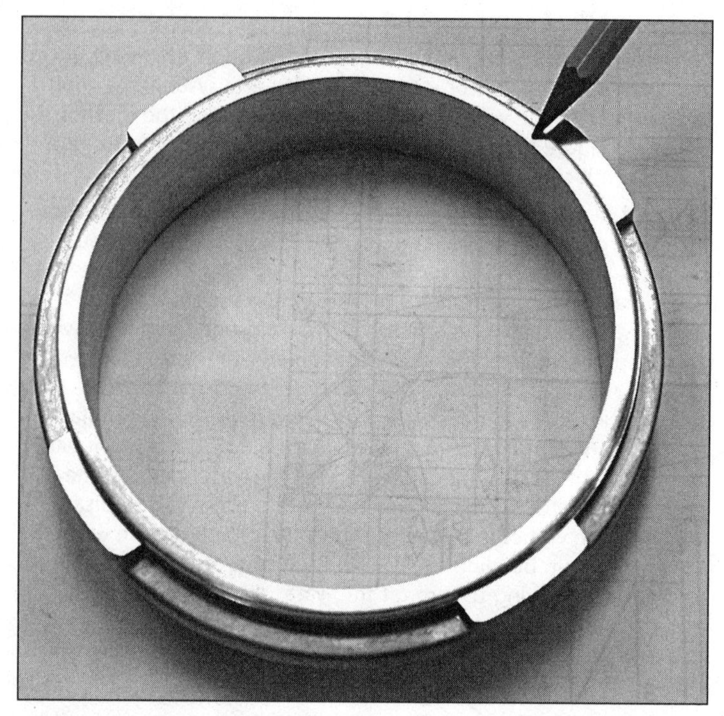

FIGURE 6.32 Check valve seat for wire draw, pitting, dirt, breaks, or erosion.

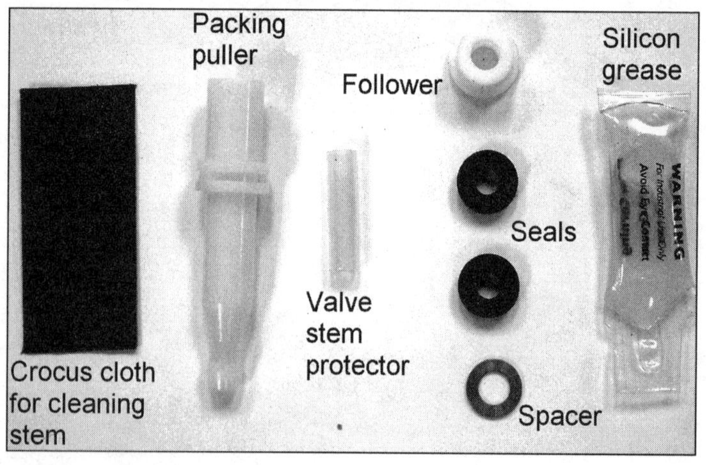

FIGURE 6.33 Valve repacking kit.

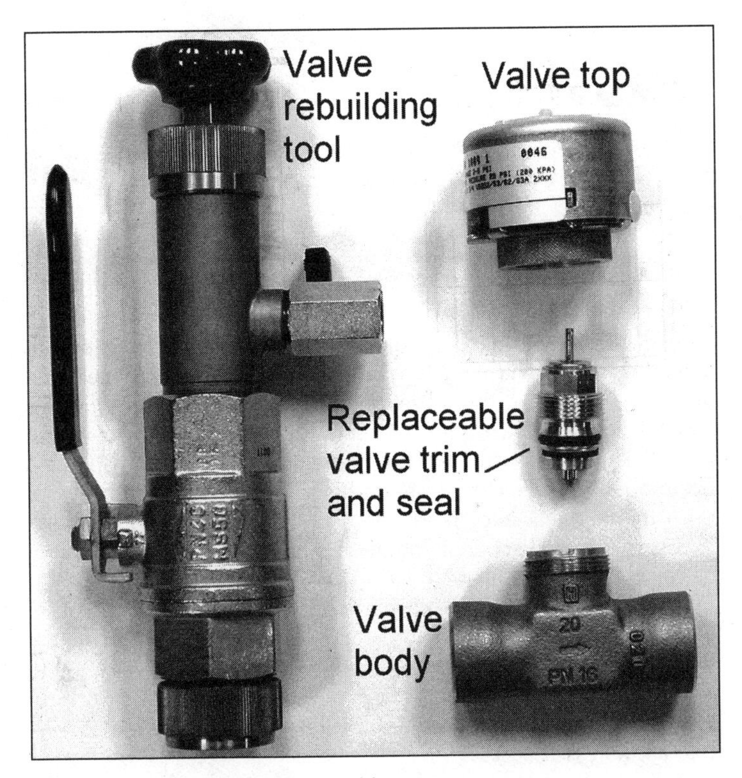

FIGURE 6.34 *In-situ* valve disassembler.

Note 13: If the valve and actuator components seem okay, you most likely have some dirt or rust in the system that is getting trapped between the seat and the disk, preventing the valve from making a tight seal. You may need to periodically flush out the valve or replace it altogether.

PRACTICAL CONSIDERATIONS

The crucial design consideration for pumps and valves is to make sure everything is sized properly for the application. Valves that are too large are difficult to control and valves that are too small don't

fastfacts

The smaller the valve, the more likely it is to get gummed up by dirt. Also, dirt tends to build up more in valves on the lower floors of buildings than on the higher floors.

have enough capacity to handle the loads. Obviously your pumps should be sized to have enough capacity without wasting energy. In addition, you should try to size the pump so that it operates near its best efficiency point. Figure 6.35 shows a typical pump curve with the efficiency lines drawn. In this figure, the pump can be operated up to about 87 percent efficiency. If you operate the pump too far from the best efficiency point then you may have a problem with excessive shaft deflection or vibration. This, in turn, leads to bearing damage and motor failure.

Also keep maintenance in mind when selecting pumps. Close-coupled pumps have the motor and the pump together on the same shaft and are generally harder to work on because you have to remove both the motor and pump housing to get to the packing, seals, etc. Pumps that are separate from the motor have a coupler that allows easier access.

Pump Bearings

Typical close-coupled pump motors use sleeve bearings, ball bearings, and roller bearings. Separate pump motors usually use sleeve bearings, roller bearings, or tapered bearings. If there are any unusual sounds or vibrations you should check the conditions of the bearings. If they get a little rough they'll start to wiggle around and

fastfacts

You should not operate a pump more than 10 percent away from the best efficiency point.

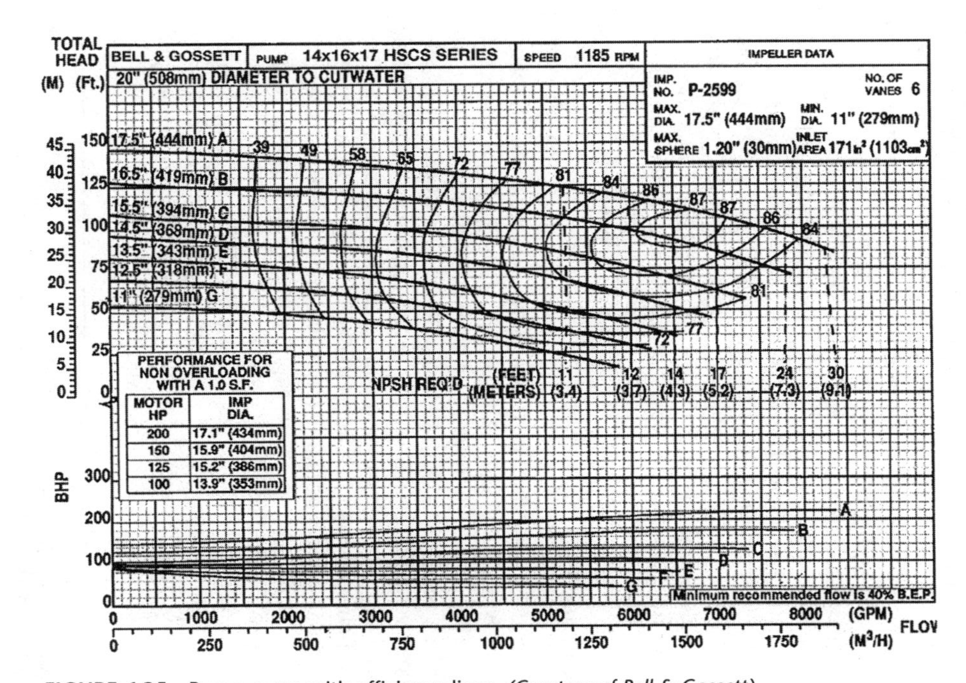

FIGURE 6.35 Pump curve with efficiency lines. *(Courtesy of Bell & Gossett)*

can knock the seals out of the motor or pump. It is important to make sure they're lubricated—some larger pumps have automatic lubrication systems, whereas most smaller pumps require manual lubrication every year.

Most pumps are held in place by steel or rubber mounting brackets. Check these brackets frequently to make sure that they are in good condition. Failed brackets can put tremendous stress on the motor shaft and bearings.

The orientation of a pump must be paid attention to. Some pumps are made to be mounted on their side, while others should be mounted vertically. This is generally more important on close-coupled pumps. On a vertical close-coupled pump there are bearings that hold up the armature. If you take a pump that should be on its side and put it vertically, the bearings can rip out since they are not made to take the weight of the impeller in the armature.

Pump Seals

One of the most critical parts of a pump is the seal. Watch the seal and make sure it doesn't leak. Once it starts leaking it doesn't get better. It is unusual to have a seal blow out—usually they start leaking first. In systems with a lot of dirt, the seals can simply grind themselves to pieces. This can happen repeatedly. This is another good reason to keep rust and dirt out of the water system.

If a pump is used in a hot water application, make sure that the pump is specified for high temperature duty and that the seals are meant for high temperature applications. General pump seals are usually made of some kind of ceramic, but these won't last very long in hot water applications. For hot water pumps you should specify seals made from tungsten or some other appropriate material. However, if you run the pump over 200°F without sufficient system pres-

fastfacts

Be aware that some new pumps are packaged and shipped without oil. They need lubrication before operating. Do not assume the pump is oiled unless the installation instructions specifically state so.

fastfacts

Pump seals may last longer if the pumps are "bumped" for a few seconds in the off-season. This keeps the seal contact area from sticking together on seasonal startup.

sure, the seals won't last long no matter what kind of material is used. The reason for this is that if the pump runs hot enough the water turns to steam right at the seal and doesn't lubricate the seal very well. This usually results in a torn-out seal. Seal coolers are available for these applications.

Sometimes if a pump is leaking, technicians wrap rags around the seals as a temporary solution. This is a bad idea all around. If the pump is having problems, something may go wrong unexpectedly and it may get quite hot. Placing flammable materials around the pump is simply not a good idea.

Pump Motors

If you replace a pump motor, make sure it is going the right direction once it is rewired. Usually the pump wiring panel is clearly labeled, but on a three-phase motor, is it easy to reverse two of the phases. This causes the motor to turn in the wrong direction. Sometimes the wiring problem can occur at an electrical panel that has been replaced. A pump that is going backwards can cavitate, causing damage to the impeller and seals.

Single-phase motors can have two sets of windings: the start winding is used to initially get the pump spinning while the run winding is used once the pump has started. Sometimes a single-phase motor can get stuck in the start winding without moving over to the run windings. This causes the motor to run hot.

If there is no flow through the pump but the shaft is spinning, turn off the pump and make sure the shaft is spinning the right way. Do not try to touch the pump shaft to determine which way it is turning. You can easily lose a finger this way. An easy way to identify the rotation of a pump is to watch the shaft carefully when the power is removed. As the pump quickly comes to a stop the rotation direction should be obvious.

fastfacts

If you find one pump going backwards and the pump wiring appears to be okay, the problem is probably at the electrical panel. If this happens, look around because there may be other pumps that are also going backwards.

If the pump trips out, there could be a bad winding, loose winding, or a lost phase on a three-phase motor. Try to identify the problem and fix it instead of just turning the power back on.

Control Valves

Plug and seat valves are used almost exclusively in HVAC applications. To ensure that these keep operating correctly, it is extremely important to prevent the build-up of dirt inside the valve body and on the plug. Dirt, rust, and other goo in the water stream can keep it from closing. In a hydronic application, sometimes build-up is so bad that it requires a valve replacement. The list of items that can clog valves is not limited to dirt. Other unanticipated objects in pipes can include a surprising assortment of things such as paper and plastic sacks, solder, welding rods, slag, dirt, and rocks. While their method of entry into a hydronic system is unknown, they usually wind up in valves because that is the smallest diameter in systems.

Stainless steel valve stems seem to last longer than brass valve stems. Most today are stainless steel, and if you need to replace a valve stem, try to find one in stainless steel. On some smaller valves, the plug is molded onto the valve, so it's necessary to replace both the seat and the stem.

When given the opportunity, replace the valve stem seals. If you have the valve apart, go ahead and replace what you can since you will need to do it eventually. Stem seals typically last between 5 and 10 years under normal circumstances, but if the valve is doing lots of travel it can cut the life down to maybe 2 or 3 years. If the water is particularly dirty, a seal may last only 6 months or so before it starts leaking. That said, it is also possible for the valve stem seals to last for 20 years. It all depends on the usage. Rubber seals can get hard and brittle, especially in hot water systems, and also have a tendency to shrink up a little bit in chilled water valves. Any glycol in the system can actually attack the stem seal causing it to begin leaking almost right away, so make sure the valves being used are compatible with glycol.

Control valves should be properly sized. Valves that are too large or too small result in poor control, occupant discomfort, and, in extreme cases, equipment damage. In particular, modulating control valves should not be oversized. When sizing valves, you should calculate the desired C_v value and make sure you specify a valve that matches this value. The only exception is that in smaller valves (less that ¾ inch) you may want to use a slightly higher C_v rating because this tends to reduce the amount of dirt build-up in the valve.

Pneumatic actuators for large valves (1 inch and greater) usually come with positive positioners. These allow for closer control of the valve and make it easy to stage valves for specific control sequences. For example, in an air-handling unit you may have a hot and chilled water coil controlled off the same thermostat. You can then sequence the positioners. For example, a 8-13 psi signal operates the reverse acting chilled water valve and the 3-6 range operates the direct acting hot water valve.

Glycols

Ethylene glycol and propylene glycol are sometimes used for freeze protection in piping systems. If there is a potential of accidental draining to the environment (for example, in an open basement without spill catchers), it's best to use propylene glycol to minimize environmental damage.

fastfacts

In one retrofit, about 200 reheat coils were installed with ½ inch, three-way valves. For dirt control, the valves were specified using a C_v value about four times higher than the normal calculated value. Even after several years of use, there were no problems with either the control or dirt build-up.

Both propylene and ethylene glycols have a tendency to seek out and leak through gaps in pipe connections, valve seals, and pump packing. Propylene glycol is slightly more problematic than ethylene glycol in this respect. Valve seals seem to lubricate and seal better with ethylene glycol.

chapter 7

WATER DISTRIBUTION SYSTEMS

Water distribution systems are the combinations of pumps, pipes, valves, and other apparatus that move hot and chilled water throughout a building. Details about the operation of pumps and piping are given in Chapters 3 and 6. This section discusses the issues related to the pumping of water over large distances between two pieces of equipment.

THEORY OF OPERATION

The basic operation of a water distribution loop is illustrated in Figure 7.1. The boiler or chiller provides heating or cooling to the *primary loop*. The temperature in the primary loop is controlled by the boiler or chiller. The three-way valve allows some of the water in the primary loop to enter the secondary loop. The valve modulates to maintain the temperature in the secondary loop. A *nulling loop* (sometimes called a *bridle* or *decoupler*) exists so that if the three-way valve is completely closed there is still circulation in the secondary loop. The figure shows two loads, but in practice there can be dozens of loads. The loads are the heating or cooling coils or any other equipment that uses the hot or cold water.

fastfacts

The primary loop allows for good temperature control and equipment safety, but has additional equipment costs. Probably half of existing HVAC systems do not use primary loops, but instead run the distribution water directly through the chiller or boiler.

Water distribution loops are pressurized to about 10 to 15 psi. Recall from Chapter 6 that the net positive suction head (NPSH) of a pump is the minimum static pressure at the pump inlet required to prevent cavitation. The NPSH dictates the minimum pressure to which the water distribution system should be charged.

There are many different kinds of water distribution systems. Some of these are listed here:

- Low temperature water (*LTW*) systems have maximum temperatures of 250°F and pressures of 150 PSI.
- Medium temperature water (*MTW*) systems have upper limits of 325°F and 150 PSI.

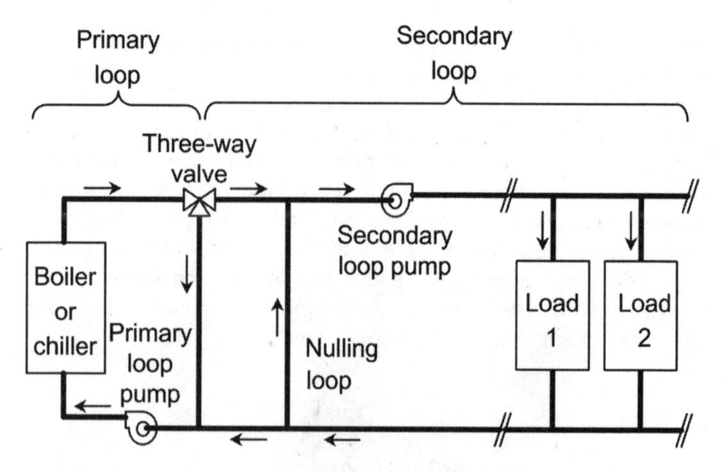

FIGURE 7.1 Basic water distribution loop.

- High temperature water (*HTW*) systems have upper limits of 450°F and 300 PSI
- Chilled water systems (*CHW*) are typically at 30-50°F and 50 to 200 PSI

Other acronyms in water temperature distribution are dual-temperature water (*DTW*) systems, condenser water supply (*CWS*), and condenser water return (*CWR*).

There are different ways that the piping can be configured in a water distribution system. The top system in Figure 7.2 shows a *direct-return* system. In the direct-return system, the first coil after the secondary loop pump sees the greatest pressure drop, while the coil farthest from the pump sees a small pressure drop. This means that the first coil gets the most flow and the last coil sees little, if any, flow. Direct return systems therefore require the use of balancing valves on all coils that are adjusted to ensure that the total pressure drop across each coil is the same.

The bottom system in Figure 7.2 shows a *reverse-return* system. In the reverse-return system, an extra length of plumbing is used on the return side so that the first coil on the supply side is the last coil on the return side. This is a natural way of getting about the same pressure drop across each coil. In reverse-return systems, there is generally no need for balancing valves but there is an extra initial cost to run the extra length of pipe. Also, the total required pumping capacity is greater in reverse-return systems due to this extra length of pipe.

Two and Four Pipe Systems

In some systems, significant piping can be saved by the use of a two-pipe distribution system. Figure 7.3 shows a typical *two-pipe system*

fastfacts

From a maintenance standpoint, the reverse return system is hard to troubleshoot because it can be difficult and confusing to trace the water lines, particularly in large systems. It's a nice way to pipe but it also makes it hard to make modifications.

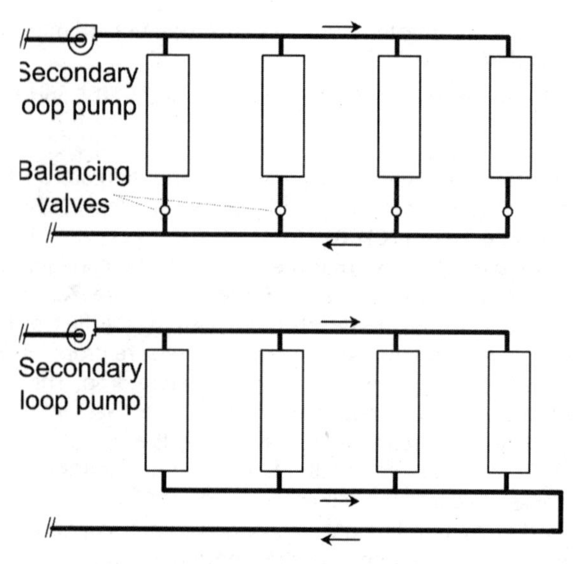

FIGURE 7.2 Direct and reverse-return systems.

that serves fan-coil units in the zones. Only one set of supply and return pipes are used to circulate both hot and cold water. The assumption is that the heating and cooling loads in the zones are strongly dependent on the season and that there will be never be a need for simultaneous heating in one zone and cooling in the other. Two pipe systems are used when there is a need to reduce initial costs, but they are a poor choice for good comfort control.

An alternative to the two pipe system is the *four-pipe system* as shown in Figure 7.4. The two sets of supply and return pipes are dedicated to their respective heating and cooling systems. These kinds of systems are common in large buildings in moderate-to-cool climates where simultaneous heating and cooling is a likely situation. A variation on this system is the *three-pipe system* in which there are separate and dedicated supply branches for the heating and cooling system, but a single, common return line. Three-pipe systems usually do not work too well and can create big control problems.

Parallel, Series, and Variable-Speed Pumping

The demand for water flow in a water distribution system changes over the course of a day as the zone loads change. Often, different

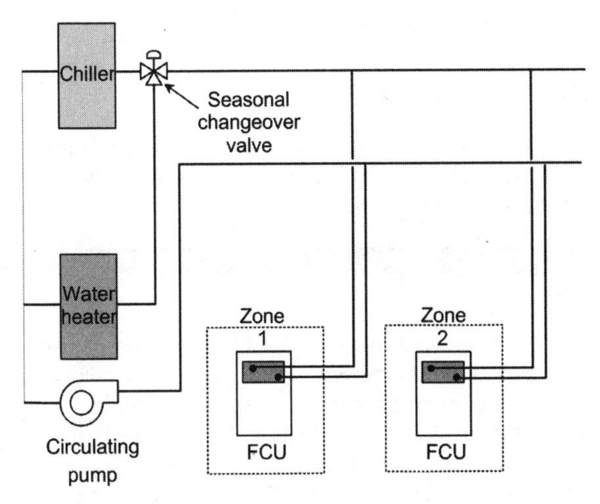

FIGURE 7.3 Two-pipe air-water system.

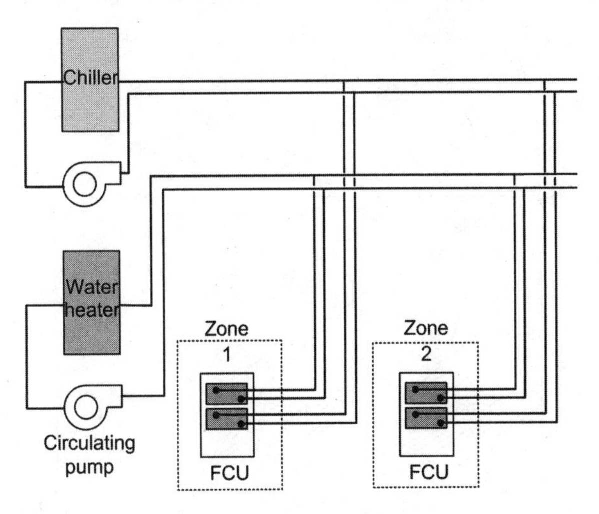

FIGURE 7.4 Four-pipe air-water system.

fastfacts

Water systems should be designed so that the pressure drop across the modulating valves is at least 50 percent of the pressure drop across the entire system.

pump configurations are used to vary the flow. Pumps can be configured to operate by themselves, in series, or in parallel with other pumps. In a series configuration the two pump curves are additive as shown in Figure 7.5. This kind of configuration is useful where there are high system pressure requirements. Parallel pumps, such as the arrangement shown in Figure 7.6, are useful for systems that have high flowrate requirements.

These days it is common to find variable speed drives in water distribution systems. They are particularly useful when working with systems that have strongly variable water requirements such as smaller HVAC systems in moderate climates. Pumps with variable speed drives

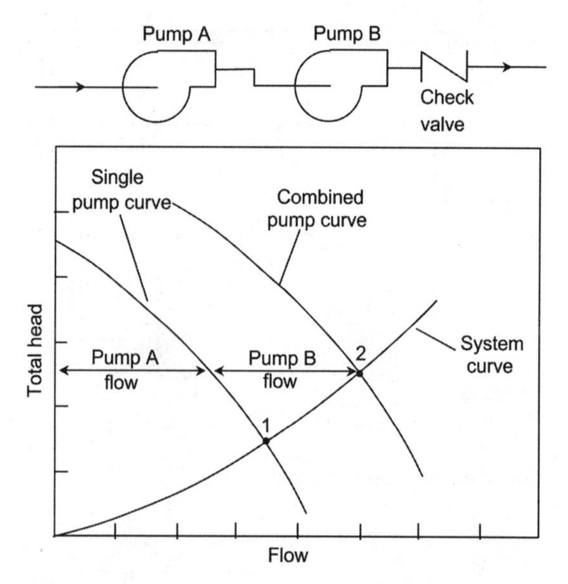

FIGURE 7.5 Schematic and pump curves for series pumps.

have pump curves similar to series pumps (Figure 7.7) and can achieve significant energy savings under low demand. Recall that the pump laws say that the power consumption is proportional to the third power of the flow rate. So a variable speed driven pump running at half the rated speed uses $(1/2)^3 = 1/8$ the rated power.

CONTROL

There are four main goals when controlling water distribution systems:

- Maintain controllable pressure across control valves. The expected flow rate through valves at a given opening will be achieved only if the rated pressure exists across the valve. If the pressure is too low or too high, so is the flow rate.

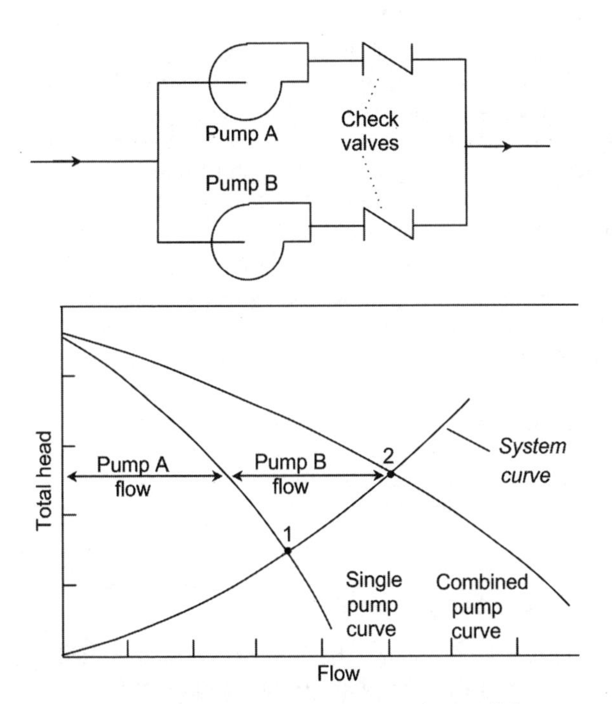

FIGURE 7.6 Schematic and pump curves for parallel pumps.

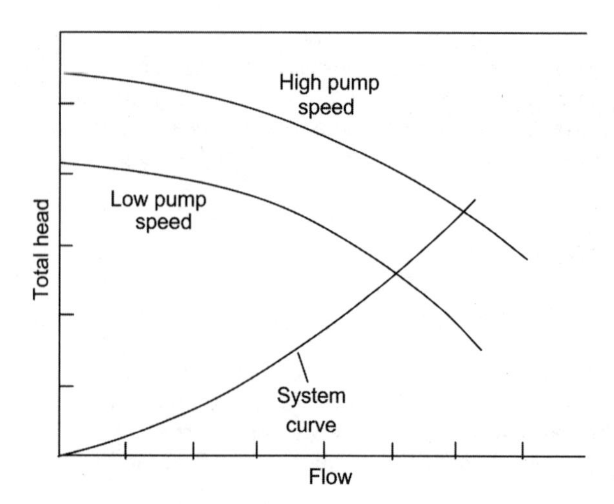

FIGURE 7.7 Pump curve for variable speed pump.

- Maintain required flow through heating or cooling source. Equipment like boilers and chillers must have flow through them when they are operating. If a boiler flow stops, there is the danger of flashing the water inside the heat exchanger and an associated explosion. Likewise, if water stops flowing through a chiller evaporator there is the possibility of freezing the water inside the evaporator barrel.
- Maintain desired water temperature to terminal units. The valves between the primary and secondary loops must modulate properly so that the supply water to the load is at the expected temperature.
- Maintain minimum flow through pumps. Pumps, like heating and cooling equipment, must have some minimum flow through them in order to prevent cavitation and the associated pump damage. It follows also that the control system must properly stage multi-pump systems.

Primary/Secondary Loop

In primary/secondary loop systems, the *primary loop* is often used to obtain rough control of the water temperature and the valve to the *secondary loop* is used to maintain more precise temperature control of the distributed water. Figure 7.8 illustrates a primary/secondary arrangement for a chilled water system. The primary loop is maintained about 5° to 10°F below the desired temperature in the secondary loop. As the cooling load increases, the temperature of the return water in the secondary loop also increases. The temperature sensor records this as a rise in the secondary supply water temperature, and the temperature controller opens the three-way valve to allow more primary loop water into the secondary loop. As the three-way valve allows more and more of the secondary return water into the primary loop, the chiller stages up until the capacity of the chiller is met.

Secondary/Tertiary Loop

Water from the secondary loop needs to flow into the coils in air-handling units, baseboard heaters, reheat coils, and other loads. These coils are branched off the secondary loop and are sometimes called the *tertiary loops*. The simplest way of controlling the amount of heating or cooling is to use a two-way control valve to vary the water flow. Figure 7.9 shows what a typical coil arrangement would look like.

However, use of a two-way valve on the coil can lead to other problems. Since the water is supplied from the secondary loop something must be done with all the water circulating in the secondary loop so that the pressure does not build up. Also, in two-way

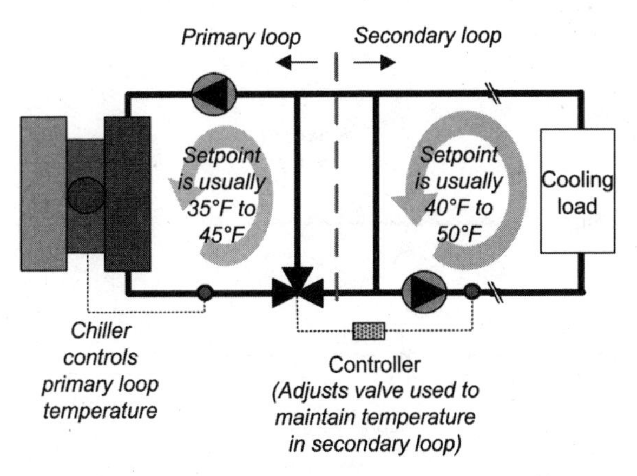

FIGURE 7.8 Example of primary/secondary loop piping.

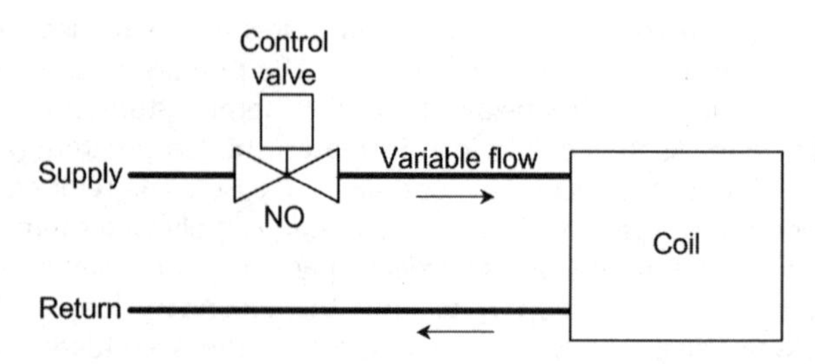

FIGURE 7.9 Two-way valve on coil.

valve systems, there can be a sudden and significant amount of heating or cooling that occurs as soon as the valve opens. For larger coils, this can make it difficult to maintain the coil air outlet temperature when the loads are low.

One way of handling the pressure problems in the secondary loop is to use a three-way valve in the coil with a bypass loop. Figure 7.10 shows how a three-way mixing valve is used to maintain constant flow through a coil branch, while Figure 7.11 shows how it is done with a three-way diverting valve. In both cases a balancing valve gives the bypass loop the same pressure drop as the coil. This makes it easier to maintain proper differential pressure in the secondary loop. In some cases, you may need balancing valves in both the bypass and the coil to make sure that the coil loop flow is the same across the entire travel of the valve. In general, three-way mixing valves offer better temperature control than three-way diverting valves.

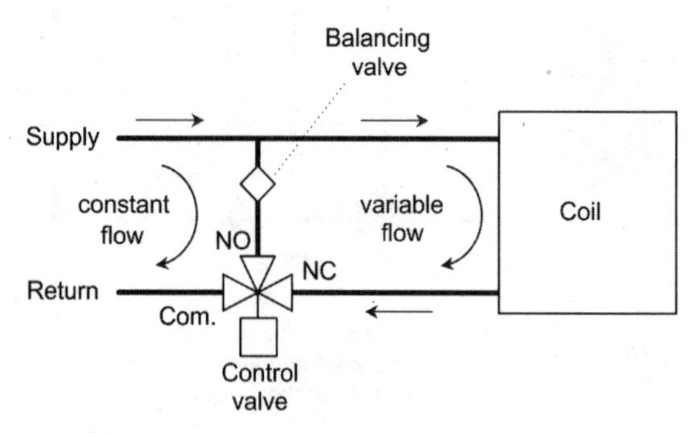

FIGURE 7.10 Three-way mixing valve on coil.

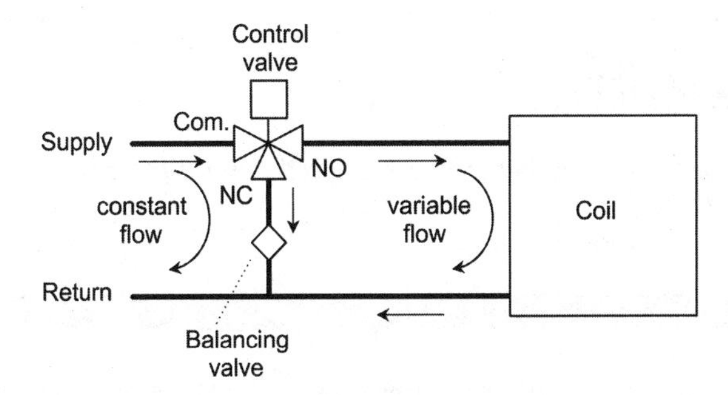

FIGURE 7.11 Three-way diverting valve on coil.

The problem of controlling the coil leaving air temperature when there are small loads can be solved by using a small circulating pump on the coil itself (see Figure 7.12). This is called a *pumped coil* and has the advantage that there is always water flowing through the coil. The three-way valve is used to change the temperature of the coil water loop. As with the primary/secondary system, you need to install a nulling loop to ensure coil flow even when the three-way valve is fully closed to the supply water. Pumped coils offer much closer control of the discharge air temperature than non-pumped coils.

You should size your control valves so that under normal heating and cooling conditions the valves operate right in the middle of their throttling range. In other words, you want your valves to work right around the 50 percent mark. This allows nice, predictable control of the valve and makes the whole water distribution system work better.

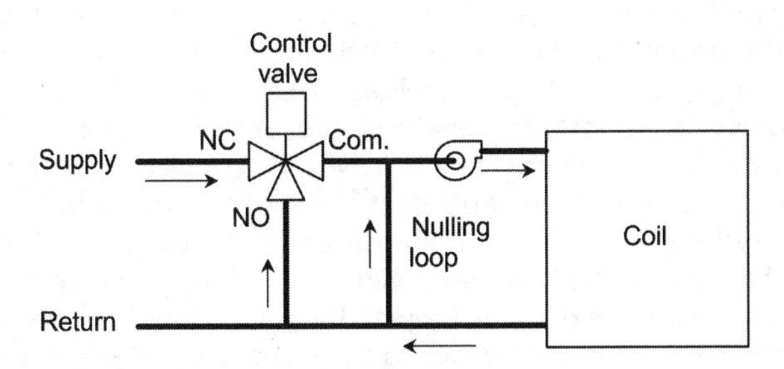

FIGURE 7.12 Three-way mixing valve on coil with circulating pump.

fastfacts

Pumped coils offer better freeze protection since there is always flow through the coil, but the initial cost is higher because of the extra hardware and pump.

Control valves on coils should be equal percentage valves (see Chapter 6). By using equal percentage valves you can counteract the rapid heating and cooling coil response under low flow and end up with a useful linear control characteristic illustrated in Figure 7.13. This lets you choose one set of control constants for the coil outlet temperature feedback loop without have the system become unstable at very low or very high loads.

Differential Pressure Control

The pressure across the supply and return lines of a water distribution system must be maintained above a minimum threshold in order for the control valves in that system to work properly. Consider the extremes: if there was no pressure drop in the system (i.e., no water flow) the valves would do absolutely nothing, while if the pressure across the valve is too high, there would be a great deal of water flow as soon as the valve was cracked open. The problem is compounded if three-way valves are not used on the system loads and the pressure control must be accomplished through other means.

The pressure control should rely on a pressure transmitter located between the supply and return piping located about two-thirds of the length down the run. The pressure signal goes back to a differential pressure control valve (in smaller applications) or a variable frequency drive (in larger applications). Unfortunately, a lot of designers do not include pressure control in their systems.

In particular, two-way reheat valves in terminal units or baseboards need some type of differential pressure control either in the main system or at the coil. These valves need a 10 to 13 psi differential to work properly; if the pressure is too high you run into noise problems and have difficulty keeping the valves closed. A lot of overheating complaint calls are caused by bad differential pressure control. In addition, too high of a pressure difference can start tearing the discs out of valves, leading to big maintenance costs.

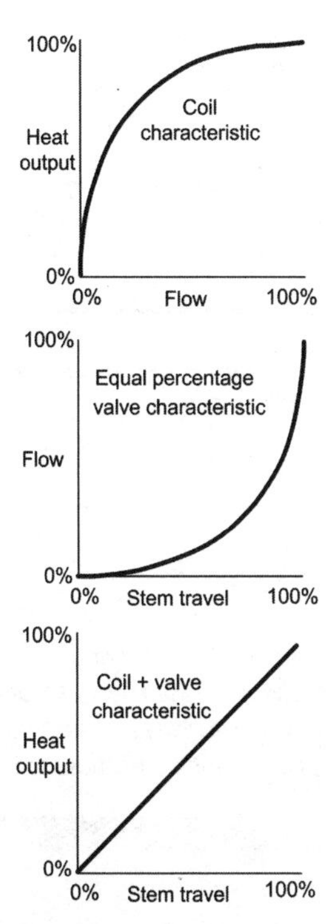

FIGURE 7.13 Combining characteristics of coil and valve for linear response.

One method of accomplishing pressure control is illustrated in Figure 7.14, where a direct-return system uses a bypass loop and a control valve in order to maintain the system pressure differential. With a fixed-speed pump and widely varying loads with two-way valves, the system pressure drop is maintained using a bypass valve controlled by a differential pressure sensor. Use care when choosing the gains for the control loop, however, since water is incompressible and small changes in the valve position can lead to large changes in the system pressure drop. It helps considerably to use double-seated valves for this kind of pressure control—twice as much water can flow through such a valve for a given pipe size. Note that the pressure sensor and bypass can exist anywhere in the system, not just at the end as in Figure 7.14. For example, the pressure control bypass loop in Figure 7.15 is located adjacent to the pump.

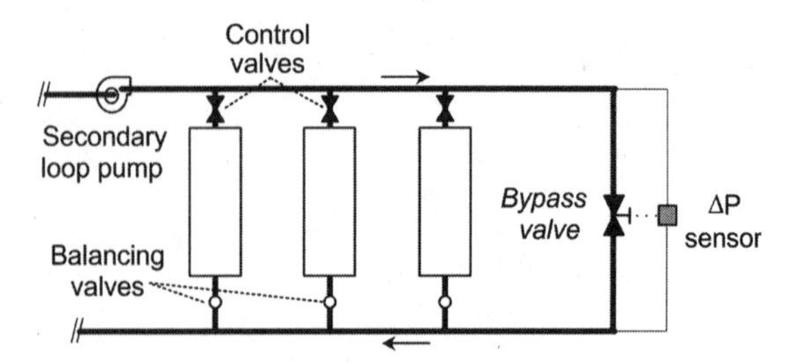

FIGURE 7.14 Direct-return system with flow bypass pressure control.

fastfacts

Always use the line size when specifying a differential pressure control valve. You can always close the valve down, but you can't open it more than necessary. This is because you're only trying to control differential pressure, not water flow rate or temperature.

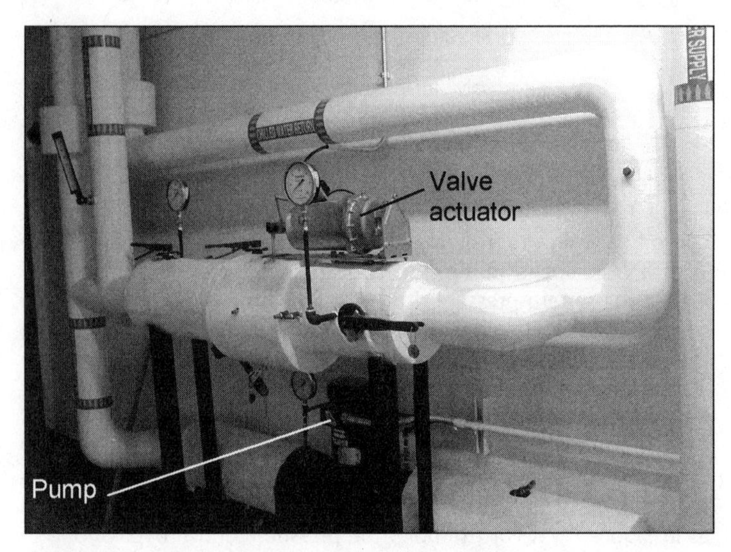

FIGURE 7.15 Pressure control bridle on CHW distribution system.

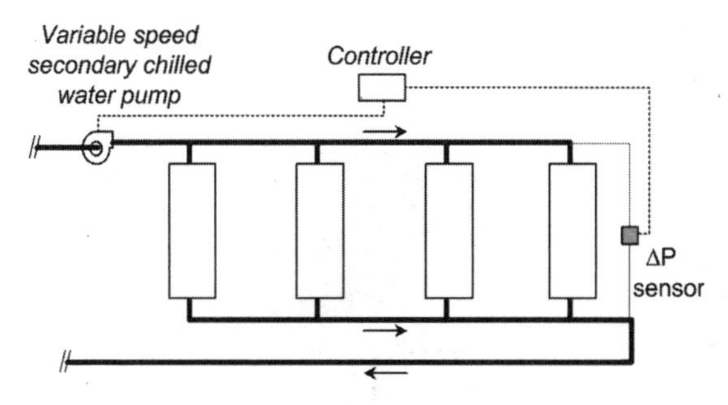

FIGURE 7.16 Reverse-return system with variable-speed pump.

A variable-speed pump system, such as shown in Figure 7.16, is used to provide water only as necessary. As the load varies, the pump speed changes to maintain a fixed pressure drop across the supply and return branches. Significant power savings can be accomplished through this kind of system. As with the bypass loop, however, care must be taken not to set the control loop gains too high because excessive system pressures can occur rapidly.

Static Pressure Control

As the water in a pipe heats up or cools down, it expands or contracts slightly. If the water system was completely closed, even very slight changes in the water volume would lead to extremely high changes in the system pressure. If the pressure goes too high in a heating system you can blow out valve seals and packing, cause leaks at flanged and gasketed pipe fittings, and destroy your pressure gauges. If the water pressure goes too low in a cooling system, you can suck air into the system, causing water quality problems. In most water distribution systems, the static pressure is passively controlled by using an expansion tank (Figure 7.17). These tanks are typically located on the return side upstream of the pump. While the tank can just contain air, they usually are bladder tanks that have a flexible diaphragm between the water and air. The air pocket above the bladder is charged (usually to between 10 and 15 psi) with compressed air at the charging port.

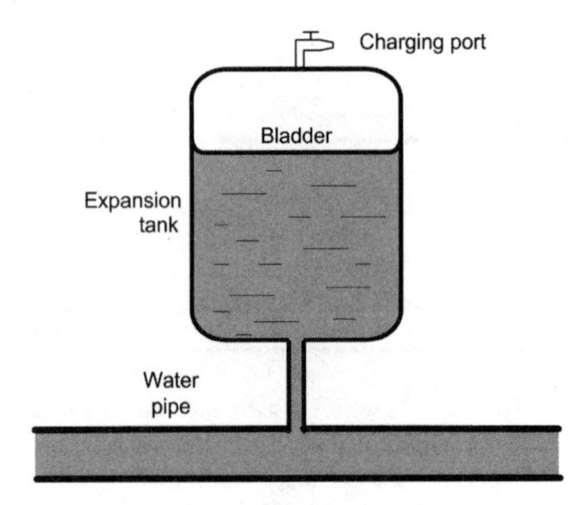

FIGURE 7.17 Expansion tank on water system.

Bladder tanks seem to work better than tanks that have just an air cushion. If there is no separation between the tank and the air, over time the air leaks out of the tank through the site glass seals and the entire tank fills up with water. Once the tank is waterlogged, there is no more capacity in the tank and the system static pressure can rapidly shoot up very high.

The static pressure in a water system depends on the height of the building. In single story buildings you should pressurize the systems to about 10 psi. In multiple story buildings the height of the water in vertical pipes creates higher static pressures on the bottom floors. Above about 80 psi, problems begin to manifest in the system, such as valve packing blowing out, pump seals ripping loose, and other undesirables. As a rule, most taller buildings have mechanical rooms every ten floors or so to help prevent these problems.

fastfacts

Be careful if you install an isolation valve between the water system and an expansion tank. If it is accidentally closed, you lose your passive static pressure control. From a practical standpoint, it is nice to have a valve to isolate the tank if necessary, but make sure this valve is well labeled.

fastfacts

The flow through a water distribution system can depend on the system static pressure. If there is not enough static pressure, you can cavitate pumps and have flow control problems. A good rule of thumb is to charge to about 10 psi over the designers calculated value (which is based on building height).

SAFETY

Water distribution systems are generally pretty safe compared with other HVAC systems. Of course, there are extreme temperatures, high voltage pumps, and fairly high pressures that you must be aware of.

It is a good idea to have *all* pipes insulated. Chilled water pipes collect condensation and drip. You will be mighty displeased if you are working with an electrical system and water drips into the open circuitry. Also, the condensation on chilled water pipes will eventually corrode the pipes and unions, making them more prone to leaks and failure. Uninsulated hot water pipes can heat up the mechanical room and the equipment in the mechanical rooms. It is not unheard of to fry out telephone equipment, fire alarms, building management systems, digital controllers, and other electronic equipment that does not like to get too hot. In addition, exposed hot water lines can melt holes through wires and pneumatic tubing that it is in contact with. Usually this happens in the plenum space above a zone where the telephone wires, control wires, and control tubing are often laid haphazardly.

Be careful when working around old pipe insulation. Most insulation installed after 1980 is pretty safe, but insulation installed before then probably contains asbestos. Don't work around asbestos insulation without the proper protection. If asbestos is discovered on a job site, stop and call an abatement firm. The asbestos clean up can easily cost more than the cost of the original job. You may have to use a glove bag containment or possibly go with full asbestos removal. Once the local authorities, EPA, and liability issues are involved, the whole asbestos issue rapidly gets complicated.

Before you begin working on a water distribution system, always make sure isolation valves are shut off and tagged out before you begin disassembling the piping. The last thing you want is a shot of

fastfacts

Uninsulated hot water pipes can melt pneumatic tubing and wire insulation even if the tubing or wires are in rigid or flexible conduit.

hot water coming down an open length of pipe. The same goes for working on piping in confined areas. If you cannot escape quickly and there's a big leak, you can get severely scalded by hot water.

In systems with faulty expansion tanks, the piping can blow up if the pressure gets too high. Always check the system pressure and use high pressure safety alarms where necessary. Check the boiler and converter safety valves for proper operation. It is safe practice to always depressurize piping before working on it. If the isolation valves are not working, drain down the system and replace the valves.

Be careful working around lines with pressure gauges and taps. The gauge is usually attached to a ⅛ inch or ¼ impulse line that is connected to the main pipe by a weldlet. These can snap off or break off easily, particularly if you are swinging around with a ladder or other big object. If you break it off below the shut-off valve, there's a scalding danger on top of the leak you now have to deal with.

Be aware that piping can pull apart, especially if it is not supported correctly. Before taking the pipe apart, trace back along the piping system and make sure that pipe supports and hangers are in enough places to take the full weight of the pipe even if a section is removed. The pipe hangers should be plentiful, frequent, and in proper tension. Iron and steel pipe is quite heavy and you don't want pieces of it falling on you. Nor do you want other sections of the piping system to shear off—this can cause pretty serious flooding and equipment damage in a very short period of time.

Wear skin and eye protection when working with glycol. You don't want to ingest any or absorb some through your skin. The glycols can be quite poisonous and if they don't kill you they may make you sick enough to wish you were dead.

In some installation and maintenance jobs, it is necessary to have welding equipment on hand to cut pipe lengths, attach flanges, etc. Be very careful when using cutting torches, especially in confined spaces. Not only is there the danger to other workers, but it is pretty easy to accidentally pass the torch across conduit. This can melt the insulation on wiring or pneumatic tubing that may be inside the conduit. Of course, since these utilities are inside the conduit, you won't notice a problem until the attached equipment starts behaving strangely. In the worst case, you could melt insulation on a high voltage line and end up with a charged conduit. When combined with an open-flame welding torch, the results can be unpleasant.

TROUBLESHOOTING

Water distribution systems range from the very simple to the mind-boggling complex. For example, if the system spans several floors, tracing the actual starting and end point of a reverse-return system can be a formidable problem. System plans are useful, but only if they are clear and have been updated to show any modifications to the system from the initial installation.

More often than not, troubleshooting a water distribution system takes time and patience. There is rarely a standard recipe that can be followed to identify problems. There are no troubleshooting flow charts in this chapter, but instead descriptions of general problems and possible causes.

Low Flow in System

It happens every winter: the calls begin when occupants are complaining that the rooms are too cold. You check and find that none of the reheat coils are getting enough flow. A number of things can cause this.

If you have a parallel pump system, make sure there is a check valve on the discharge of both pumps. If you run just one pump and

fastfacts

If you are using an arc welder on an existing water distribution system, make sure the ground does not pass through controllers or valves. Take the time to find a suitable location for the ground so that you do not fry any existing equipment.

use the other as a back-up, you don't want to have back flow through the back-up pump. This just circulates water around the two pumps without sending it out into the distribution system. Worse than that, the back-up pump can start turning backwards, and then as soon as you turn on the back-up pump, you can twist the coupling out or twist the armature off the motor shaft or even snap the shaft. Even if there is a check valve, check to see if the back-up pump is going backwards: you may have a stuck check valve.

Check the pump static pressure on the suction side. If the pumping system is running with low static pressure, this can lead to cavitation, especially on hot water systems as the temperature goes up. Once this happens the water can turn to steam and lock the system where there's no flow. Not only do you have no flow but this can make the pump run hot enough to destroy the seals.

In both hot and chilled water systems, check to make sure that a reset controller has not set the hot water temperature down or the chilled water temperature up. If water temperature has been reset too far, then each coil may demand a lot of water to satisfy the load. Once enough coils have opened up, the entire system pressure drops and there is simply not enough pumping capacity to handle the full flow requested by all the coils. If this happens, return the temperature setpoint to its original value so that some of the coil valves close up. This returns the system differential pressure to its normal value and hopefully brings the system demand back within the capacity of the pump.

Another thing to check is if the system pump has been worked on recently. When maintenance is performed on any pump, whether it be a main circulator or coil pump, make sure the balancing valve is put back to the same position is was in before the maintenance started. Usually there is no shut-off on the downstream side of the pump. While there is most likely a shut-off on the suction side (such as a ball valve), there will be no equivalent on the discharge

fastfacts

In many cases, the balancing valve will be in an obvious location as in Figure 7.18. In fact, it will be so obvious that you can easily forget to check it.

FIGURE 7.18 CHW circulation pump showing differential pressure switch, strainers, and balancing valve.

side. Maintenance personnel usually just shut off the balancing valve to isolate the pump. When the pump is returned to service, the balancing valve should be set back to its original position. If the balancing valve is left in its closed position, you can deadhead the pump when there is no coil flow. If the balancing valve is set all the way open, there will be too much flow in the piping system, leading to an over-amping of the motor and subsequent motor trip-outs.

Many large valves are preceded by a strainer (as illustrated in Figure 7.19). This is a mixed blessing. Strainers do keep the dirt out of the valve but are also a maintenance item. They can trap big pieces of dirt but in doing so they can clog up so that there is no flow. Where the water is very dirty it is worth it to have the strainer, but otherwise it may not be necessary. Make sure all pump and valve strainers are cleaned out. If there is no pump strainer on the suction side of the pump, there may be big chunks of plastic, cardboard, and other stuff that can get into the impeller. Every now and then a big piece of something (like a welding rod) can tear the impeller out of the pump.

Some pumps have suction diffusers that contain a strainer. During the original system installation a construction strainer (with a

FIGURE 7.19 Strainer upstream of hot water valve.

very small screen hole size) is installed in the diffuser to protect the pump. It is not unusual to take the suction diffuser off and find the construction strainer still in place. This causes a low-flow condition and can cause the pump to cavitate.

Low flow and pump cavitation can also happen if you try to throttle the pump on the suction side rather than the discharge side. This is the same effect as having a clogged upstream strainer.

If everything else appears fine and you still have low flow in the system, you may need to take the system apart and see if there are any obstructions in the pipe itself. An amazing variety of strange objects can find its way into the water distribution system. Not surprisingly, pumps don't like to see a lot of dirt, wooden boards, construction materials, welding blankets, cloth, rags, plastic bags, clumps of solder, etc. If there is clogged flow or unusual behavior of pumps and valves, do not be surprised if you take the piping apart and find things you never suspected nor have any idea how they came to be in the system.

Water Temperature Not at Setpoint

The usual suspect for improper water temperatures is a setpoint reset at the chiller, boiler, or converter. Check the thermostats on these devices, along with the sensors to make sure that they are installed and working properly. Usually a quick once-over of the system can

fastfacts

A large building was constructed in the early 1960s with the chilled water coil on the air-handling unit piped backwards. The system ran fine for three decades and was only identified when a contractor connected another coil to the same line during a building addition. The pipe markings showed the coil to be connected the right way, but in fact the chilled water on the new coil was trying to go backwards through the coil pump and three-way valve.

identify an obvious solution to the temperature problem. Also check for things like closed isolation valves, non-operating pumps, disconnected equipment, etc.

If the problem persists, it may be that there is a problem with the piping system itself. This happens more often than you might think, and unfortunately these problems can be difficult to track down. What you need to look for is an improperly piped system, such as an accidental cross connection between the heating and cooling systems. In some cases the cross connection can be located underneath the insulation, making it even more difficult to find.

The problem with improperly piped or designed systems is that they can run fine under some conditions, but not in others. This makes the problems difficult to track down. For example, an undersized pump is not recognized until the peak heating or cooling load occurs.

fastfacts

A building heating system worked fine for 15 years until a couple of very cold winters. At the high loads, the heating system just quit working. The hot water pump checked out fine and showed plenty of flow, but the water was not hot enough. On close inspection, it was discovered that the steam converter had been piped in parallel rather than in series with the pump. This worked fine until the piping system load was so great that the converter could not keep up. All the piping had to be torn out and the converter reinstalled in series with the pump.

Control Valves Not Working or Too Noisy

If a valve appears to be stuck open and the actuator is okay it may be that the stem plug has gotten stuck down inside the seat. This can happen if the plug has encountered a piece of dirt or solder that has traveled through the pipe. This requires a mechanical "adjustment" of removing the actuator and forcing the stem back up. Once the plug is freed it usually continues to work properly.

Most hydronic systems run at about 10 to 13 psi differential pressure across the valve. If the differential pressure goes over about 15 psi, there can be noise problems from the water turbulence as it passes between the plug and seat. As the plug comes down onto the seat, there can be some significant noise problems, especially in baseboard units. Specifically, the baseboard can start rattling, shaking, and banging like it is possessed. Sometimes the water hammer can be so bad that it knocks the ends out of coils or knocks piping loose. Hammering problems should be fixed promptly to avoid both occupant complaints and equipment damage.

If there are complaints of baseboard or reheat valves hammering, ask the occupant how long the noises last. If the hammering goes on for two to three minutes and then quits, it could be that the valve is installed backwards. As water runs backwards through the valve it starts to hammer as the valve nears the full closed position. That is, at certain times and operating conditions the valve starts hammering because the valve is in a certain position and the reverse water pressure slams the valve shut. At that point, the water flow stops and spring pressure brings it back up again, only to be slammed shut by the flowing water. This hammering occurs for a few minutes until the thermostat either puts the valve all the way closed or opens it more. The only solution is to isolate the valve and reinstall it properly.

fastfacts

A contractor was hired for the retrofit of several floors of a large office building. Unfortunately, the contractor managed to pipe every single reheat coil backwards and there were tremendous noise problems when the system was commissioned. Once the problem was identified, it was deemed easier to turn all the valves around rather than to re-pipe the system.

fastfacts

If you identify one piping system or valve being backwards in a system, there is a good chance that other nearby valves are also installed incorrectly.

One way to test if a valve is piped backwards is to start by taking the actuator off. With the system operating, manually press on the valve stem. You should feel water pressure pushing on the valve stem (that is, pushing the valve stem back up) as the valve nears the closed position. If the valve is piped backwards, the water pressure will pull the stem out of your hand as you near the closed position. If you experience the latter case, it is a sure sign that valve is piped backwards, or possibly that the whole coil assembly and valve is piped backwards. This can be a problem because some of the newer design balancing valves have movable pieces to control flow and will not operate very well in backwards flow. Note that the older balancing cock valves work well in these circumstances and so perhaps just the valve will need to be reversed.

For hot water systems, there is another, perhaps easier, way to test if the flow is backwards through a valve. Manually shut off the water flow ahead of the valve and then force the valve open. For a pneumatic valve this may require disconnecting the compressed air or using a squeeze bulb, while on an electric valve this may require disconnecting the control signal wires or applying an artificial signal. Wait for the baseboard or reheat valve to cool down. Once the temperature on both sides of the valve is about the same, put your hands on the pipe, open the valve and see which side heats up first.

Always remember that dirt is the enemy of the water valve. It ruins seats, plugs and can prevent the valve from achieving a tight shut-off. When the water is dirty with rust, the oxide particles can get up in between the stem and valve packing where it will cut scores in the valve stem, causing leaks.

Reheat Coils Not Working

Baseboards and reheat coils are low-flow applications. When the valves just barely crack open, the entrained dirt settles out and the valves and baseboard quickly fill up with a mud-like substance. This is iron oxide from the pipes. If you take the valve bonnet off and find excessive amounts of mud, you should flush out the whole length of pipe and valve.

If you think a pneumatic plug and seat valve on a reheat coil is not working, check the balancing valve. These can also fill up with dirt. If sand or mud goes into the coil or valve, it tends to pack in and must be blown out backwards. If you blow out in the normal direction, you will pack dirt in even more. You may find it necessary to try to dislodge the clog by first running it through with a brush or wire. Sometimes it's best just to take the reheat coil out of the system and hit it against the ground a few times to knock the dirt loose. Of course, in a baseboard the valve can be difficult to remove. Sometimes you need to saw it out, clean it, and solder it back in. Also check the balancing valve on the coil (if one exists). Sometimes these get shut off if room was once too hot but then never returned to the correct position to allow flow through the coil.

PRACTICAL CONSIDERATIONS

This section provides some information that you may find useful. While not directly related to troubleshooting or system design, you will hopefully find some suggestions here that reduce your headaches and maintenance costs.

Piping

There is usually one main supply line in the water distribution system that feeds different coil and reheat loops. The take-offs can come off the main line in any direction, up, down, or sideways. If the take-off goes sideways or up then there will not be a lot of dirt migration down the take-off. If the take-off goes down, however, dirt will drop in from main supply and lead to a lot of gunk build-up in any valves at the end of the line. If the take-off does come from the bottom, it is a good idea to have a drip leg ahead of the take-off with a valve so that it can be blown out easily.

Any air that enters the water distribution system rises to the highest part of system. To help prevent air lock of the system and associated no-flow, place automatic air vents at the highest part of system. If it is not possible to install such a vent, at least put in an automatic bleed valve. Even though the average life of a bleed valve

fastfacts

Make sure your system static pressure is high enough. Too low of a static pressure leads to cavitation in the pumps and can cause flashing at the pumps in hot water systems.

is only about two to three years, it is cheaper to replace the valve than it is to send someone up to keep bleeding air out.

Avoid or replace any inverted U-shape piping loops anywhere in the system. These can create air locks that prevent water from flowing. Also, don't put pipes in or near exterior walls. The insulation space is generally limited and you run the risk of freezing the pipes in the winter.

There is some concern about connecting pipes of two dissimilar materials (such as iron and copper) together. An electric potential develops between the two pipes that some system designers believe causes deterioration of the pipes. While it is true that two metals produce a small but constant current when in contact, the seriousness of the problem is not certain. If you still have some concern about this, use a dielectric union to make the connection, although be conscious of the fact that these tend to leak. Teflon waterways can also be used to connect different pipe types.

Pumps and Valves

Water systems should be designed so that the pressure drop from modulating valves is at least 50 percent of the total system pressure drop. This helps insure that the valves provide good control rather than being at the whim of the system frictional pressure drops.

Normally, valves are situated so that the actuator sits upright. If you are working with a hot water system, however, and especially a system over 180°F, it helps to rotate the valve by 90° so that the actuator sticks out to the side. Since convection heating from the pipe rises straight up, the valve actuator will be out of the heat and will last longer. In fact, it may be possible that the convective flow will help draw cool air up over the valve and counteract any conductive heat flow along the actuator bracket. Before you do this, however, check to make sure that the actuator you are working with works on its side. Some actuators are hydraulic, and some electric actuators have an oil reservoir that needs to stay in a particular orientation to ensure that the gears remain lubricated. Pneumatic actuators, on the other hand, work in any orientation.

After reading that last paragraph, you may be tempted to swing the valve around a full 180° so that it sits underneath the pipe. This is actually not such a good idea. If the valve packing starts leaking the water will drip down into the actuator, killing any electronics and rusting out springs.

Leaks in valve and pump seals allow air to enter the water system. This can oxygenate the system and cause the pipes to deteriorate very fast. In some cases, rusted pipes can occur within in a month, and this leads to a whole variety of other problems.

Water Treatment

There are conflicting opinions about water treatment. Some people swear by it and others insist it is not necessary. Most water distribution systems are pressurized and closed systems, and if the system remains truly closed and was filled with city water, there will generally be no corrosion problems once the oxygen has been driven out of the water. It's true that steel piping rusts, but once it has a coating of iron oxide the rusting stops if there is no further oxygen introduced into the water. However, any leaks or make-up water will add oxygen. If you have a system that leaks a lot, you probably want to use water treatment. On the other hand, if there is a leak in your system, the water treatment may not do any good because there will always be make-up water and the treatment will just leak out anyway.

If you take apart a valve and see just a thin black coating around the edges of valve, you can relax. That's a sign of a healthy water system. If you see red rust, however, then it is time to start worrying. Red water indicates that the system is under attack from rust and

fastfacts

Repair hydronic system leaks as soon as they are found. Any oxygen that enters the system through the make-up water will quickly rust steel piping and lead to expensive repairs.

fastfacts

Glycols lower the heat carrying capacity of water by about 20 percent. Glycols are made by different companies. In most cases you should stick with one manufacturer and not mix different brands together. Each company seems to use different additives and this can create problems and clog systems due to the chemical interactions.

means you have some real problems. Valves can get packed full of rust and not seat, and pipes can be clogged with rust.

One treatment that is difficult to argue with is using some form of glycol for freeze protection. However, glycol tends to create all kinds of leaks in systems that were previously tight. Some valve stem seals do not hold glycol well. Once the glycol begins reacting with the seal material, the seals can shrink up a little bit and the glycol starts leaking out. Keep in mind that glycol can ruin what it touches (ceiling tile, carpet, sheetrock, etc).

In glycol systems, there is usually a glycol feed tank with a pump and pressure switch, where the glycol is pumped into the system to keep the system pressurized. It is a good idea to have any overflow or pressure relief systems piped back to the water treatment plant to avoid losing glycol or other treatment chemicals. If you are using city supply for make-up water, most codes require that you have backflow preventers on the makeup so that treated water does not go back into the domestic water system.

Water treatment is often necessary to prevent biological or chemical deterioration of a water distribution system. Treatment can consist of adding biocides, corrosion inhibitors, and glycols into the water using pressure feeds and timers as shown in Figure 7.20. Failure to properly treat water can lead to valve failure from rusting or corrosion of the valve components. The corrosion can be due to dissolved solids such as calcium or magnesium or from dissolved gases such as CO_2 and oxygen. Usually a corroded component has a characteristic color that can be used to identify the suspect material (for example, Figure 7.21). Use Table 7-1 to help identify the possible corrosion culprit for different types of piping. Remember that the corrosion on some pipes can come from water condensing on the outside of the pipe and that the pipe itself is not actually leaking.

fastfacts

On chilled water systems, keep the lines insulated to keep them from sweating. If the pipes are steel, it is possible to rust them through from the outside in.

FIGURE 7.20 Water treatment system for open condenser water system.

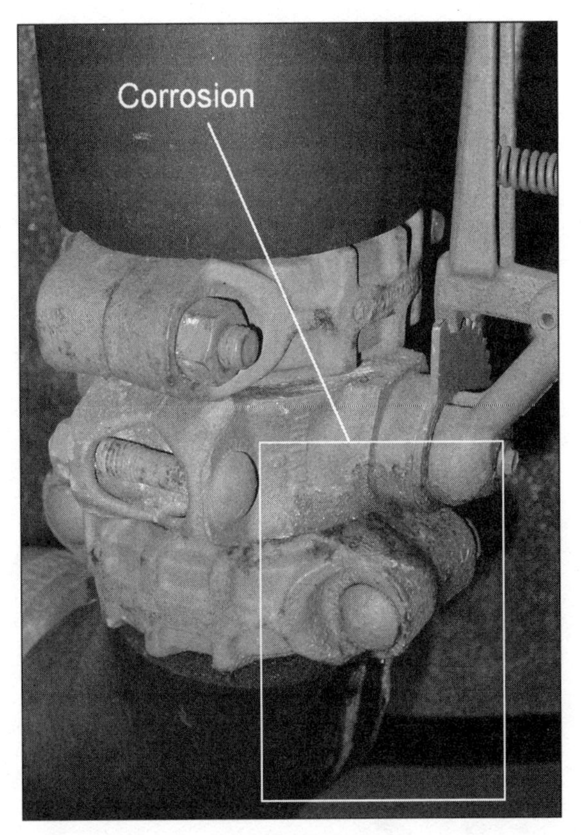

FIGURE 7.21 Corrosion on shut-off valve from magnesium in the water.

TABLE 7.1 Characteristic Corrosion Colors *(From Honeywell, 1988)*

From Honeywell(1988)

brass/ bronze piping

Corrosion color	Possible cause
Blue or dark blue	Ammonia
White	Calcium
Dark blue-green	Carbonates
Light blue-green	Chloride
Black (gas)	Hydrogen Sulfide
Rust	Iron
White	Magnesium
Black (water)	Oxides

iron/steel piping

Corrosion color	Possible cause
White	Calcium
Rust	Iron
White	Magnesium

chapter 8

CHILLERS

Chillers are pieces of equipment used, appropriately enough, for making things cold. Most commercial and industrial buildings employ a number of chillers for cooling water and air. These chillers are integrated into packaged stand-alone units or as one of the main components of a central cooling plant.

THEORY OF OPERATION

There are two basic kinds of chillers used in HVAC systems. *Vapor compression* chillers use a special fluid called a refrigerant that circulates between a high pressure and low pressure side of the cooling cycle. Heat is absorbed on the low pressure side and rejected on the high pressure side. A *compressor* is used to move the fluid between the two sides and to maintain the pressure differential. *Absorption* chillers replace the compressor with a small pump and then use heat to drive the refrigeration cycle. The refrigerants used in absorption cycles are very different from those in vapor compression cycles.

The rated chiller capacity is the maximum amount of cooling the chiller can achieve under the best of circumstances. This number is given in tons of cooling where 1 ton = 12,000 Btu/hour.

Vapor Compression Cycle

A vapor compression chiller has four basic parts as shown in Figure 8.1.

- The *compressor* (points 1 to 2) is the pump that drives the refrigerant through the cycle and maintains the pressure difference. It increases the temperature and pressure of the refrigerant so that heat can be rejected from the cycle.
- The *condenser* (points 2 to 3) rejects the heat that has been gained through the evaporator and the compressor. It is a heat exchanger that is cooled by air or water.
- The *expansion valve* (points 3 to 4) holds the pressure difference across the cycle and is where the cooling capacity is modulated.
- The *evaporator* (points 4 to 1) is the part that does the cooling. It is a heat exchanger in which the refrigerant evaporates, cooling the heat exchanger and whatever fluid is on the other side.

This cycle is commonly used in large chillers in central cooling plants, in direct expansion (DX) units on packaged air-handling equipment, and in heat pumps. The vapor compression cycle uses the properties of refrigerants to move energy (in the form of heat)

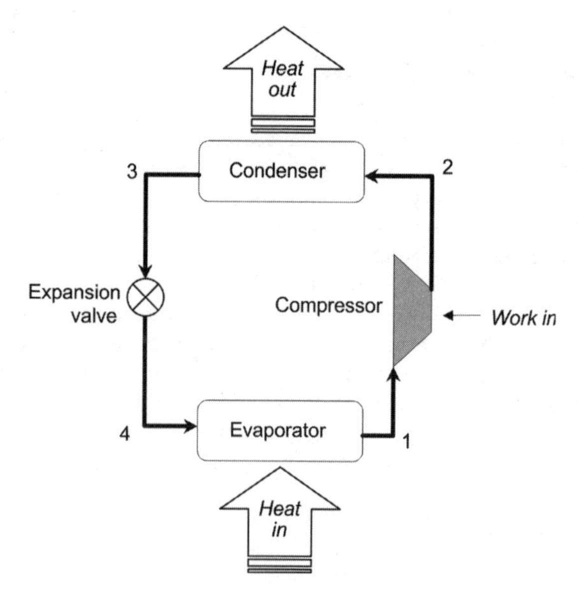

FIGURE 8.1 Components of vapor compression cycle.

from one location to another. Figure 8.2 shows the vapor compression cycle on a refrigerant *pressure-enthalpy* diagram. The *enthalpy* of a substance tells how much energy is stored in a given weight. For example, the enthalpy of a refrigerant can be given in Btu per pound. In the pressure enthalpy diagram, the enthalpy of the refrigerant increases as it passes through the evaporator. This means that the refrigerant has gained energy from something else—this something else is the fluid cooled by the chiller. The refrigerant then passes through the compressor, increasing in both enthalpy and pressure, and then into the condenser. The condenser removes the heat that has been added from both the evaporator and the compressor. (Chapter 9 discusses air-cooled condensers and cooling towers, two different kinds of techniques that people use for condensers.) Finally, the refrigerant passes across the expansion valve and loses both pressure and temperature.

The amount of cooling achieved by the chiller is given by the enthalpy difference across the evaporator (that is, between points 4 and 1) and is expressed as Btu of cooling per pound of refrigerant that passes through the evaporator. For example, if the evaporator

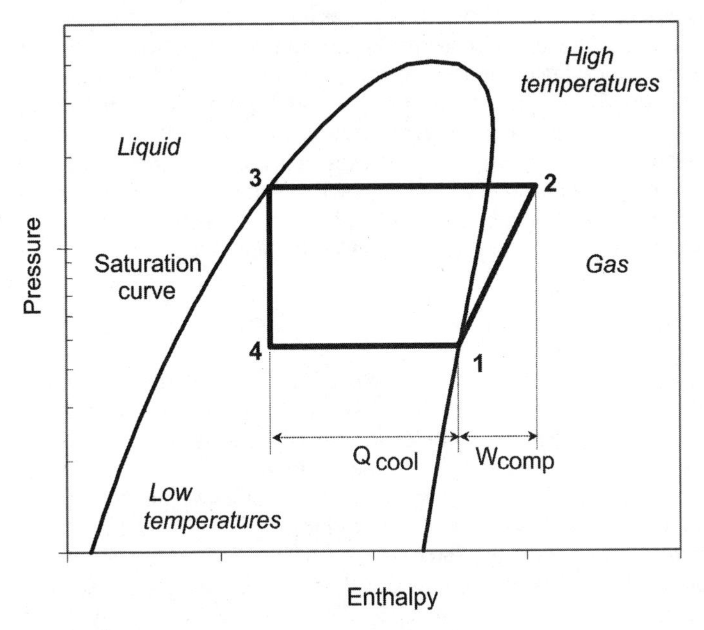

FIGURE 8.2 Vapor compression cycle on pressure-enthalpy diagram.

inlet enthalpy (at point 4) is 250 Btu per pound and the outlet (at point 1) is 275 Btu per pound, then the refrigerant has gained 25 Btu per pound. This is labeled Q_{cool} in Figure 8.2. Suppose the compressor can move 400 pounds of refrigerant through the evaporator each minute. Then the chiller would provide:

$$25\frac{\text{Btu}}{\text{lb}} \times 400\frac{\text{lb}}{\text{min}} \times 60\frac{\text{min}}{\text{hr}} = 600{,}000\frac{\text{Btu}}{\text{lb}}$$

This is about 50 tons of cooling. The amount of work that needs to be put into the compressor is given by the enthalpy difference across the compressor (that is, between points 1 and 2). The *coefficient of performance* (COP) of the chiller is the ratio of the cooling energy to the compressor energy. If the compressor in the above example uses 10 Btu per pound (W_{comp} in Figure 8.2), then the COP is 25 Btu/lb ÷ 10 Btu/lb = 2.5. This is typical for a vapor compression cycle.

In some cases a *heat-recovery* condenser uses the heat rejected from the chiller to satisfy other heating loads in the building. There is usually a *receiver tank* between the condenser and the expansion valve. The purpose of the receiver is to act as a buffer so that the expansion valve does not experience strong pressure surges as the condenser temperatures fluctuate or the compressor stages up and down.

Many diagnostics of chiller performance are made by measuring the pressures and temperatures around the refrigeration loop. The pressure just upstream of the compressor is called the *suction pressure* or *suction line pressure,* while just downstream of the compressor it is called the *discharge pressure.* After the condenser, the refrigerant is mostly liquid and the pressure is called the *liquid-line pressure.*

Compressors

Several different methods can be used to compress the refrigerant. A *reciprocating compressor* behaves like a car engine in reverse. Refrigerant enters the cylinders and is compressed by pistons that are driven from a central shaft (see Figure 8.3). Valves at the top of the cylinders allow the refrigerant to enter or leave the cylinders at the proper stages of the cycle. To reduce the capacity of the chiller, some of the valves are forced open so that no compression takes place. These types of compressors therefore have very distinct operating capacities (for example, 0, 25%, 50%, 75%, and 100% of rated peak capacity).

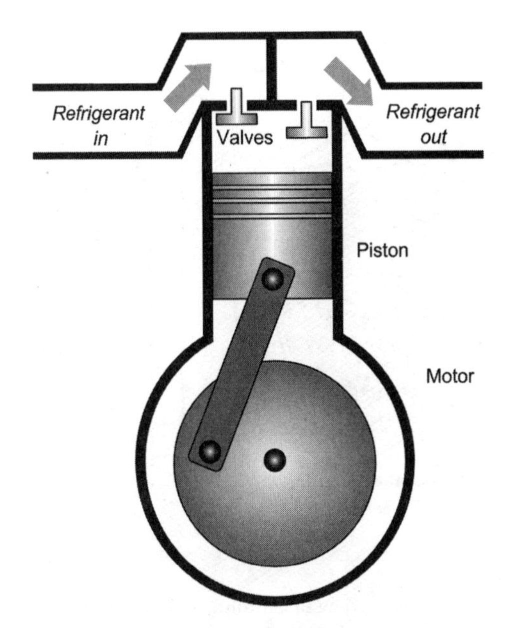

FIGURE 8.3 Breakaway view of reciprocating compressor.

A *centrifugal compressor* is like a centrifugal pump. The refrigerant enters the compressor and is subject to the centrifugal forces of an impeller (see Figure 8.4). This increases both the velocity and the pressure of the refrigerant. Centrifugal compressors can also use inlet vanes to pre-swirl the refrigerant or to impede the refrigerant flow. The chiller controller operates a pneumatic or electric actuator to reposition the inlet vanes as a function of the chiller water temperature. If speed control is available, the controller sequences the motor or rotor speed with the position of the inlet vanes.

fastfacts

A typical chiller is sized for one ton of cooling per 500 square feet of occupied floor space.

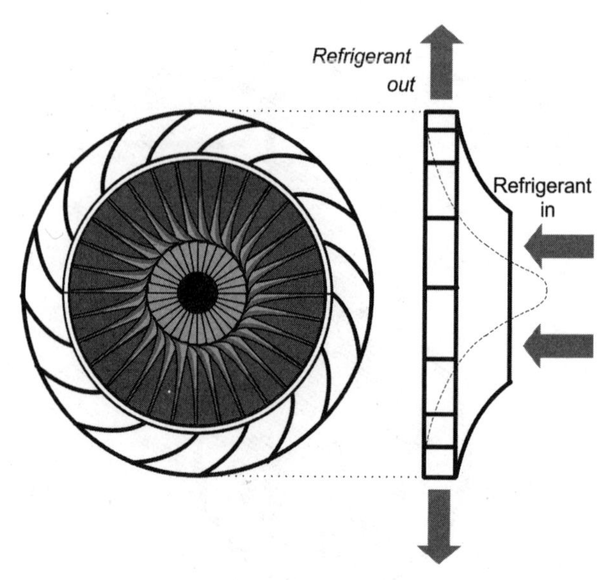

FIGURE 8.4 Breakaway view of centrifugal compressor.

A *screw-type* compressor forces the refrigerant into a series of interlocking screws. Figure 8.5 shows what this kind of compressor looks like from the side. Only one screw is shown for clarity. The second screw would be on the other side of the screw shown and the refrigerant would enter and exit from the spaces in between the two. The farther along the mesh the refrigerant travels, the greater the compression. Modulation of capacity occurs with a pneumatic or electric actuator used to position a sliding bypass valve. Figure 8.6 shows a small screw compressor with the slider valve unloading and loading relays. The valve allows refrigerant to leave the mesh at various points along the screws. As a result, these kinds of compressors have good modulating characteristics.

Scroll compressors are used in small systems of less than 10 tons capacity. Their primary use is in residential systems and heat pumps. Figure 8.7a shows a schematic of the two scrolls that make up such a compressor. The scrolls are like rolled-up sheets of metal that interweave with each other. One scroll is fixed and the other rotates in small orbits as shown. Figure 8.7b shows how the refrigerant enters at the perimeter (left side of figure), is compressed by the orbiting

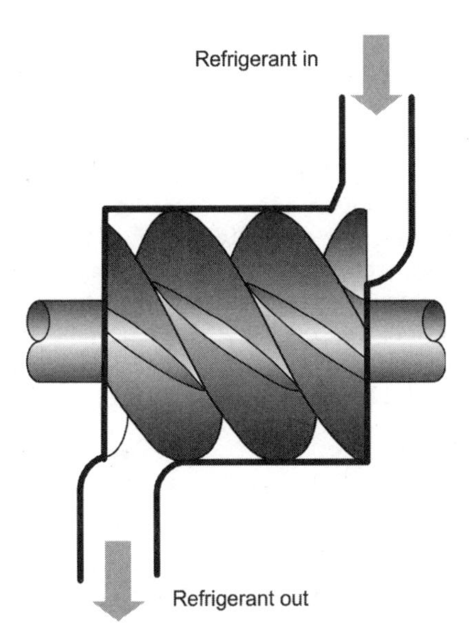

FIGURE 8.5 Schematic side view of screw compressor.

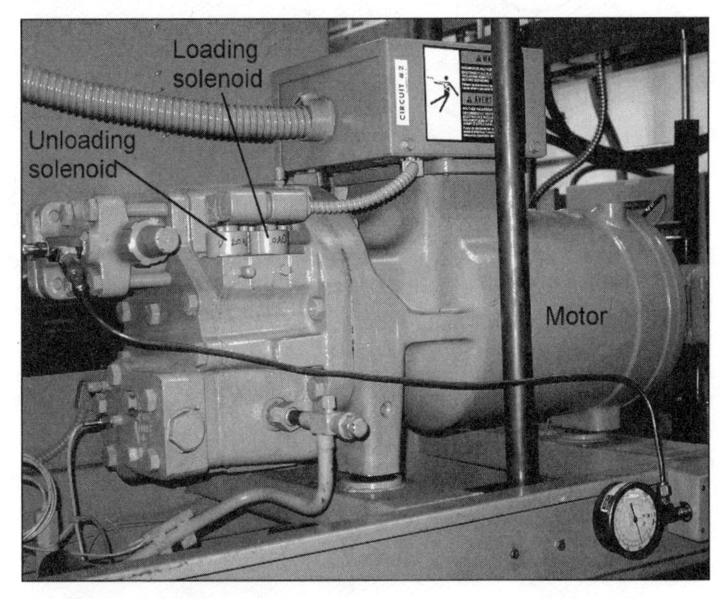

FIGURE 8.6 A 35-ton screw compressor.

motion of the moving scroll, and then is forced to the center (right side of figure) where it exits the compressor. During each orbit of the scroll there are several pockets of refrigerant being compressed, so the output of the compressor is essentially continuous.

Rotary compressors are typically used in low temperature applications. In this kind of compressor, a series of blades moves in and out of slots as the rotor turns. The gap between the blade and the slot is large where the refrigerant enters the compressor. The gaps then become successively smaller as the rotor turns, as shown in Figure 8.8. Rotary compressors can be started with low-starting-torque motors, are relatively lightweight, and have good part-load characteristics down to about 20% of rated capacity.

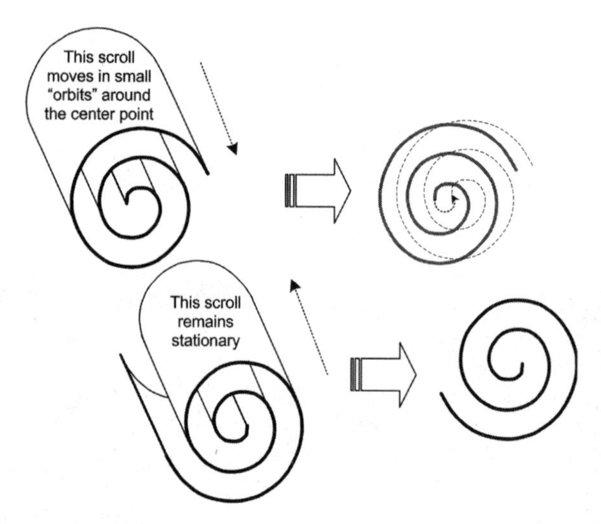

FIGURE 8.7a Scrolls in a scroll compressor.

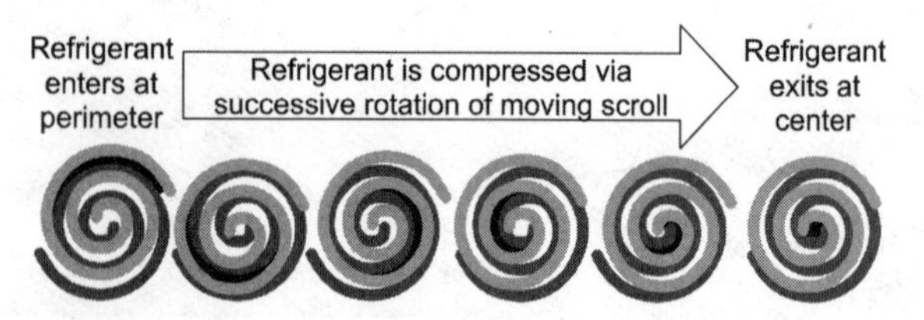

FIGURE 8.7b Compression of refrigerant in a scroll compressor.

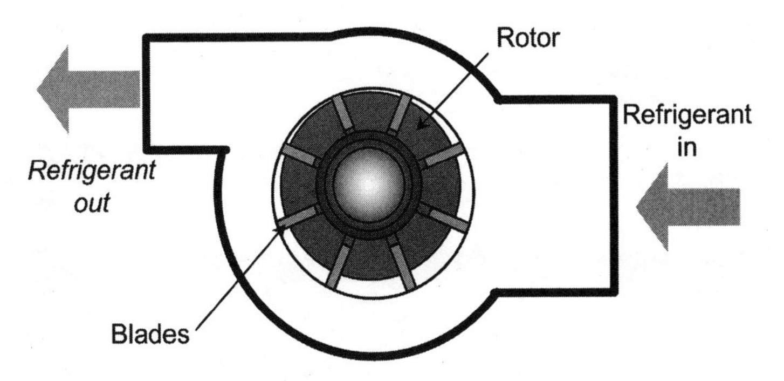

FIGURE 8.8 Rotary compressor.

The choice of compressor depends on required chiller capacity and part-load performance characteristics. Some compressor types are best suited for smaller chillers. The performance of a chiller is often cited in kW of electricity per ton of cooling. It is also given as the COP of the chiller, where the COP is the ratio of the cooling energy provided by the chiller to the electrical work input to the compressor as defined earlier. To use either value you need to calculate the amount of cooling coming from the chiller. The cooling rate in Btu per hour is given by

$$\text{Cooling rate} = 500 \times (\text{water flow rate}) \\ \times (\text{inlet temperature} \; - \text{outlet temperature}$$

where the water flow rate is in gallons per minute and the temperatures are in Fahrenheit. The COP is calculated from the ratio of the cooling rate to the power input,

$$\text{COP} = \text{Cooling rate} \div (\text{Chiller kW} \times 3412)$$

The other way of measuring systems performance is to find the kW per ton of cooling,

$$\text{kW per ton} = \frac{\text{Chiller kW}}{\text{Cooling rate} \div 1200}$$

fastfacts

Chiller power consumption will range from about 0.5 kW per ton of cooling up to about 1.0 kW per ton. The lower the number, the better.

The COP should be above 2 and the kW per ton should be between 0.5 and 0.75. Lower COP values or higher kW per ton values indicate that the chiller is not operating very efficiently and that something may be wrong.

Other Chiller Components

Evaporators are usually shell-and-tube heat exchangers where the water, flowing in the tubes, transfers energy into the refrigerant and in the process boils the refrigerant. The cooling capacity of the chiller is determined by how much energy the evaporator can absorb and is expressed in tons of cooling.

The cooling in the cycle comes at the *thermal expansion valve* (TEV) between states 3 and 4 in Figure 8.2. When the refrigerant passes through the TEV, the refrigerant temperature drops so that it can absorb heat in the evaporator. The amount of expansion is modulated to maintain the temperature of the refrigerant leaving the evaporator. Often there is a *receiver tank* located between the condenser and the expansion valve to store excess liquid refrigerant and to smooth out refrigerant pressure surges.

Refrigeration cycle condensers can be air-cooled or water-cooled. In an *air-cooled condenser* the refrigerant travels from the refrigerant compressor to an outdoor heat exchanger. Fans are used to drive air across the heat exchanger coils and to cool the refrigerant. In some cases the compressors are integrated into the air-cooled condenser in order to shorten the refrigerant runs.

In *water-cooled condensers,* the refrigerant is cooled by water that is itself cooled by either ground water or in a cooling tower. Heat recovery condensers combine several heat exchangers, one or more of which is used to transfer heat back to processes that require it. Some heat recovery condensers are *double-bundle heat exchangers,* meaning that there are two coils in the shell to allow the chiller to dissipate heat either to a cooling tower or a heating load. Chapter 9 has more details on air-cooled condensers and cooling towers.

Direct Expansion

In a direct expansion system, the fluid to be cooled (usually air) passes directly through the evaporator coil. Figure 8.9 shows a schematic of a typical direct expansion system.

Absorption Cycle

The absorption refrigeration cycle is similar to the vapor compression cycle with two important differences: first, the compressor is replaced by a *generator/absorber* combination, and the working fluids are at pressures much lower than atmospheric. The other principle difference is the materials used in the cycle. Instead of a single refrigerant, there is a combination of two fluids that circulate, one is the refrigerant and the other is the *absorbent*.

Figure 8.10 shows a schematic of the absorption cycle. Only the refrigerant travels from the generator through the condenser, expansion valve, and evaporator and back to the absorber. The absorbent

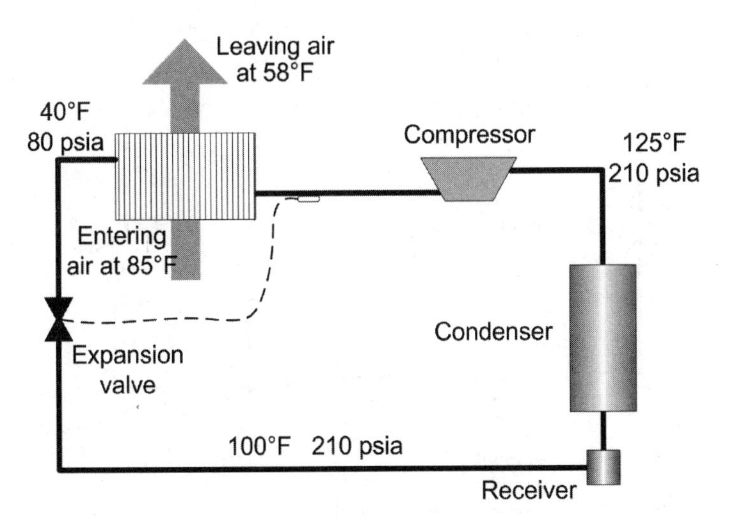

FIGURE 8.9 DX system with typical temperature and pressures for R-22 system.

remains in the absorber/generator combination, traveling through points 1, 2, and 7 in the figure. The refrigerant is mixed with—and absorbed by—the absorbent at low pressure in the absorber. The mixture of the two fluids has the interesting property of having a lower enthalpy than either of the two components taken alone. In order to accomplish this, energy must be removed from the absorber. The resulting mixture is then pumped to a higher pressure in the generator. Here heat is added to force the refrigerant out of the mixture. The refrigerant then completes the cycle as it would in a conventional refrigeration cycle, except that it returns back to the absorber rather than to a compressor.

Absorption cycles can be used where waste process heat (such as steam) is available or with solar heat. This energy is used to "power" the generator, where the refrigerant is separated from the absorbent. These chillers generally have lower operating costs than their more conventional counterparts since there is no need for the large compressor. The pump is used mostly to maintain the pressure differential across the refrigeration cycle, and since the pump acts

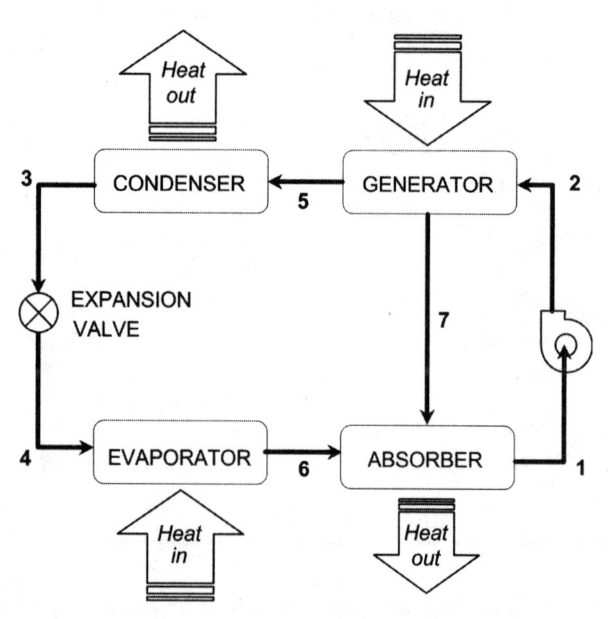

FIGURE 8.10 Schematic of absorption cycle.

on a liquid (rather than a gas) to force it to high pressure, the power required for the pump is much less than what would be required for a vapor compression cycle compressor.

A commonly used set of fluids is lithium bromide (LiBr) and water, where LiBr is the absorbent and water is the refrigerant. Obviously in this case the cooling temperature cannot go below the freezing point of water. Another common set of fluids is the water-ammonia mixture, but in this case water is the absorbent and ammonia is the refrigerant.

Figure 8.11 gives the properties of a LiBr-water brine solution. This diagram shows how the enthalpy of the brine changes as a function of the LiBr concentration. As the amount of water is absorber by the LiBr, the enthalpy increases.

CONTROLS

Because of their complexity and costs, the controls on chillers are essential to both the proper operation and the safety of personnel. Some chillers work with a primary loop and primary loop pump with a bridle between the two (see Chapter 7 for more information about primary and secondary loops) while others just plumb the distribution water directly into the chiller evaporator. A primary loop is

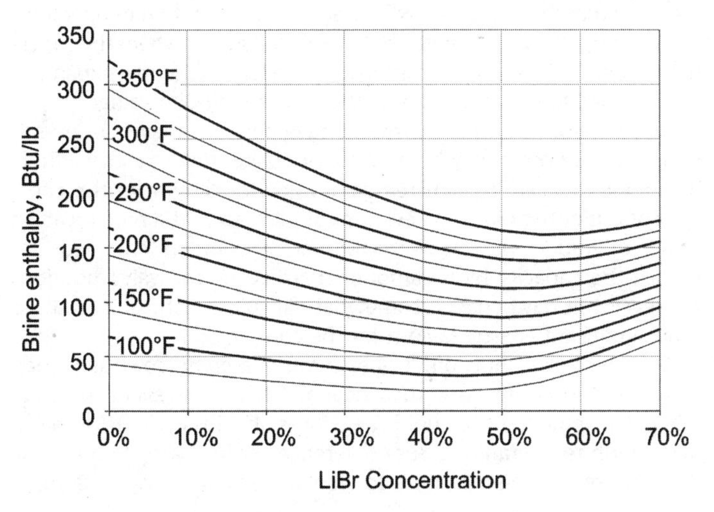

FIGURE 8.11 Properties of lithium bromide.

always a good idea with a chiller. It makes it much easier to control the chiller and helps ensure that the evaporator always has sufficient flow to prevent freezing water inside the evaporator barrel.

Most chillers are tied into a number of safety flow switches (like that shown in Figure 5-19) that confirm water flow through the evaporator and condenser before the chiller is allowed to turn on. This is because a running chiller that does not have water flow can quickly get too cold, with catastrophic results. Figure 8.12 shows a ladder diagram for a typical chiller startup with a cooling tower used for heat rejection from the condenser. In this diagram, the chiller start signal turns on the chilled water pumps in the primary and secondary loops. When the flow switch has verified the water flow, the condenser water pump is turned on. If the thermostat registers a call for cooling, the refrigerant flow switch is opened and the chilled is turned on. Finally, the cooling tower fan is cycled to maintain the condenser water temperature. The chiller start-up controls are usually incorporated into the chiller control panel (Figure 8.13).

It is best to avoid having any form of liquid refrigerant enter the compressor. Small drops of liquid traveling at high speed can pit blades of centrifugal compressors and can damage the seals and valves of reciprocating compressors. This is called slugging the compressor. To avoid this, chillers usually operate with some amount of *superheat* where the evaporator outlet temperature is slightly outside the saturation curve. A typical value for superheat is about 10ºF, meaning that the refrigerant is 10ºF above the saturation temperature at the compressor inlet pressure. Figure 8.14 shows where the superheat is applied in the vapor compression cycle. The thermal expansion valve is used to maintain the superheat at the setpoint value.

Since reciprocating compressors have only fixed levels of operation, *hot gas bypass* may be used to modulate their capacity. In this technique, a valve lets hot refrigerant flow directly from the compressor outlet back into the evaporator inlet, so that a reciprocating chiller operating with a given number of cylinders engaged can reduce the capacity by an amount specified by the valve position. However, hot gas bypass is somewhat difficult to control, and there is no significant energy savings for the compressor.

When a chiller shuts off, it goes through a *pump-down* sequence in which the compressor runs near full load for several seconds before the refrigerant solenoid valve shuts off. This cycle helps maintain a high refrigerant pressure difference across the system so that the compressor does not have to work as hard at the next startup.

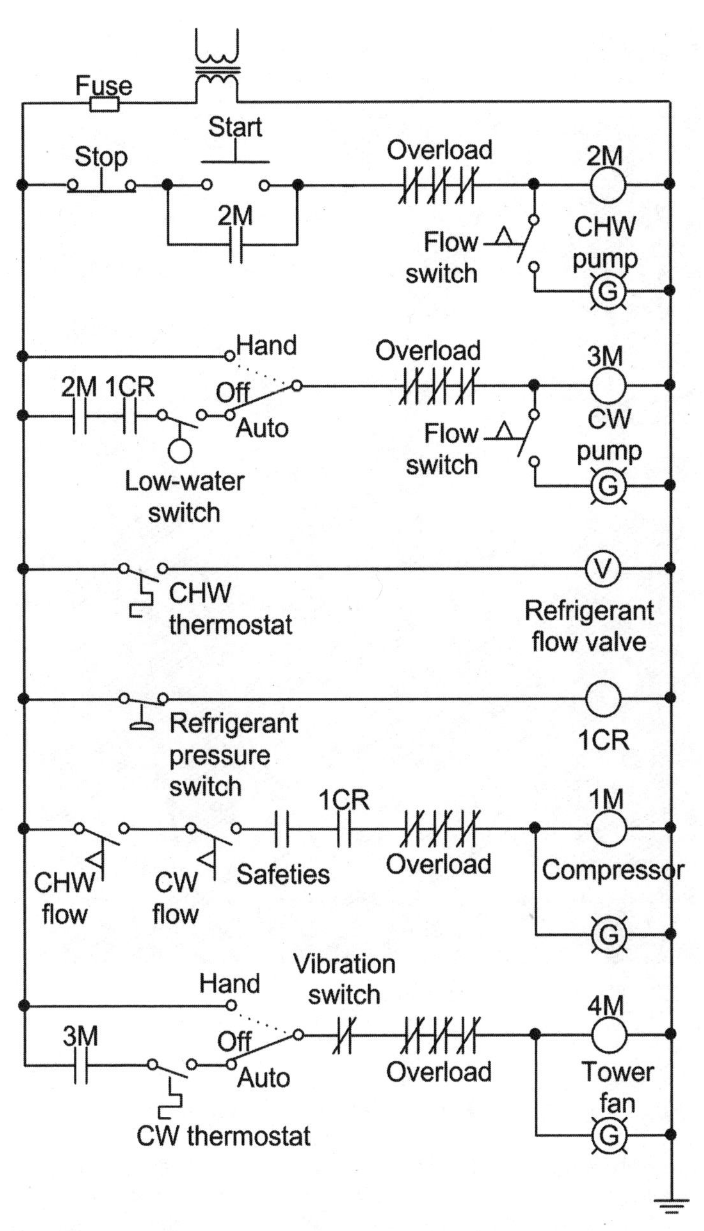

FIGURE 8.12 Ladder diagram for chiller and condenser start-up.

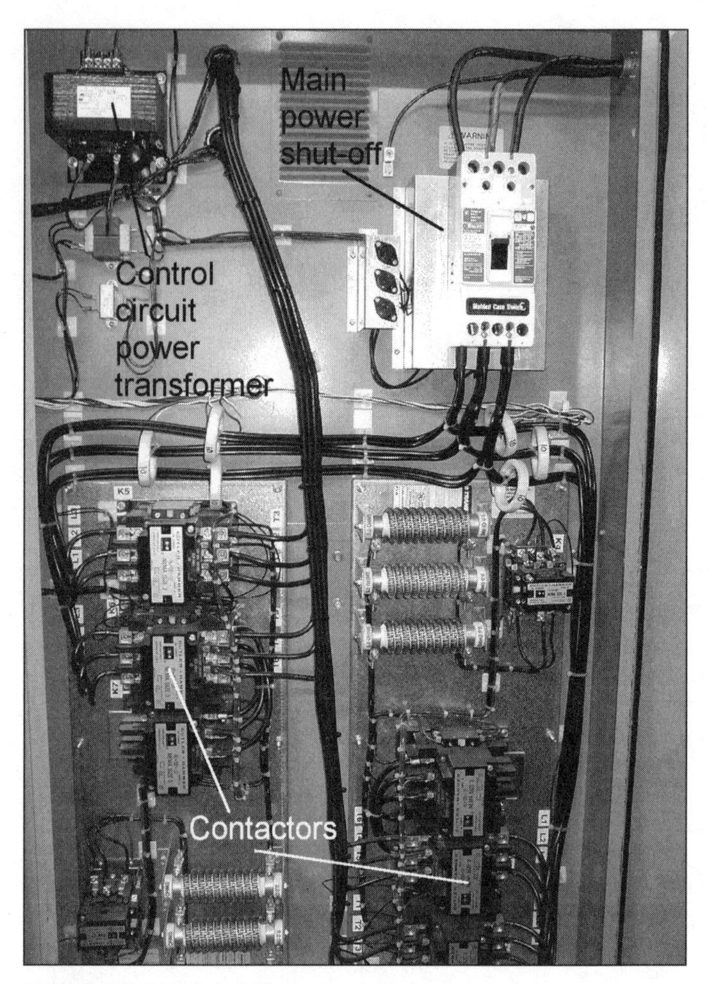

FIGURE 8.13 Typical chiller start-up circuit.

Chiller Safeties and Flow Interlocks

There are a number of safety controls on a chiller designed to protect it from freezing or high refrigerant pressures. Some of these safeties are listed here. Most, if not all, of these safeties cause the chiller to shut off and not reset automatically; human intervention is usually required.

fastfacts

In normal operation, it is a good idea to use the chilled water and condenser water flow switches to turn on the chiller. Use the control system to start the chilled water pump and then let the flow from the pump start the chiller.

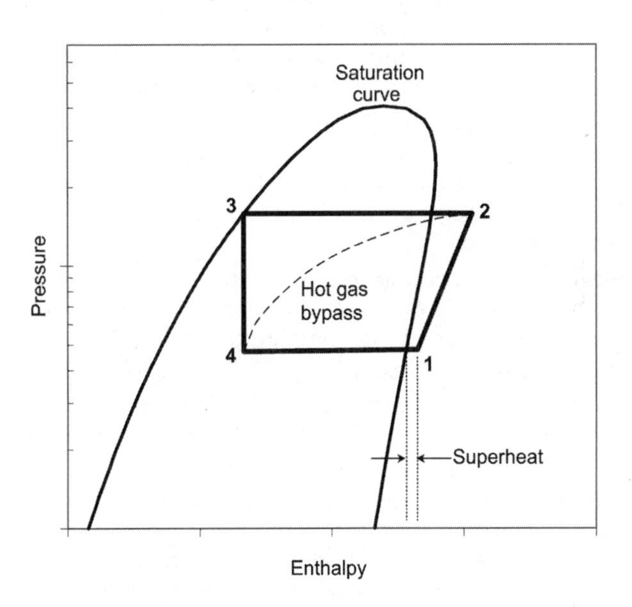

FIGURE 8.14 Superheat and hot gas bypass on the pressure-enthalpy diagram.

- *Low refrigerant pressure* is measured at the evaporator barrel and indicates that the evaporator may be close to freezing. If you blow a tube in the evaporator, you can have water mixing with the refrigerant and the compressor will be very unhappy. Low refrigerant pressure can also mean that the refrigerant charge is getting low. Since the refrigerant is used to carry oil around the system, low refrigerant levels can also spell doom for the compressor if the bearings are not lubricated.

- *High refrigerant pressure* measured at the compressor discharge shuts off the compressor in case the chiller "runs away" or in the event of a condenser failure.
- *High motor temperature* is used to indicate that the compressor is working very hard and needs to take a rest. You do not want to burn out the motor since most compressor motors are hermetically sealed in the compressor and repairs can be incredibly expensive.
- *Motor current overload* usually means the same thing as high motor temperature, just measured differently. Inside the chiller controls there are current transducers that measure how much electricity is going to the compressor motor. If the chiller trips out on current overload, the chiller may be under too high of a load or may have lost a phase.
- *Low oil sump temperature* means that the compressor oil is too cold and may be too viscous to properly lubricate the motor. Some large chillers have an oil heater that keeps the oil from getting too cold.
- *High oil sump temperature* means that the oil cannot help cool the motor and you may be leading to a compressor failure. Check to make sure the oil levels are correct.
- The *evaporator water flow interlock* ensures that there is water flow through the evaporator before the chiller starts running. If this safety fails, the low refrigerant pressure safety should shut off the chiller before the evaporator freezes.
- The *condenser water flow interlock* shuts off the chiller if there is no flow through a water-cooled condenser. Since the condenser water is used to reject the heat removed from the evaporator (plus heat put into the refrigerant from the compressor), failure of the condenser can lead to a very hot condenser and a very high refrigerant pressure. If this safety fails, the refrigerant high pressure cutout should shut off the chiller.

DX Control

Direct expansion air cooling is used where precise temperature control is not that important. The compressor generally does not modulate, or if modulation is required, there are simply two or more small compressors rather than one big one. Figure 8.15 shows the control points for a DX controller. When the zone thermostat goes

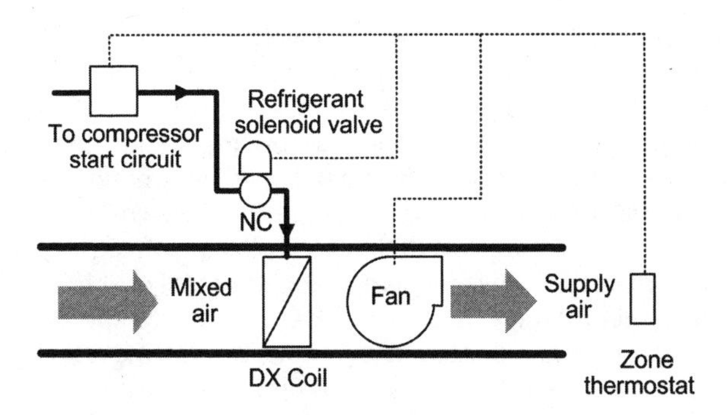

FIGURE 8.15 DX controls.

above the setpoint, the two-position refrigerant valve is opened and at the same time the compressor is energized.

Current Limiting

Some chillers have built-in control circuitry that limits the current draw. This is useful if one is trying to manage the utility demand charges. Under current limiting control the chiller unloads capacity rather than trying to meet the setpoint if the current limit is reached. Some variable speed drives also have circuitry that automatically controls the maximum output current of the drive. This is necessary to protect the current carrying components. Typically, a drive's rating is at 100 percent current. Constant torque drives typically have a maximum current limit of 150 percent and variable torque drives have a maximum current limit of 100 to 115 percent.

SAFETY

Chillers are big, heavy, and loud. In addition, they work at large temperature extremes, high refrigerant pressures, and fatal voltages and currents. That alone should be enough to make you respect chillers. If you keep your chillers in good working order, they will be safe and provide for the comfort of the building occupants. Perform all manufacturer-specified maintenance and check refrigerant and oil levels frequently. A scattered compressor is not only inconvenient—it can be extremely dangerous.

Remember that high pressures are involved in both the refrigerant line and the water lines. The refrigerant can run at several hundred psi. A ruptured or leaking refrigerant pipe can cut through skin very quickly. Refrigerant discharge lines can be hot. Avoid contact since they can and will burn you. Be careful working around refrigerant piping so that you do not accidentally trip and grab a refrigerant pipe to catch yourself. Never cover refrigerant piping with rags, papers or any other materials. While the refrigerant is generally not hot enough to ignite things, there is a certain amount of temperature loss through the piping, especially in long runs. You can make things easier for the condensers by not covering up the refrigerant piping.

On the other hand, the evaporator refrigerant inlet piping can be very cold—cold enough to raise frost. Usually the temperatures are not cold enough to cause frostbite, but in some systems (such as refrigerated warehouses and supermarket freezers) the refrigerant piping can get cold enough to cause your skin to stick to the piping.

Most mechanical rooms are not very well lit. Exercise caution when moving around chillers and any HVAC equipment. Pipe weldlets, gauges, sensors, and other small items will be sticking out of the side of the chiller just waiting for you to catch an arm or eye. In the best case you end up with a small scratch, in the worst case you have a serious injury and a broken sensor or gauge. Also be very careful when working on the chiller control systems. There are many hot wires and exposed circuits. If in doubt, shut the chiller off and lock out and tag out the motor control center. Most large chillers run at an unfriendly 460 V.

Too much time around large chillers can lead to permanent ear damage. There are OSHA standards for hearing protection around loud equipment. Heed them. If you will be working in a chiller room for an extended period of time, wear earplugs and take frequent breaks to make sure your hearing is not being affected. A constant ringing in your ears after spending too much time in the mechanical room should be treated as a warning sign.

The chiller plant should always have refrigerant leak detectors both in and outside the room containing the chiller. The alarms should sound outside of the chiller room. This is extremely important, since the refrigerant can displace the oxygen in a room and if you must enter the room to investigate an alarm, you can suffocate.

Never put a chiller and a boiler or furnace in the same room. If Freon encounters open flame, phosgene gas is produced. This is a killer. Most local codes prohibit refrigeration equipment and combustion equipment in the same room no matter how much makeup air is available.

TROUBLESHOOTING

Chillers come in sizes from small DX units to large central plant components. While the sizes and compressor technology can vary significantly, there are some common threads to the operation of all chillers that can be used when troubleshooting. Some of the basic problems with chillers include the chilled water distribution system being too hot, too cold or fluctuating between the two. In many cases the root cause lies within the water distribution system and not the chiller. Figure 8.16 shows a troubleshooting flow chart for vapor

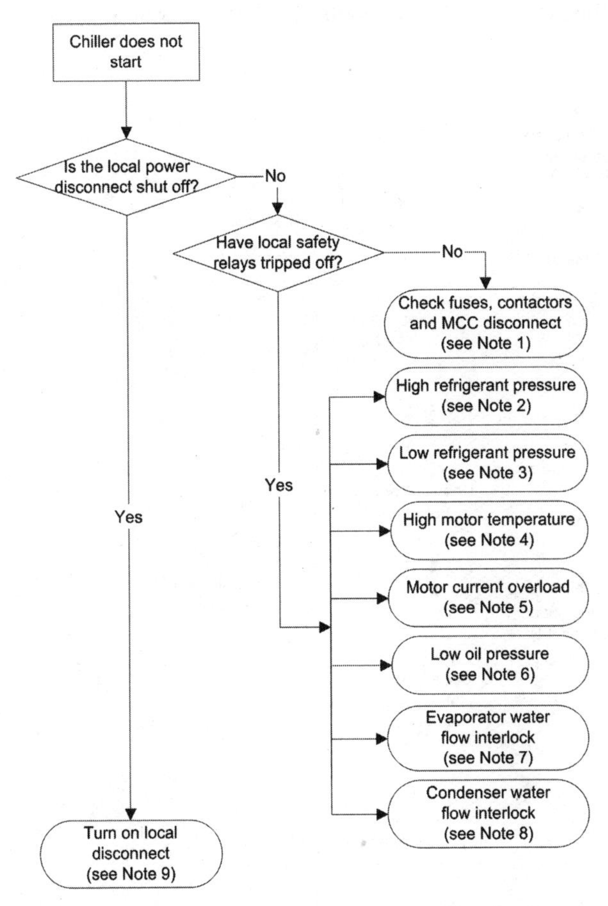

FIGURE 8.16 Chiller start-up troubleshooting flowchart.

compression cycle chillers that do not start. If the chiller starts but does not maintain the right temperature, refer to the flowchart in Figure 8.17. If your water temperature problem does not appear on this flow chart or does not seem to apply, look for similar material in the section on water distribution systems (Chapter 7).

Note 1: If the electrical connections are all valid between the motor control center and the chiller motor, then there is a good chance the problem may be a blown fuse or that the power is shut off at the motor control center. You can check for fuse continuity using a standard ohmmeter. Make sure the motor control center power is turned off before accessing the fuses. Do not attempt to override the safety latch on the motor control center access.

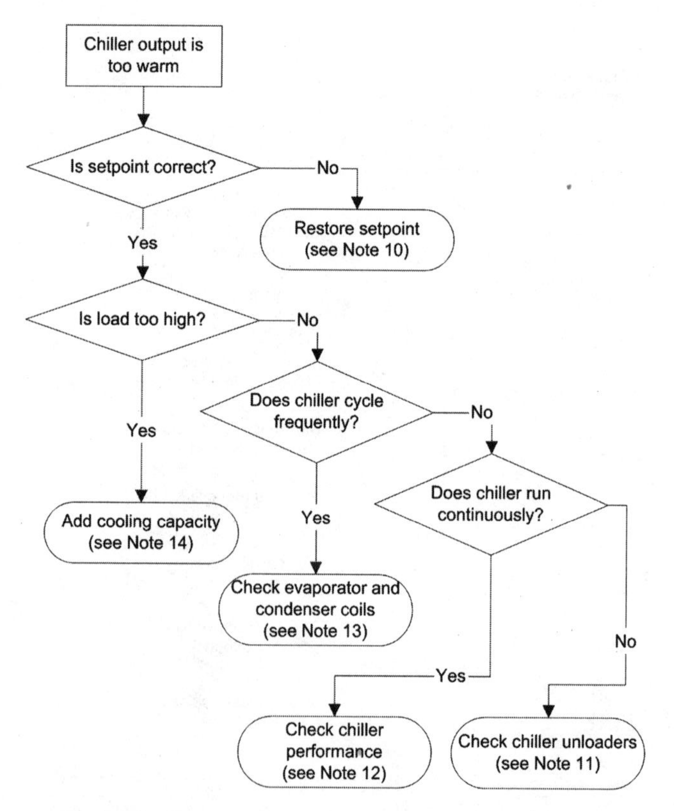

FIGURE 8.17 Chiller temperature troubleshooting flowchart.

> **fast**facts
>
> *Never reapply power at the motor control center if you do not know why the power was turned off. Normally the motor control center is tagged and locked for maintenance or other service reasons, but this step is sometimes overlooked. Never energize a circuit until you have verified that there is no threat to personnel or equipment.*

While at the motor control center, make sure that the chiller supply voltage is within about 10 percent of the nameplate value. You may also want to check this at the chiller itself to see if there are any substantial resistive losses in the connecting wires. The resistive losses only show up if there is current flow, so you have to take these measurements while the chiller is running.

Also check the contactors on the chiller. A pitted, dirty, or malfunctioning contactor can prevent current from going to the motor. Remember to lock and tag out the shut-off on the motor control center before you test the contactors.

Note 2: High refrigerant pressure on the condenser side can be caused by a number of things. The first thing to do is check to make sure there is proper water flow through the condenser. In the condenser the refrigerant pressure and temperature follow each other. If the condenser is too hot then the pressure will be too high. If it's an air-cooled condenser, make sure the condenser coil fins are not dirty or clogged and that the fans are operating as they should. If the water flow is okay, then you may have dirt or lime build-up on the inside of the condenser tubes. This requires taking the condenser apart to clean. If you are reasonably sure that the condenser is okay, then there may be air or some other gas in the refrigerant system. If the foreign gas is non-condensing, it won't return to liquid form and reduce the pressure. You should purge the foreign gas from the condenser. Finally, you may have an overcharged refrigerant system. Check the refrigerant charge and discharge refrigerant as necessary.

If the high pressure trip out occurs on the suction side, you should first check to see that the chiller load is too high. This is the easiest to diagnose although "fixing" it may mean adding

fastfacts

Never vent refrigerant directly into the atmosphere. This wastes money and can contribute to the destruction of the ozone layer. It's also illegal.

additional cooling capacity. Otherwise, check the thermal expansion valve and sensor bulb location to make sure these are working correctly. If they appear to be fine, then check the superheat setting to make sure it has not been changed from the manufacturer's recommendation. It should be around 10°F.

Note 3: Low refrigerant pressure on the condenser side reduces the efficiency of the chiller. If you have low refrigerant pressure you may have a condenser that is working overtime. If the chiller uses an air-cooled condenser, make sure the fans are staging correctly. The fans should cycle up and down between about 150 psi (lowest fan setting) and 250 psi (highest fan setting) discharge pressure. Different chillers and different refrigerants have different settings, so check the chiller documentation. If the chiller uses a cooling tower, check the tower controls and make sure the condenser return water temperature is between 85° and 90°F. Different chillers have different requirements, so check the documentation. If the condenser checks out okay, the problem may be that there is too much liquid refrigerant in the evaporator and this is flooding back into the condenser. Check the thermal expansion valve and the temperature sensor bulb. If the bulb has fallen out, the valve may be all the way open, allowing the backup of refrigerant.

If the low refrigerant pressure occurs on the suction side, this may be an indication that you have a refrigerant leak. Check the charge. Similarly, it could mean that your oil-to-refrigerant ratio is too high and that you need to remove some oil. Check the oil level according to the manufacturer's instructions. If the refrigerant and oil levels are fine, then see if the expansion valve is working correctly. If it is not opening all the way then it could be starving the evaporator of refrigerant. Usually you notice this by a decrease in available cooling from the chiller. If the chiller has been overhauled recently, there is a chance that the expansion valve was replaced with one that is too small. Finally, if all else seems fine, you may need to pump down the system and look for blockages in the chiller liquid line. Check the strainers to see if anything has gotten into the system.

If the problem occurs on a DX unit, check to see if the evaporator coil has frosted up. Particularly in humid climates, air flow through a DX coil can cause ice build-up to the point where there is no air-flow through the coil. Usually this happens when the defrost cycle of the coil had failed or is not frequent enough, or if the airflow has been reduced (probably by a dirty filter).

Note 4: A high motor temperature means that you are close to burning out the motor. This is bad, particularly with hermetic or semi-hermetic motors, since the motor replacement is much more difficult and expensive. The first thing to do is see if the chiller is short cycling—that is, if it switches on and off more than about 10 times per hour or if the chiller has been running continuously for a long period of time at a high load. If this is happening, then there are other problems with the chiller operation (see Notes 11-14). If not, then you should check the electrical supply of the chiller. A "hidden" problem is low voltage at the chiller from line loss. With the chiller running, check the voltage at the chiller terminals. It should be at the rated voltage plus or minus five percent. Also check the terminals to make sure there is a good mechanical connection. If all appears well, then use an ammeter to make sure the three legs have approximately the same current draw. An unbalanced motor may mean that you have lost a winding, a fuse in the motor control center, or that the starting contactor may not be closing a phase.

Note 5: Some chillers use current transducers instead of motor winding temperature sensors to detect when the motor is approaching burn-out. See Note 4 for more information. Keep in mind that high compressor current draw can indicate other serious problems, such as bad compressor bearings or lack of lubrication. If the problem persists, do not keep resetting the safety without trying to identify why the problem is occurring.

fastfacts

A lost phase on a chiller motor can quickly lead to an overloading and burning out of the remaining phases.

fastfacts

If the cooling tower sump sits outside, for example, on the roof, a cool night might cause the sump temperature to drop even if the fans are off. If there is no three-way valve to maintain the condenser return water temperature, the sitting sump water could be cold enough to trip off the compressor on a low-oil safety.

Note 6: Low oil pressure usually indicates that the condenser water is too cold. Check the condenser return water temperature to make sure it is in the manufacturer's recommended range. This should be about 85°F. If the condenser water is okay, check the oil level in the chiller and make sure the oil strainers are not clogged.

Note 7: A good chiller start-up design incorporates the chilled water flow switch directly into the start-up circuit. If the chilled water pump is not running then the chiller should not start. Check the pump to make sure it is working. If it is, check to make sure no isolation valves have shut-off flow to the evaporator. If there appears to be water flow, check the flow switch to make sure it is working correctly. Most flow switches have adjustable sensitivity. If there is a leak anywhere around the flow switch, the tension adjust spring may be rusty and not moving. Also, the blades on the flow switch can fall off. You may need to isolate the switch, drain the system and make sure it is operating correctly.

Note 8: Chillers with water-cooled condensers must confirm water flow before the chiller comes on. Many systems are designed with the condenser water flow switch in-line with the chiller start-up circuit. Check to make sure this switch is working correctly and that you truly do have condenser water flow.

Note 9: Never activate a local disconnect until you have identified why it was shut off in the first place. If there is currently any maintenance taking place on the chiller or any of its subsystems, turning on the local disconnect can cause severe damage to equipment and poses a health and life risk to any personnel. Keep in mind that the chiller can and will have interlocks with other equipment that may be out of your line of sight, such as air-cooled condensers and water towers.

Note 10: Obviously, the first thing to check if the chiller leaving temperature is too warm or too cold is to check the thermostat setpoint. The setpoint can be reset either manually or by a supervisory controller. This happens more often than you might expect.

Note 11: In reciprocating compressors, the unloader valves cycle on and off to modulate the capacity. If these valves unload and then never come back on line, the available cooling rate can be greatly diminished. Usually this occurs on individual cylinders rather than on the entire compressor.

Note 12: One of the most common reasons why chillers run continuously is that there is not enough refrigerant in the system. Check the charge and add refrigerant if necessary. If you suspect something is wrong with the chiller but can't identify any obvious problems, you may want to calculate the performance of the chiller. In the same way that the gas mileage of a car can tell if something is wrong with the engine, the performance of a chiller can tell if the cycle is operating correctly. The two common metrics for gauging chiller efficiencies are the coefficient of performance and the power consumption per ton of cooling. The equations for calculating these values are given in Chapter 8.

Note 13: A short-cycling chiller can rapidly ruin the compressor and motor. This can happen when there is a low load on the chiller and the chiller cannot unload any further. If the load is high and the chiller still cycles, there is a good chance that the condenser is not providing enough heat rejection due to scaled or fouled tubes in water cooled condensers or restricted air flow in air-cooled condensers.

If the condenser seems fine then you may have non-condensing gases in the refrigerant line, most likely air. This requires that you purge the condenser and recharge. If the compressor is providing cooling for a DX coil, make sure the coils are not iced or frosted. If so, the defrost cycle is probably not working, the coil filters are clogged, or the fans have failed.

Note 14: Under high load conditions, you may simply not have enough cooling capacity to keep up with the load. Either find some way to reduce the load (almost impossible) or find some way to increase your cooling capacity (almost as impossible). Another possibility is that the chiller has current limiting circuitry that unloads the chiller automatically to prevent the power consumption from going too high. This is done as an economic move to limit the utility demand costs. If the building activities are critical, you may wish to get permission to raise the current limit and increase the chiller capacity.

fastfacts

Your condenser outlet (supply) water temperature should be no greater than about 120°F and the inlet (return) temperature should be around 85° to 90°F.

PRACTICAL CONSIDERATIONS

This section describes some of the more practical issues to consider when designing and working with chiller plants.

Staging Multiple Chillers

Large buildings may have several chillers of different sizes that are brought on-line according to the load on the cooling plant. Since the part-load performance of a chiller may not be very good, it is often desirable to operate the chillers at 80 percent to 100 percent of their rated capacities to ensure that the kW/ton ratio is sufficiently small. A number of issues are associated with multiple chiller operation, however, particularly if the chillers are of different capacities. If the morning load is small but the afternoon load is expected to be large, it may be advantageous to bring the large chiller on-line in the morning even at a low part-load ratio. This prevents the large chiller from doing a "warm-start" in the afternoon when the start-up transient power consumption can be high enough to reset the demand load for the building.

In a multiple-chiller plant, the outlet setpoint temperature on each one should be the same. The condenser and evaporator flows through each chiller should be in proportion to the relative capacity of each chiller.

DX Units

If the airflow through the evaporator coil is too low, the cooling capacity will decrease and the unit will use more energy than necessary. If the airflow is too low, the efficiency of the unit will drop and there is a great risk of slugging the compressor. Typically the airflow should be at least 350 CFM per ton at the evaporator to guarantee that only gas refrigerant is leaving the coil, that liquid refrigerant is not evaporating in the liquid line, and that the chance of freezing the coils is minimized.

Noisy Compressors

It pays to listen to the chiller compressor and to become familiar with the way it sounds under normal operation. If your compressor suddenly sounds different or is noisier than normal, it could be a sign that there is a problem. One culprit may be a faulty thermal expansion valve that is letting too much liquid refrigerant leave the evaporator. This can lead to slugging the compressor—the noise you are hearing—and a shortened compressor life. If this happens, try increasing the superheat setting by a few degrees. In a similar circumstance, too much oil in circulation can cause a slugging noise. Check the oil level and receiver tanks frequently.

Another possible culprit is the compressor itself. As the compressor ages, the seals and bearings begin to wear out. Once the tolerances are sufficiently worn, the compressor can become unbalanced.

Keep It Clean

It is pretty important to make sure that the water systems for chillers are kept clean. Even the slightest fouling of condenser and evaporator tubes can greatly lower the efficiency of the machine. Since chillers are one of the biggest energy consumers in an HVAC system, you should try to keep the efficiency as high as possible.

Most of the problems occur in the heat rejection systems. These are usually open to the environment and consequently get dirty and

fastfacts

You may want to use vibration sensors on your large chiller as an indicator of when the compressor may be wearing down too much. It will be cheaper to perform preventative maintenance than to do a full compressor rebuild.

fastfacts

During the winter, a contractor was called in to disassemble and clean out the condenser on a large chiller. But the building manager neglected to clean out the associated cooling tower. When the condenser pumps were started for the first time at the beginning of the cooling season, a big slug of dirt came down through the pipes and fouled the condenser, requiring that they be dismantled and cleaned yet again.

fouled easily. One common problem in cooling towers is that the condenser water pump strainer gets clogged with leaves and dirt. This must be kept clean, because a clogged filter causes restricted flow, and this leads to problems maintaining the condenser water temperature. If this temperature gets too high the chiller trips out on high head pressure. If it gets too low, the chiller trips out on low oil pressure. It is a big challenge to keep cooling towers and strainers clean, but it is necessary.

Keep It Under Control

Chiller plants can become very complex, particularly when you try to get fancy with free tower cooling and staged chillers. The function of such a system depends on the design and controls. When they work, they work well and can save a lot of money in energy and maintenance. When they do not work, they generally behave very badly and can cause a lot of headaches.

For example, it takes a little more wiring, but it is a good idea to use the chilled water flow switch to start the chiller. This ensures that the water flow is on before the chiller will start. If the chiller is wired separately from the flow switch and the flow switch fails while the chiller is operating, it is possible to break a tube in the evaporator. This will take your entire chiller plant down while repairs are made.

Proper control can also keep the plant operating. If the condenser water is too cold, the chiller will go out on low oil pressure. Condenser water systems should use a tower primary loop with a three-way valve and a nulling loop. This allows water to bypass the tower while still maintaining condenser flow. The cold water does not come out of the tower sump until it is needed. The position of the three-way valve is used to control the condenser return water temperature. Of course, the tower fan is also used to maintain the water temperature—this is discussed more in Chapter 9.

AIR-COOLED CONDENSERS AND COOLING TOWERS

Air-cooled condensers and cooling towers are used to remove heat from refrigerant and water streams. Air-cooled condensers take their name from the condenser component of refrigeration cycles (see Chapter 8 for more details on chillers). Cooling towers take advantage of the cooling effect of evaporating water to provide cooling. Air-cooled condensers are usually used on smaller refrigeration systems, while cooling towers can be used for large refrigeration plants, heat rejection equipment for power plants, general evaporative cooling, etc.

THEORY OF OPERATION

Air-Cooled Condensers

In an *air-cooled condenser,* the refrigerant travels from the refrigerant compressor to an outdoor heat exchanger as shown in Figure 9.1. Fans are used to force air across the heat exchanger and in turn to cool the refrigerant. Air-cooled condensers are most often found on rooftop units and smaller refrigeration systems, typically 200 tons or less. The reason for this is that large air-cooled condensers can take up considerable space and consume a good deal of electricity to operate the fans. Figure 9.2 shows a typical air-cooled condenser.

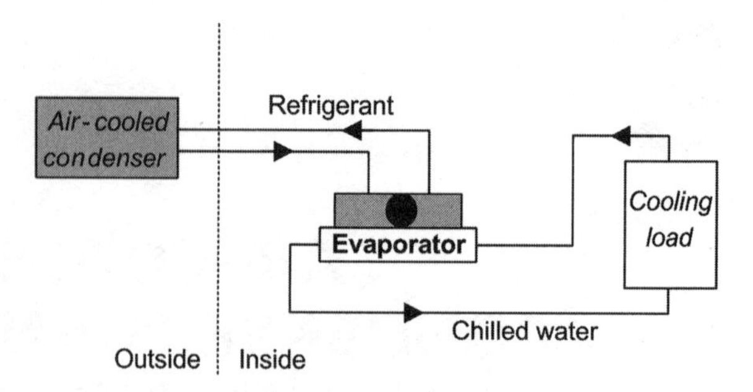

FIGURE 9.1 Air-cooled condenser piping.

Sometimes these units use filters along the sides to prevent the air side of the heat exchanger from becoming clogged with dirt, dust and other particulates.

An air-cooled condenser loads and unloads capacity by staging the fans on the unit. The fans operate in pairs to ensure that some cooling occurs even if one fan per pair fails. However, at low loads it is usually possible for just one fan to operate. On some systems, the overall capacity is increased by using evaporative cooling, either through wetted media or by spraying water directly onto the heat exchanger fins.

Occasionally the compressor for the refrigeration cycle is located within the body of the air-cooled condenser. The advantage of such an arrangement is that the amount of plumbing (and therefore total amount of refrigerant) is minimized. This also allows for faster and better control of the discharge pressure of the chiller.

A *dry-cooler* is a variation of the air-cooled condenser used in larger systems and where it is desirable to minimize the refrigerant piping length. In this type of system, the refrigerant condenser is located next to a fan-coil unit with a closed water or glycol loop

fastfacts

An air-cooled condenser must be sized so that refrigerant condensing can occur even when the ambient air temperature is hot.

FIGURE 9.2 Typical air-cooled condenser.

used to remove heat from the condenser (see Figure 9.3). These kinds of units can generally operate at higher temperatures than an air-cooled condenser because of the better heat transfer characteristics of the condenser heat exchanger. Figure 9.4 shows a dry-cooler installation with the condenser located behind the fan unit.

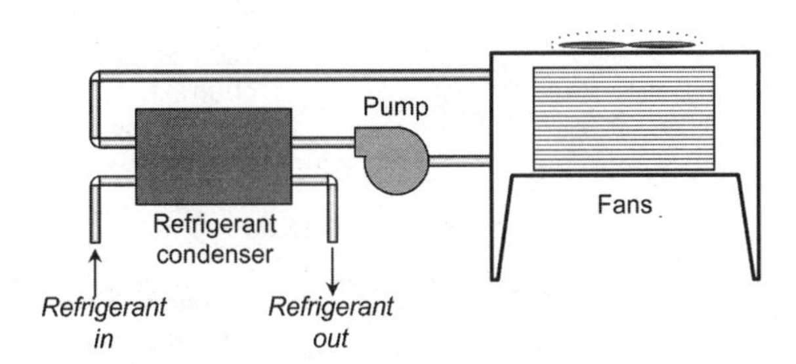

FIGURE 9.3 Schematic of a dry cooler.

FIGURE 9.4 Dry cooler.

Cooling Towers

In a cooling tower, water is sprayed into an airstream and the evaporation of the water cools the surrounding air and water. The water then falls into a basin at the bottom of the tower where it can be used for cooling. Figure 9.5 shows a cross sectional view of a forced air cooling tower that uses a fan to drive air through the tower. Water towers like these are used on large HVAC systems to reject heat from refrigerant condensers.

An *open-loop* cooling tower means that the water in the cooling tower goes directly to the condenser. The loop is open in the sense that the water is open to the atmosphere. In a *closed-loop* cooling

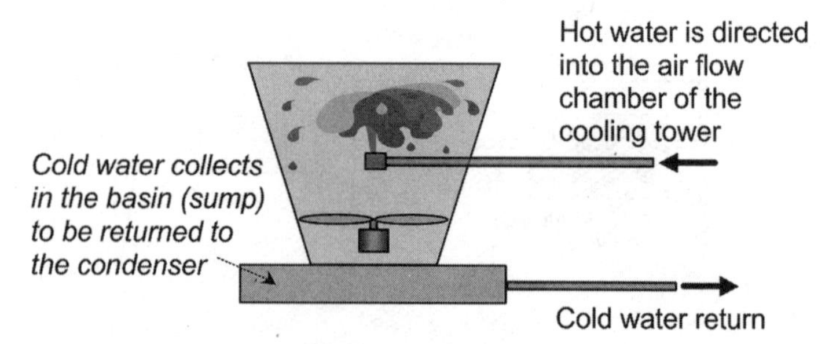

FIGURE 9.5 Schematic of forced-air cooling tower.

tower, the water circulates from the condenser to a heat exchanger. The heat exchanger is cooled by the cooling tower water. Figure 9.6 shows the plumbing schematics for open and closed loop towers. In both cases you must be careful to keep the cooling tower sump and piping free from leaves, dirt, and other objects that can clog the associated strainers and pumps. In many cases the tower uses a three-way valve and a nulling loop to provide additional control of the tower leaving water temperature. These are also shown in Figure 9.6. The valve can be modulated to return water back to the tower sump (and return condenser water back to the condenser) or to get more cooling by letting tower water go to the condenser.

There are many different variations on the basic cooling tower. An *atmospheric tower* (Figure 9.7) is considered a *passive cooling tower* where water drops in from the top of the tower. As the water falls, it helps create a natural downdraft of cooler air that in turns draws in dry air from the top of the tower. These towers are very simple in operation but don't have very high capacities.

In *forced-draft towers* a fan is used to move the air through the tower at high velocities. By doing so you can have much higher cooling capacities. Figure 9.8 shows how a centrifugal fan is used to push air into the bottom of the tower where it passes through the water stream and exits at the top of the tower. While efficient, this kind of tower can suffer from a lack of capacity if the warm, humid air leaving the top of the tower gets sucked back in by the fan. The

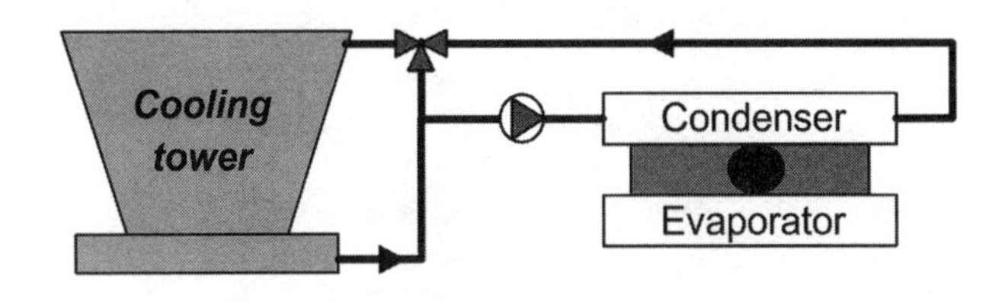

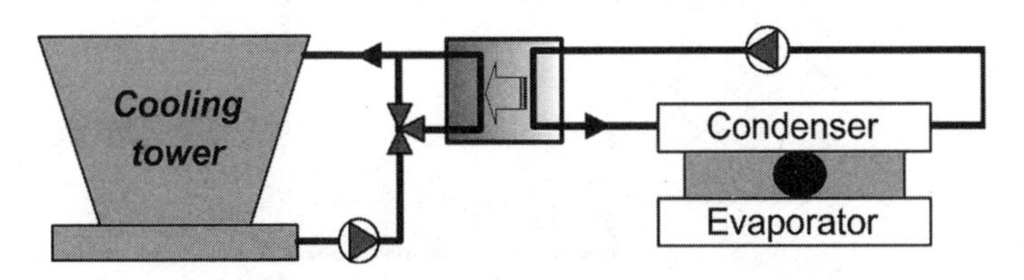

FIGURE 9.6 Open and closed loop cooling tower condenser circuits.

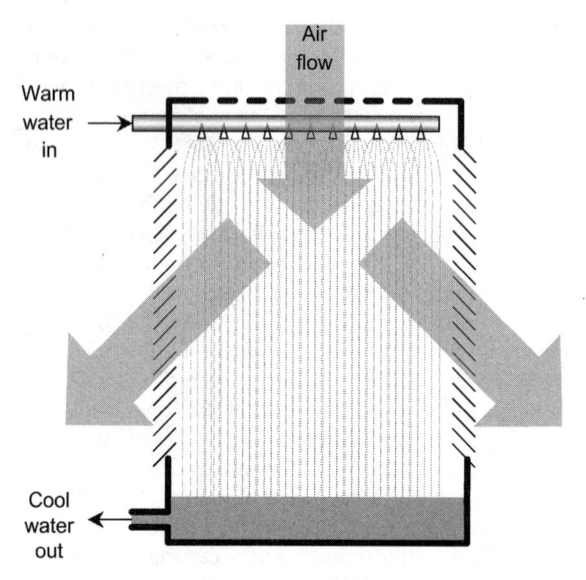

FIGURE 9.7 Schematic of basic atmospheric cooling tower.

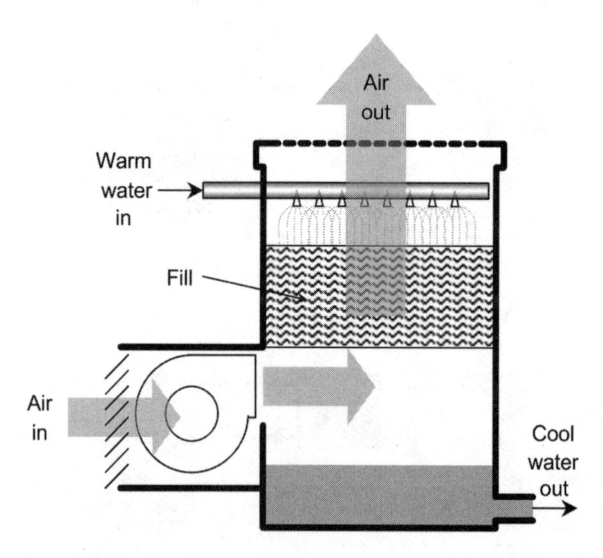

FIGURE 9.8 Schematic of forced-draft counterflow cooling tower.

short-circuiting can cause air with a high wet bulb temperature to enter the tower, decreasing the available cooling. In addition, if the weather is very cold you can have ice forming on the fan blades and an unbalancing of the fan motor. In a worst-case scenario you can throw a fan blade, damaging both the tower and the motor. The short-circuiting and freezing problems can be solved by using an *induced air tower* as shown in Figure 9.9. The air leaving the tower through the fan is generally too warm to allow ice to develop, and the direction of the fan outlet prevents the air from returning to the air inlet louvers.

CONTROLS

Air-Cooled Condenser Fan Staging

The fans on an air-cooled condenser are staged to provide a level of cooling so that the compressor outlet pressure is maintained at the setpoint. For simple on/off controls, the fans are dispatched by either the refrigerant pressure or temperature. For example, in the case of the six fan condenser in Figure 9.10 operating with R-22 system, fan 1 may come on at 150 psi discharge pressure, fan 2 at 185

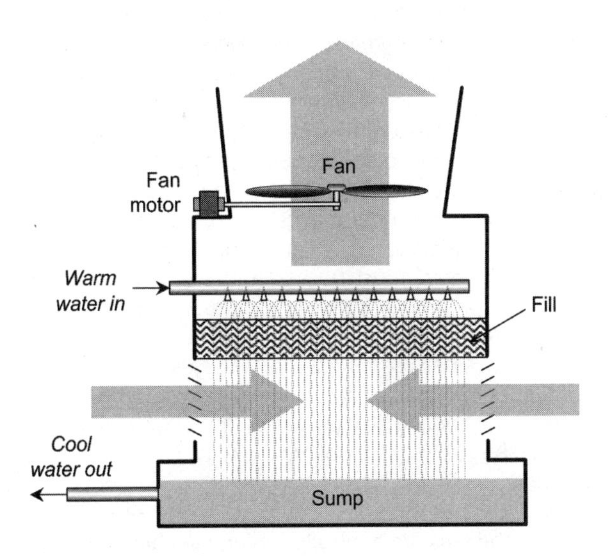

FIGURE 9.9 Schematic of induced-draft counterflow cooling tower.

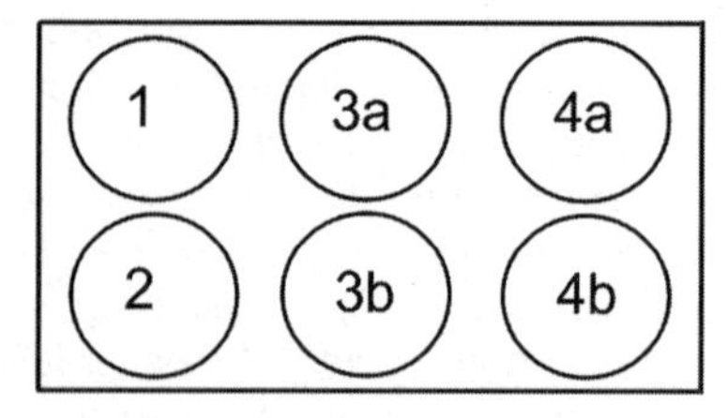

FIGURE 9.10 Fan numbering on air-cooled condenser.

psi, fans 3a and 3b at 220, and fans 4a and 4b at 255 psi. The actual values depend on the type of chiller and refrigerant used. The fans also usually have some built-in hysteresis (perhaps 10 to 20 psi) so that they do not cycle frequently when the refrigerant is hovering around certain pressures.

Cooling Tower Capacity Control

In HVAC applications where the cooling tower is used to reject heat from a cooling plant, the towers are controlled to maintain the condenser return water temperature, that is, the temperature of the water that goes back into the chiller condensers. This is done by either staging multiple cooling tower fans or by varying the fan speeds with variable frequency drives. Since cooling towers take advantage of evaporative cooling, the setpoint of the cooling tower can be no lower than the wet bulb temperature of the ambient air. Thus, ambient temperature data is needed both in the design and control of cooling towers. Also note that cooling towers tend to create their own microclimates so any measurement of the ambient conditions for tower control must be made in the vicinity of the cooling tower. The *approach temperature* of a cooling tower is an indication of how effective the tower is at cooling the water. It is defined as the difference between the outlet water temperature and the ambient air wet bulb temperature. As the approach temperature decreases, the available cooling also decreases. If the approach temperature is below about 5°F the cooling capacity of the tower is effectively negligible.

If the cooling tower has multiple cells (that is, multiple fan and draft columns) and the fans are controlled using variable frequency drives then the fans in all cells should be operated at the same

fastfacts

A cooling tower cannot produce water colder than the wet bulb temperature of the ambient air.

speed. This provides the desired amount of cooling at the minimum fan energy cost. If the cooling tower has multiple cells and the fans are controlled with multiple speed fans then the lowest speed fans should be put on-line first as additional capacity is needed.

A cooling tower controller (such as that shown in Figure 9.11) must maintain the sump temperature through staging of the fans. If there is a three-way valve in the condenser return water line, the controller also varies the valve position to maintain a desired water temperature.

The controller also manages *tower cooling* or *free cooling*. Free cooling is used when the wet bulb temperature is below the desired chilled water temperature and the cooling load can be completely met using the cooling tower. For this to work, the system must be

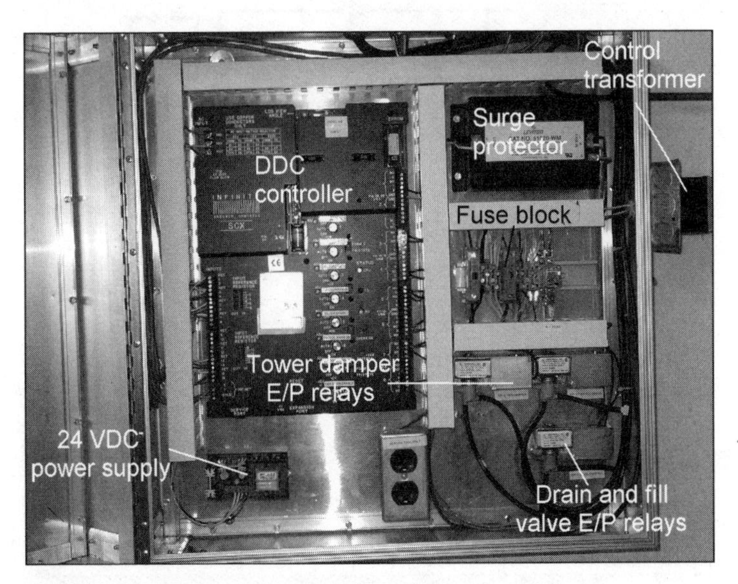

FIGURE 9.11 Cooling tower control panel.

piped similarly to that shown in Figure 9.12 where the cooling water can be diverted to the secondary water distribution system. There is usually a large heat exchanger used to separate the tower water and the chilled water loops so that dirt from the tower loop does not enter the secondary water distribution system. When the wet bulb temperature is close to the desired chilled water circulation temperature, the controller switches the valve so that the chiller can be turned off and only the tower fans run. This ends up saving a tremendous amount of energy and reduces the wear and tear on the chiller compressor.

The water level in a cooling tower is often not that easy to measure since the water is always jumping around in the air drafts. Even a simple float switch will be bounced and possibly broken due to the constantly changing water levels. Yet it is important to get a decent water level reading because the amount of make-up water added to the system affects your chemical treatment and sump level. One way of filtering the water level signal is to use a stilling tank like that shown in Figure 9.13. This is a small tank located next to the cooling tower that is connected to the tower pan. Since the connecting tube

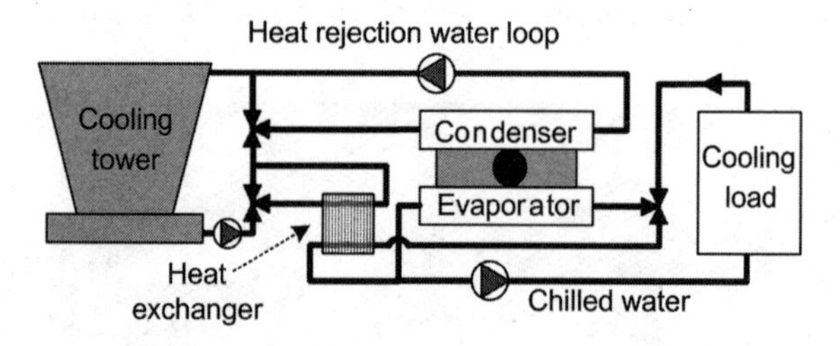

FIGURE 9.12 Schematic of piping for free cooling.

fastfacts

The cooling tower sump temperature is a few degrees higher than the outdoor air wet bulb temperature. You can take advantage of free cooling if the wet bulb temperature is close to your desired chilled water circulation temperature.

is small, the level in the stilling tank does not experience the big variations caused by the draft, but still is a good indicator of the tower water level. The level sensor or float switch is located in the stilling tank, ensuring that make-up water is added at the proper time.

Heat Recovery

Some chiller models allow for the recovery of condenser heat for relatively low-temperature (~120°F) hot water applications. Heat from the condenser can be used to provide some or all of the heating energy for hot water distribution loops in the range of 100°F to 130°F.

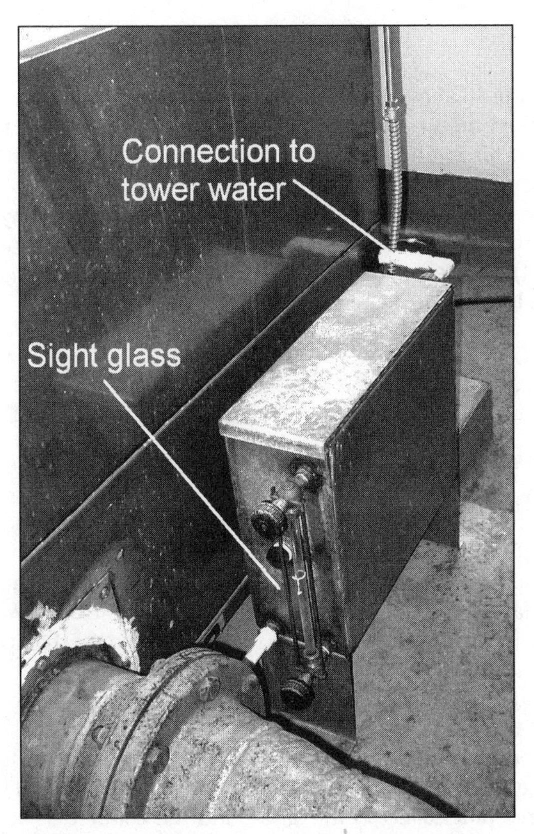

FIGURE 9.13 Cooling tower stilling tank.

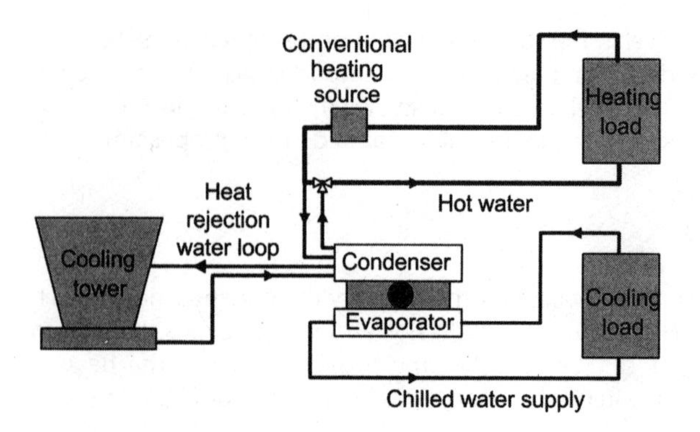

FIGURE 9.14 Piping configuration for condenser heat recovery.

Temperatures above this are not advisable since the condenser would then be too hot for efficient chiller operation. The temperature controls for heat recovery loops are usually installed by the manufacturer. The condenser heat recovery can be placed before or after the conventional heating source. Note that it is always necessary to have conventional refrigerant cooling equipment (as in Figure 9.14) in case the chiller is operating when the demand for hot water is low.

SAFETY

Cooling towers and condensers operate at high temperatures and voltages and have large exposed fans. Always keep in mind that cooling towers have the delightful combination of water and electricity (208/460 V) and are, by nature, slippery. A little caution can go a long way in preventing injury and equipment damage.

The fans in air-cooled condensers can start at any time. Since they are located outside and away from the main mechanical room, you usually have little warning as to when they are about to come on. If you must work on the fans or remove the protective grills, lock out and tag out the fan shut off.

Air-cooled condensers are filled with refrigerant. Or, at least, you hope they are filled with refrigerant and that there are no leaks letting refrigerant collect in the air around the condenser. Remember that Freon and open flame together can produce phosgene gas, the same stuff used in trench warfare in WWI. In other words, don't smoke around the condensers.

The fins on air-cooled condensers can be thin and very sharp. Be careful when cleaning the grill and don't let your fingers or arms slide up against the coil. The wounds are painful and deep.

In cooling towers, the fans can also start without notice, even if the chiller is not running. Unfortunately, since people are generally not expected to be working in cooling towers when they are operating, the fans are often unguarded. To make matters worse, the fans in forced air towers can often be found right at head height if you are standing in the pan. Not even decent head protection will stop a 5 HP fan blade.

Some older cooling towers use redwood slats along the outside of the tower. Despite the presence of a lot of water, these towers can sometimes catch fire. If working in a wooden tower, be careful when sweating pipes or doing any welding. When the tower is not operating, the wood can dry out very fast and catch fire very easily. Newer towers use fiberglass slats that are much less likely to ignite.

Many towers use chemical treatment to prevent the growth of mold, mildew, and fungus on the cooling tower media. This treatment is necessary considering the presence of humid air and water at 80° to 100° that is ideal for growing all kinds of unpleasant things. However, these chemicals can be hazardous to your health as well. Do not spend too much time in the sump of a cooling tower, and never stick your face in the discharge of a forced draft tower.

Many condensers and cooling towers are located on roofs. These are not the safest places given that they are high up in the air and exposed to the elements. Never go up on a cooling tower during a wind storm or a thunderstorm. For some reason, lightning likes cooling towers. In addition, the lighting is generally not too good at night so it's a good idea to bring a flashlight if you have to check out the tower.

fastfacts

Remember that if the cooling tower is slippery when climbing up, it will be even worse coming down. If appropriate, wear a harness and tie in to the railing so that a minor slip does not become a great fall.

Be careful on cooling towers when it is cold outside. The air around a cooling tower is usually saturated with water vapor and this can adhere to and freeze on any exposed surface. The bigger cooling towers have stairways that can become very slippery or can even freeze up and become dangerous. It is easy to slip or slide on the tower and fall off.

Another danger of winter operation of cooling towers is that the tower can actually freeze solid. If the tower starts to ice up, the water can begin to pump over the side and turn into ice as well. As the sump water collects as ice on the side of the tower, the make-up valve keeps operating to maintain the water level in sump. This is a dangerous cycle that can eventually lead to enough ice on the side of the tower to collapse the tower or even to have the tower crash through the roof. Special towers are available that are made to work in cold weather. If you operate your tower through the winter, you should probably spend the extra money to purchase a freeze-tolerant tower.

TROUBLESHOOTING

This section provides some troubleshooting tips for air-cooled condensers and cooling towers. With air-cooled condensers and dry-coolers, the problems are usually quite obvious. A coil that is blocked by dirt or leaves can have a reduced air flow and a reduced cooling capacity. Usually the cure is a backwashing of the coils with water.

Cooling towers, on the other hand, can call for a heavy maintenance schedule between the chemical treatment and the frequent cleaning. The trade-off is the efficiency of the cooling tower over the air-cooled condenser. On large buildings, the cooling tower can pay for itself just from the increased efficiency.

fastfacts

The efficiency of a cooling tower drops off after a while if the fill starts falling apart. Wooden fill has to be replaced about every ten years, but it's not used much any more. Check the fill annually and replace it if necessary.

Cooling Towers

Cooling towers are relatively simple in operation. Many problems with cooling towers can be found by inspecting for obvious problems, such as low sump water level, non-operating fans, or excess dirt and debris in the airflow stream. Figure 9.15 shows a troubleshooting chart that can be used to track problems that make the tower return water temperature too warm.

Note 1: If the electrical connections are all valid between the motor control center (MCC) and the fan motor, then there is a good chance the problem may be a blown fuse or that the power is shut off at the MCC. You can check for fuse continuity using a standard ohmmeter. Make sure the MCC power is turned off before accessing the fuses. Do not attempt to override the safety latch on the MCC access.

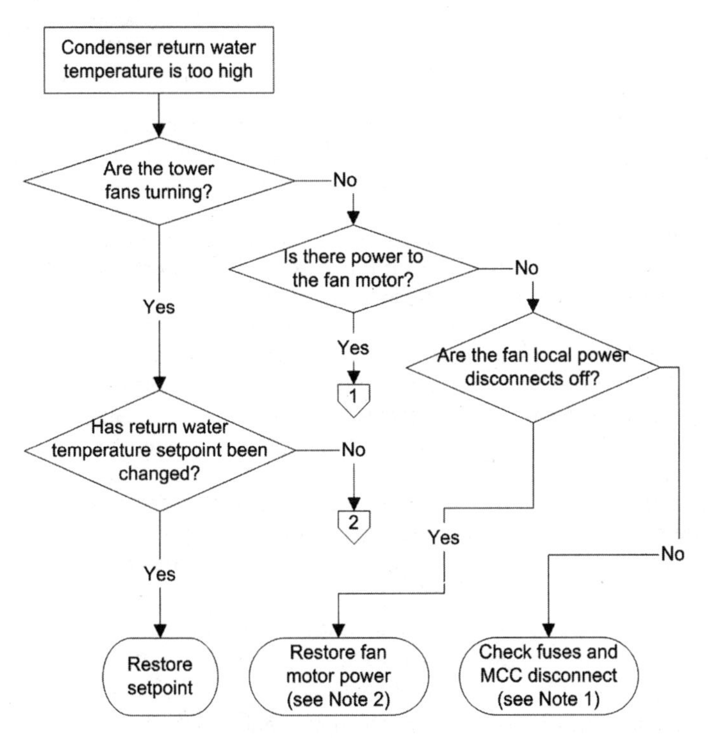

FIGURE 9.15 Troubleshooting flow chart for cooling tower.

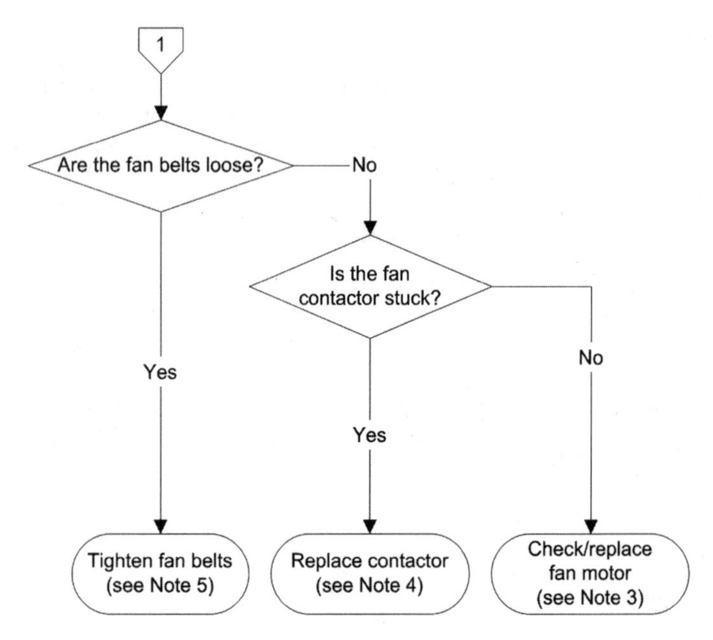

FIGURE 9.15 *(continued)* Troubleshooting flow chart for cooling tower.

If the problem recurs after replacing a fuse, there may be a short somewhere in the line between the MCC and the fan motor. Check the connections in the local disconnect as well as the wire nut connections within the motor housing that connect the motor to the line voltage. If these connections are mechanically and electrically secure, then a constantly blowing fuse may indicate that the motor is nearing failure. If the MCC uses thermal breakers, check for the presence of variable speed drives anywhere on the system (not necessarily the fan motor). In some cases, VSDs can introduce harmonics into the main line voltage that can cause thermal breakers to trip repeatedly. This usually happens at certain VSD frequencies. If possible, identify those frequencies and set the VSD controller to avoid them.

Note 2: Never activate a local disconnect until you have identified why it was shut off in the first place.

Note 3: Check the windings of the motor for obvious problems. If possible, remove the belt from the motor and check to see if the motor has seized. If so, it is necessary to replace the fan motor.

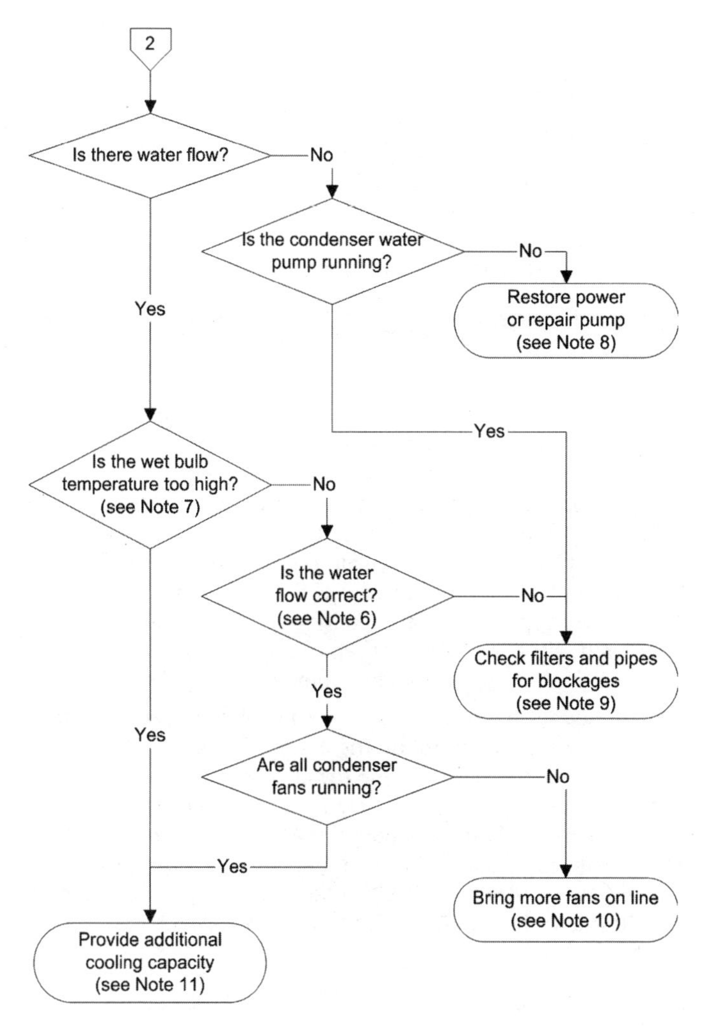

FIGURE 9.15 *(continued)* Troubleshooting flow chart for cooling tower.

Note 4: The fan contactor can be stuck in the off position. This is not as common as a contactor being welded in an on position, but excessive pitting of the contactor surfaces or just excess use of the contactor can cause it to fail. Check the contactor operation and replace if necessary. Remember to always shut off, lock, and tag the local power disconnect when performing electrical repairs.

fastfacts

Never reapply power at the MCC if you do not know why the power was turned off. Normally the MCC would be tagged and locked for maintenance or other service reasons, but this step is sometimes overlooked. Never energize a circuit until you have verified that there is no threat to personnel or equipment.

Note 5: Over time, the fan belts can stretch and come loose. If the belt is not too loose or the belt tension adjustment is already at the extreme, it may be possible to extend the life of the belt using a belt spray or equivalent adhesive substance. Remember that this is only a temporary repair; the fan belt has to be replaced eventually. Another place to check is the fan pulley. If an adjustable pulley is used, it may be that the set screw of the pulley has come loose and the pulley gap has opened, relieving tension in the belt. If this has occurred, either re-tighten or replace the pulley. Make sure the local power is off and locked before removing the pulley.

Note 6: An easy way to see if you have a cooling tower problem is to check the water flow through the condenser water piping. If excess dirt has accumulated in the tower or in the sump, it is possible that the piping has become clogged and the constricted water flow is decreasing the amount of available cooling from the tower. However, unless you have a sensor that directly measures the water flow, it may be difficult to determine how much water is flowing through the pipes. If you do not have a direct measure-

fastfacts

Belts on fans can come loose without warning. A fast-moving belt that comes loose can whip around causing serious injury. Always use caution when examining such systems. Always remove power to the motor before testing the belt tension.

ment, check the associated water pressures upstream and downstream of the condenser water pumps. High downstream pressure may imply that filters or strainers may need service.

Note 7: Recall that the cooling tower outlet water temperature can go no lower than the ambient wet bulb temperature. If you know the ambient temperature and relative humidity (as found in most building management systems) you can look up the associated wet bulb in the appendices. For example, if the outdoor air is at 90°F and 60% relative humidity at sea level, the wet bulb temperature is 78°F, meaning that the cooling tower leaving water temperature can be no lower than 78°F and is most likely a few degrees higher than that. If you are not getting enough capacity from your cooling tower and the pumps and fans seem to be working correctly, it may simply be too humid to have effective evaporative cooling.

Note 8: After restoring the pump operation, check for flow both at the supply and return water side as well as inside of the tower. If the tower has spray pumps, make sure that these are working correctly.

Note 9: Since cooling towers are outside, they are naturally susceptible to accumulation of dirt, dust, pollen, leaves, etc. Check for the presence of blockages in the system filters and strainers.

Note 10: Even single-fan cooling towers usually have multi-speed fans or variable speed drives on the fan motors. Check to make sure that the fan speed is staging properly. If there are multiple fans in the cooling tower, make sure that the fans are coming on as needed. If certain fans are not activating when they should, check the fan motor along with the fan contactor (see Note 4).

Note 11: If everything is operating normally but the condenser water is still too warm, it may simply be a case of having a condenser water load that is too high for the cooling tower. If this is the case you may have to add cooling capacity by either installing more cooling towers or by increasing the fan power or water flow of the existing towers.

PRACTICAL CONSIDERATIONS

Dirt is the enemy of condensers and cooling towers. Try to keep the fill and coils as clean as possible. This may mean daily cleaning in some areas, but it is worth it when compared to maintenance and energy costs.

Maintenance

The fill material on cooling towers can get clogged with debris like leaves. This decreases the air flow and reduces the tower capacity. Sometimes the fill material just falls apart and tumbles down inside the tower. Usually the water feed pans on top of the tower have holes where the water filters down over fill material. These holes can also get clogged, causing water to spill over the top of the pan and not onto the fill, further lowering cooling tower efficiency. Quite often there is a screen at the bottom of the tower that's supposed to catch dirt and leaves before they fall into the sump. This screen itself can become clogged.

The bottom line is that you should try to stay ahead of the dirt accumulation. A regular maintenance schedule helps you do this. Some things to consider are:

- The basin of the tower and the sump should be flushed out every year. If large clumps of dirt are allowed to accumulate, they can come loose and foul the condenser and plug the pump strainer.
- Water treatment should be checked every week. This is necessary because the tower is an open system. Use biocides, chlorine, copper sulfate, corrosion inhibitors, and whatever else is necessary within EPA guidelines to prevent clogging up the tower.
- Check the pump strainer every week. Because so much outside air passes through the tower, the water tends to gather dirt and pollen and deposits it in the filters and strainers. If the tower is in a particularly dirty area, you may want to check the strainers even more frequently.
- Check the fan motors every month and keep them lubricated. If there is a gear assembly (reducer) on the fan attachment, make sure that it is also maintained with oil. If not, the oil reservoir can fill up with condensate and water. In the winter it will freeze and break and all the oil will drain out.
- Some towers have steam or HW coils in the sump to prevent freezing in the winter. You should check these at the start of each heating season to make sure they work correctly.
- Check the float level in the tower to make sure it is providing the right amount of make-up water. Too little water lowers the tower efficiency; too much can overflow the pan, wasting water and treatment chemicals.

- Make sure that the bleed works and that you have good control of the total dissolved solids in cooling towers. These can accumulate on and ruin the fill and can foul condenser heat exchangers.
- Make sure the fan drive shaft is working and that any flexible connections between the motor and the gear box are tight.
- Check to make sure the fan does not go out of balance. Usually there is a vibration sensor on the tower that turns off the fan if it becomes too unbalanced. An unbalanced fan can throw a blade, causing damage to the tower and further damage to the fan motor.

Some air-cooled condensers use evaporative cooling to lower the temperature of the air reaching the condenser coils. This can significantly increase the heat-rejection capability of the condenser. Other systems just use sprayers to wet the exterior surface of the coils. In both cases, you need to make sure that the water is treated to prevent scale and lime build-up on the air side of the condenser coil. This is pretty difficult to remove and lowers the heat transfer rate of coil, defeating the whole purpose for using evaporative cooling in the first place.

Free Cooling

If you want to use free cooling, remember that the heat exchanger used to separate the tower water from the chilled water distribution

fastfacts

A very large office building complex has a chiller plant with ten chillers and a total rated capacity of 15,000 tons of cooling. Water from a nearby river is used to cool the condensers through a series of flat plate heat exchangers. Unfortunately, the river water contains far more silt than originally anticipated and the actual capacity of the plant is less than 90 percent of the rated capacity because of fouled condensers and high condenser return water temperatures. The loss of capacity is over 1500 tons of cooling, or the equivalent of one chiller.

loop must be quite large since it must substitute for the capacity of the chiller. These heat exchangers can take up a lot of room in the plant (see Figure 9.16) and must be planned carefully. Before installing such a system, check the local weather data to see if it in fact makes sense. These work when the outdoor temperature is less than about 70° but it really depends on the wet bulb temperature. Also make sure that the control system for free cooling properly controls the valves and pumps. In particular, make sure that you have sufficient strainers and filters on the tower side of the heat exchanger. You don't want to take the heat exchanger apart to clean it—this is costly and labor intensive.

Controls

If you are working with two or three towers and only one supply and return pipe, it can be difficult to get the water distributed evenly among the towers. You may need to install equalization lines across the sumps so that the levels are the same in the different towers.

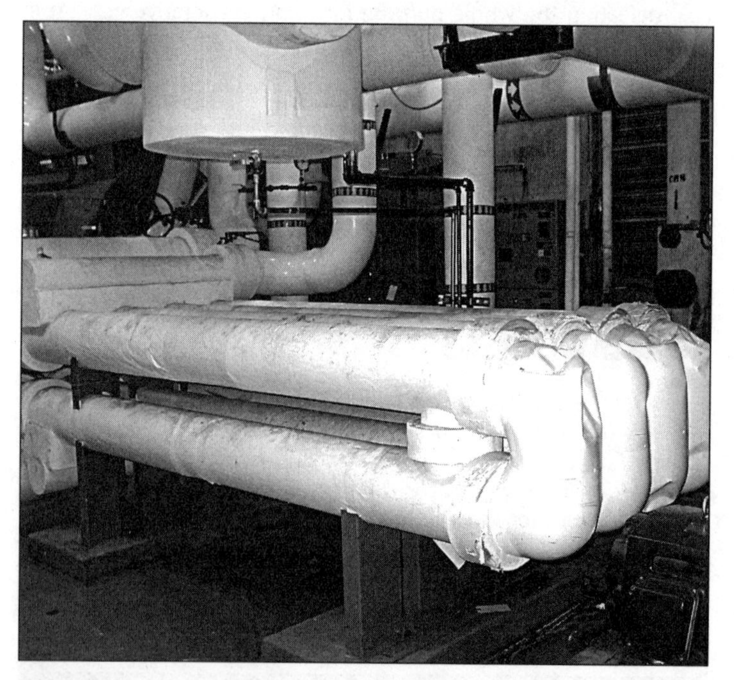

FIGURE 9.16 Shell and tube heat exchanger for free cooling.

Keep in mind that this works only when the towers are next to each other or otherwise at the same elevation. In some cases you may have to install automatic valves on the sump outlet in order to get an equal draw from all.

If you do have parallel cooling towers with cycling fans, you need to be careful in the winter. One tower can ice up before the other tower, causing loss of that tower. When the fans are cycled to maintain the condenser water return temperature, it is usually the tower in operation that freezes up first. If this continues to be a problem, you probably need to install a pan or sump heating system.

In freezing climates it is not unusual to find an open-top sump inside the building. The benefit of this kind of system is that you avoid weather problems like frozen towers and sumps that are too cold. In general you can start the chiller right away without having to warm the sump water first. Some towers use a drain and fill cycle at nighttime when it goes below freezing. Keep in mind, however, that every time you drain the pan or sump you lose the chemical treatment. Some of the treatment chemicals can cost around $1,000 per 55 gallon barrel, so it's easy to go through a lot of money in the winter if you use this technique.

chapter 10

THERMAL ENERGY STORAGE SYSTEMS

Thermal energy storage (TES) systems are becoming more popular as the cost of energy keeps increasing. The term *thermal energy storage* can apply to both heating and cooling energy, although it is quite rare for commercial buildings to store heating energy. More common (but certainly not prevalent) are thermal energy storage systems that use ice or water tanks to store cooling energy at night for use during the next day.

THEORY OF OPERATION

Cooling Storage

Cooling energy storage systems are used to shed daytime peaking loads to the evening hours when overall electrical demand is low and the cost of electricity is usually cheaper. The "coolth" is stored either by creating ice or by cooling a large volume of water. Figure 10.1 shows a chilled water primary loop system with thermal energy attached. The dark lines indicate the flow of water under normal chiller operation.

At night or during any periods when there is no cooling load, it is possible to use the chiller to "charge" the storage as shown in Figure 10.2. Note that some TES systems have small chillers devoted to charging the storage. During normal discharge, the chiller bypass

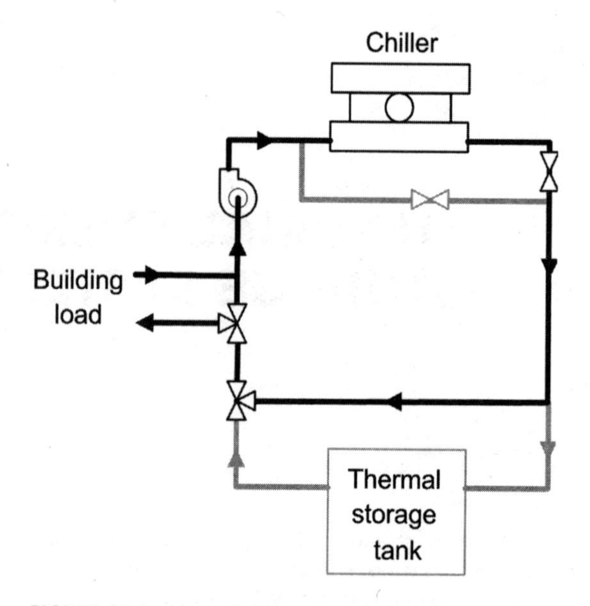

FIGURE 10.1 Normal chiller operation.

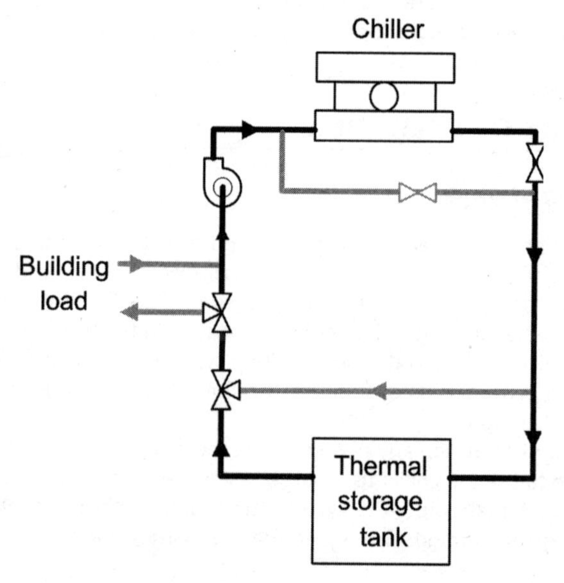

FIGURE 10.2 Charging of cooling thermal stage.

loop can be opened and the chiller shut off (Figure 10.3). It is also possible to run part or all of the primary loop flow through the chiller and simply unload the chiller to reduce the energy consumption.

In cooling applications, energy is stored by either direct water storage or by some kind of phase change process such as ice creation. An advantage of direct water storage is that the chilled water system does not operate below the freezing point so there is no need for glycol treatment. The main disadvantage of direct water storage is the need for a large insulated volume for water storage. Every 100 GPM of chilled water flow is equivalent to almost 50,000 gallons of daily storage requirements.

Ice storage, on the other hand, uses the heat of formation when water freezes to store the cooling energy. These kinds of systems are much more compact, but are also more complex. There are several methods used to make ice in an ice storage system. In systems such as those shown in Figure 10.1 through Figure 10.3, the glycol-treated chilled water system can be operated at temperatures below freezing, generally about 20 to 25°F, and circulated through the storage tank to make ice. The thermal storage tank is a large, insulated

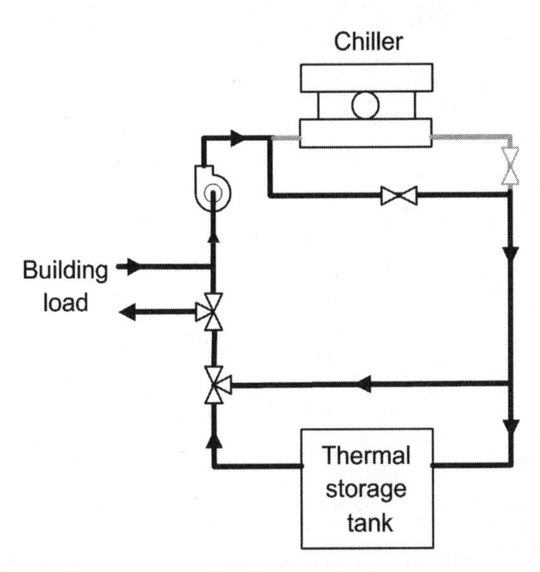

FIGURE 10.3 Discharging of cooling thermal storage.

tank that contains a closed coil of pipe immersed in water. This is diagrammed in Figure 10.4. The inlet and outlet pipes are attached to manifold headers that distribute the water through many different coiled pipes in an effort to reduce the total pressure drop across this closed system of piping.

A variation on this theme uses an open system in which the cold glycol solution freely circulates around the storage media. A typical method uses a tank with water-filled polyethylene spheres as shown in Figure 10.5. These spheres are dimpled to allow for expansion of the water within the sphere as it turns into ice.

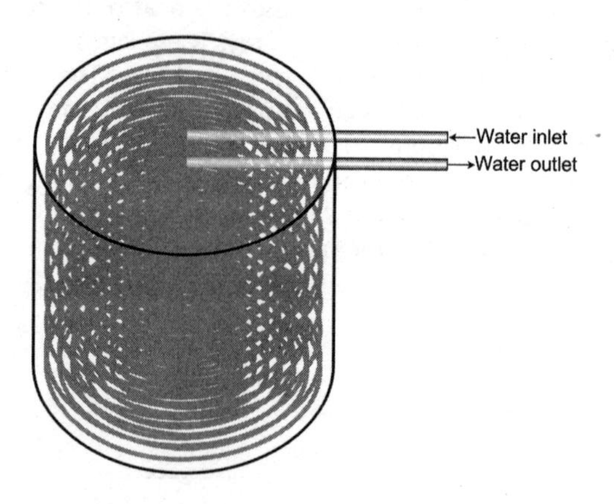

FIGURE 10.4 Cut-away view of an ice storage tank.

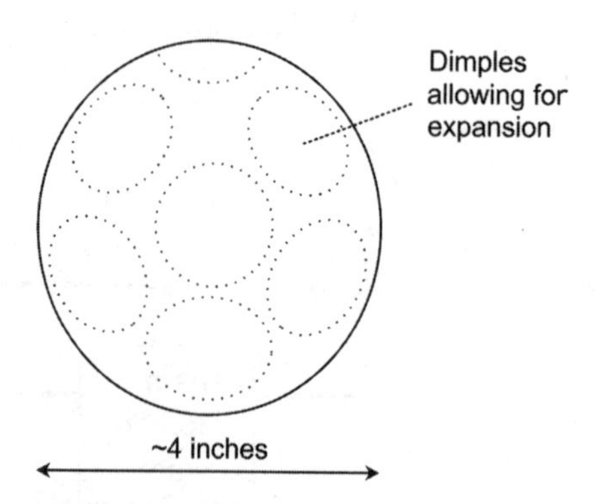

FIGURE 10.5 Storage sphere used in ice storage systems.

Another technique is to use an *ice harvester* that behaves very much like an ice-cube maker. This system uses a devoted chiller to produce ice shavings that are deposited around the water circulation piping. The chiller evaporator is a plate on which water flows and freezes. As the ice reaches a certain thickness, it is released into the storage tank either through a mechanical scraper or by using hot gas refrigerant from the vapor compression process.

The storage capacity of cooling TES systems is measured in ton-hours as a surrogate for tons of refrigeration. One ton-hour is therefore equal to 12,000 Btu. Since ice can store much more energy than water alone, the required volume for water systems can be five to ten times that for ice systems. In any case, when designing a thermal energy storage system, you should pay attention to the tank size. Even in a relatively small application, the tanks can occupy a decent amount of floor space in the mechanical room. Figure 10.6 shows the size of a 150 ton-hour ice storage tank. It is about ten feet across and

FIGURE 10.6 150 ton-hour ice thermal storage tank.

stretches from floor to ceiling and provides enough storage to cut about half the daily peak load of the accompanying 50-ton chiller.

Heating Storage

Since demand charges for thermal energy are exceedingly rare, there is not much call for heating storage. The economics for thermal storage of heating arise in cases where the contractor or architect wishes to reduce the capacity of the heating equipment and to supplement the smaller capacity with energy from storage. The storage medium in this case can be water, rock beds, or the earth itself. Water storage is similar to that for cooling energy storage: a large volume of water is heated using energy recovered from the exhaust air system or from heat recovery condensers on cooling systems. This water is then used to provide perimeter zone reheat during the occupied hours of the day.

Rock beds are large volumes of crushed rock and gravel through which air is circulated. During the charging of the storage, air at 90° to 120°F is passed over the rocks for a number of hours. The warm air comes from building exhaust air or from solar collectors. During the hours when there is a requirement for heat, building air is circulated through the rock bed and introduced directly into the building air distribution system. Since rock beds are inherently dusty, the building air must be carefully filtered.

Ground-coupled heat pumps are basically residential heat pumps that have the outdoor condenser/evaporator as many hundreds of feet of buried piping. During the summer, the heat removed from the building is stored in the ground. This energy is then used in the winter to heat the space. Such systems are mostly used for residential air conditioning in the southeast states. Large-scale commercial applications are rare.

CONTROL

Cooling Storage

The typical control points of a cooling TES system are shown in Figure 10.7. At night the three-way mixing valve CV-2 is closed to force all the water from the chiller back into the primary loop. If ice is used as the thermal storage medium, then the chiller setpoint $T_{CHW, SUP}$ is set somewhere around 25°F to 30°F. The water is used

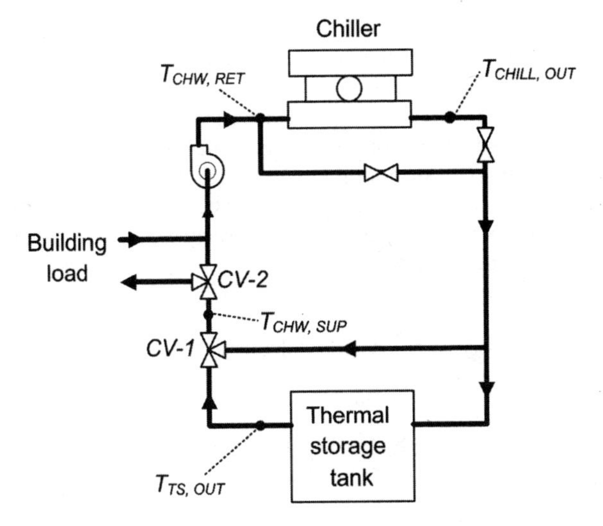

FIGURE 10.7 Control points in a cooling TES system.

to cool the thermal storage tank. During the daytime the CV-2 modulates to maintain a desired setpoint temperature in the secondary loop. To retrieve the stored energy one of the following methods is usually used.

The simplest method for thermal storage control is the *chiller-priority* strategy. The chiller is sized for a capacity less than the building peak load. During the evenings, the chiller is used to make chilled water or ice. During the day the chiller operates at full capacity and the difference between the chiller output and the building load is made up with the cold storage. This technique maintains a high chiller efficiency and provides for a known demand curve in case demand limiting is implemented. The main disadvantage of this method is that the chiller still operates during the daytime hours and incurs both consumption and demand charges.

Storage-priority control attempts to use as much of the stored energy as possible and to minimize the chiller operation during the hours when electricity is most expensive. The clear advantage of this strategy is that the chiller can often be turned off during these on-peak hours leading to significant cost savings. The main disadvantage is that a control scheme is required that can predict the next-day building loads and generate enough ice to handle that load. If the storage is depleted during the day then the chiller may need to be placed on line during periods of high energy cost.

Constant proportion control of thermal storage refers to a technique where the ratio of the cooling provided by the chiller and the storage is a constant. In other words, the proportion of cooling provided by either device remains constant throughout the day regardless of the load. This combines the simplicity of chiller-priority control with the demand limiting potential of storage-priority control. When using constant proportion control the outlet temperatures of the chiller and storage tank remain constant. The chiller outlet temperature setpoint is given as

$$T_{CHILL\ OUT} = T_{CHW,\ RET} - f \cdot \left(T_{CHW,\ RET} - T_{SUP}\right)$$

where f is the fraction of cooling that is to be handled by the chiller.

SAFETY

Most thermal energy storage systems are relatively safe. Normal safety precautions for piping and compressors are sufficient. There is one aspect of contained ice storage that must be mentioned, however. Recall that ice is less dense than water; ice cubes float in drinks and the ice in a closed TES tank also wants to float. However, these containers are usually sealed at the top to aid in the insulation of the tank. Under no circumstances should the bolts holding the top of the tank be loosened while there is ice in the tank! The upward force of the ice against the top can be many hundreds of pounds, enough to lift a technician and equipment. The top would then be an unstable surface and the danger of falling or being caught between the tank top and the room ceiling is high.

It is important to know what the level is in the tank for a number of reasons. The first is to determine the state of charge of the tank. Since the ice expands on freezing, you can estimate how much ice is in the tank by the water level. Another reason to measure the level is so that you do not run the risk of overflowing the tank. A simple way to measure the level is to use a bubbler level sensor (Figure 10.8). This device sends a constant stream of air into the tank from the compressed air system. The column of water displaced in the bubbler tube is reflected in the air pressure at the gauge. In fact, if your gauge measures pressure in equivalent inches of water, then you can make a direct reading of the water level in the tank.

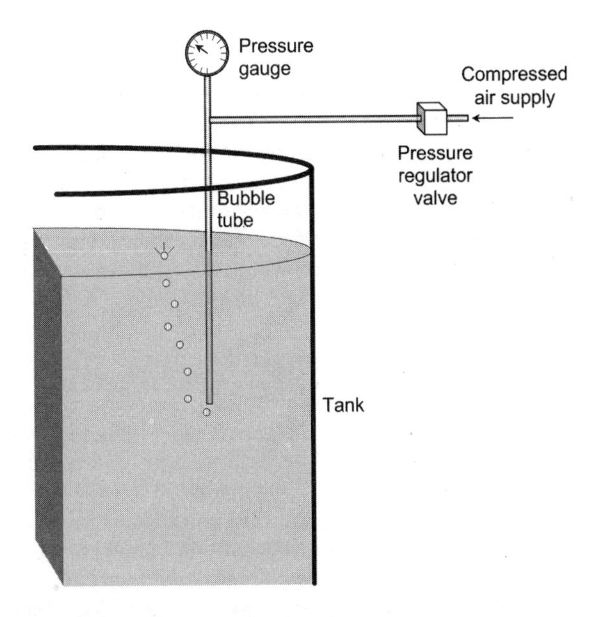

FIGURE 10.8 Bubbler level sensor.

TROUBLESHOOTING

A problem in many existing TES systems is that the design or control system is not properly implemented. This creates a distrust of these systems. Since there are few TES systems in existence, most are custom jobs and each has its own specific sets of problems. Given this situation, the troubleshooting of cooling TES systems is limited to the circumstance where the system is not providing adequate supplementary cooling and there is no significant reduction in the energy bills. If this is the case, there are a few areas that are worth investigating.

The first point to consider is the presence of a decent control system. Under the best of circumstances, the TES should finish charging at the same time that the time-of-use rates increase, and it should finish discharging just as the rate decreases. A single 15 minute period during on-peak hours in which the TES does not operate can lead to the demand charge being set for the month. The general rule of thumb is: *poor control leads to poor economics.*

Is the system properly configured for thermal energy storage? Investigate the system plans and identify any components that may prevent the system from operating correctly.

If you try to discharge ice storage systems too quickly you can end up creating a pocket of water between the water circulation tube and the ice (see Figure 10.9). The water acts as an insulating layer and can prevent the tank from providing enough cooling. Slowing the discharge rate can solve this problem, but by doing so you may need to add more capacity to achieve the required peak shaving.

PRACTICAL CONSIDERATIONS

Cooling TES systems can be used successfully in places where electricity costs more during the daytime on-peak hours than at night. Of course, one must account for the fact that the storage system will have losses, so the daytime cooling load must be large enough to make storage worthwhile. It is an interesting circumstance of TES systems that they can use more total energy but lower costs. Other instances in which cooling TES may be feasible include buildings with large daily load variations, buildings with short duration cooling loads, and utilities that offer rebates for load shifting.

The main goal of a cooling TES system is to lower the operating costs, primarily by reducing the peak daytime electricity demand. If properly designed, the conventional chiller can be undersized relative to the peak cooling load; the assumption is that the storage system will supplement the chiller and provide full coverage of the load.

The benefits from TES systems are not limited to cooling energy requirements. Since the chiller can now be reduced in size, there may be lower pump horsepower and energy consumption, smaller pipes, increased reliability, and decreased maintenance costs.

fastfacts

The ton-hour rating of many cooling thermal energy storage systems is usually optimistic. If using the manufacturer's ratings, you may want to oversize your storage tank to ensure that you meet the design demand reduction.

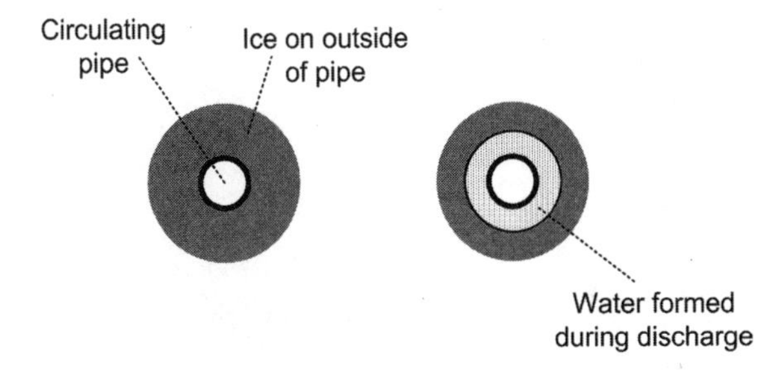

FIGURE 10.9 Insulating layer of water formed during discharge.

fastfacts

A large thermal energy storage system (three million gallons of chilled water) was installed in an office complex. The water distribution system was pressurized while the water storage tank was at atmospheric pressure. To de-couple the pressures, a heat exchanger was used between the storage and the chiller plant. However, the temperature differences across the heat exchanger were so low that the heat exchanger effectiveness was only about 10 percent. The building operator simply could not discharge the storage fast enough to obtain any kind of useful cooling from the storage. The control algorithm did not matter; the system design prevented the TES from being used to its full extent and the entire system was essentially useless.

chapter 11

BOILERS

Boilers are used to generate steam for use in larger buildings. Steam is a convenient medium since it does not require additional pumping power to circulate. The boiler produces steam and the steam expands into the distribution system. The steam then passes through heat exchangers to heat water and air where needed. Of course, there will be some condensation within the distribution system as some of the steam cools off. This condensate is removed using steam traps that feed to a condensate receiver tank. A float switch in this tank closes when the tank is full and activates a pump that returns the condensate back to the boiler.

THEORY OF OPERATION

Boilers use electricity, natural gas, oil, or coal to heat water and produce steam. Natural gas boilers are most common for large systems while electric boilers are common for smaller boilers. *Hydronic* boilers do not produce steam but rather pressurized hot water. Most of the control circuitry for boilers is installed by the manufacturer, although often the user has the ability to change the outlet pressure or temperature setpoint.

Boilers are rated by both their *heat rate* (in pounds of steam produced per hour) and by their overall heating capacity (in millions of Btu per hour). In addition, boilers are rated by their pressure range.

- Low pressure boilers produce steam in the 5 to 15 psig range or hot water at pressures up to 160 psig and temperatures of 250°F.
- Medium pressure boilers produce steam pressures up to 50 psig
- High pressure boilers operate at steam pressures greater than 50 psig

High pressure and low pressure boilers require about the same maintenance, but local codes often dictate that a high pressure boiler requires a full-time operator.

Boilers are relatively simple, with components shown in Figure 11.1. Water passes through a heat exchanger and is warmed by a combustion-based or electric heater. The water then turns to steam and exits the boiler through some piping while any combustion gases leave through a flue. There are many variations on this theme, of course. The different fuels have different combustion arrangements, while the heat exchangers can be water boxes, shell and tube, tube and fin, etc. In *cast iron* boilers the heat exchanger is composed of a number of cast iron modules joined together. The capacity of the boiler is increased by attaching additional modules. *Firetube* boilers are essentially shell-and-tube heat exchangers where the combustion and flue gases travel through the tubes. The surrounding water is then heated and converted into steam. Some firetube boilers have *external fireboxes* where the firebox is not surrounded by water. Watertube boilers are also shell-and-tube heat exchangers but in this case the water runs through the tubes while surrounded by hot flue gases within the jacket. *Modular boilers* are basically big water heaters that produce steam instead of hot water. There are even combination boilers that produce both steam and hot water.

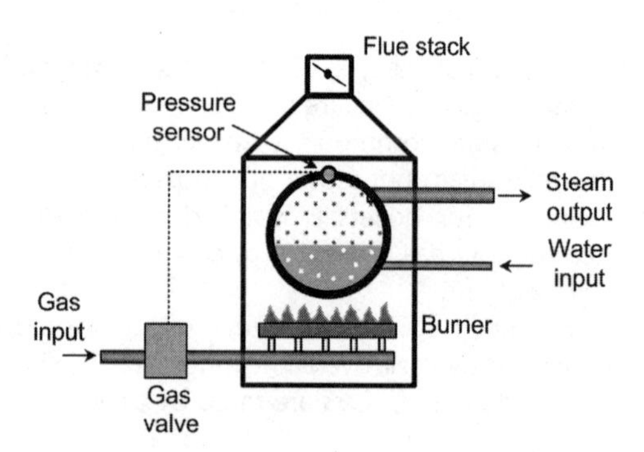

FIGURE 11.1 Basic boiler schematic.

On *natural draft* boilers, the flue flow is maintained by the buoyancy of the heated air as it rises up the exhaust stack. A *forced draft* boiler uses a fan and burner assembly. The fan is referred to as the *blower* (Figure 11.2) and aids in the combustion as well. The blower forces air into the combustion chamber and the flue damper is modulated to provide the correct amount of air for a given boiler load and corresponding fuel flow rate. The blower usually works together with some kind of carburetion device to make sure the air and fuel are well mixed before combustion. In most cases the blower is on the burner side (so it doesn't get too hot), although in some cases it may be on the stack side.

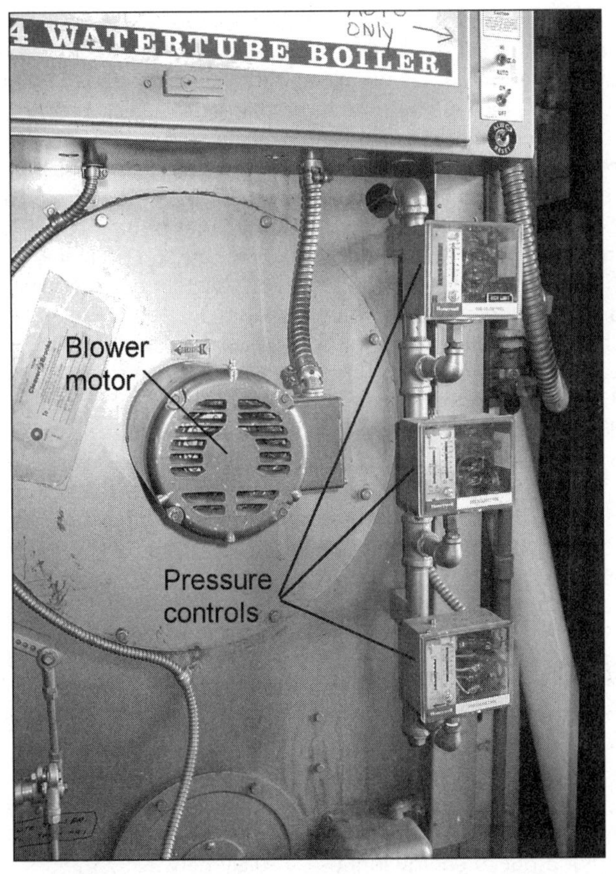

FIGURE 11.2 Boiler front showing blower fan motor and pressure controls.

Electric boilers can heat water in two different ways. *Resistance heating* elements boil water by immersing the resistive elements directly in the pressure chamber. As the elements get hot, they heat the water through direct contact. *Electrode boilers* have the actual open electric terminals exposed to the water. As the electric current travels through the water, the resistance of the water itself makes the water warm up to the boiling point. In both cases, there are generally at least five resistive elements or electrodes that can be staged independently so that a fairly continuous staging of the boiler capacity is achieved.

As water travels into the boiler, it leaves deposits of dirt, lime, scale and other dissolved solids on the inside of the heat exchanger. It is necessary to blow down the boiler occasionally to get rid of this material and maintain the dissolved solids concentration at a manageable level. Small steam boilers probably don't need to be blown down very often—perhaps no more than once a year. Large boilers, on the other hand, need to be blown down once a week or even once a day. Most steam distribution systems (see Chapter 12) include a feed pot located on the condensate or water return line just before the boiler. The purpose of this pot is for chemical treatment of the water, such as corrosion inhibitors and pH maintenance. If the water becomes too fouled, then the boiler may develop scale and the heating capacity will be diminished.

If a steam boiler is under a heavy load or is overfiring, the water can start to bounce around and jump out into the steam line along with the normal steam flow. This reduces the capacity of the lines and creates a danger of condensate hammering. This can also happen if there is too much water treatment and the boiler starts "foaming." A *Hartford loop* is a piping arrangement that creates a gravity feed between the boiler header and the return piping. This allows any water siphoned out of the boiler to drain out of the steam distribution system and back into the boiler.

Boiler Rating

In combustion boilers, the Btu per hour rating is the energy input rate to the boiler. This is what is usually shown on the nameplate. The boiler horsepower or pounds per hour of steam are the energy output rates. The energy input does not equal the energy output due to losses. The efficiency of combustion boilers—the ratio of energy output to energy input—is usually between 75 and 90 percent. The remaining energy is lost in the hot gases that leave through the flue. The heating elements are typically a flat set of burners with some kind of standing pilot or spark ignition and a flame verification safety. Natural gas boilers can use the existing gas line pressure, although in very large boilers you may find gas compressors. In oil-based boilers, the liquid fuel is converted into a fine spray that is then introduced into the air stream. This is done through nozzles or by atomizing the oil using a jet of high-pressure steam or air.

Electric boilers are rated in kilowatts and are assumed to have 100 percent efficiency. Resistance boilers have capacities up to about 2500 kW while electrode boilers can be as much as five times this amount.

The US Department of Energy rates boilers in terms of the annual fuel utilization efficiency (AFUE). The nominal efficiency is a simple, steady-state ratio of input energy to output energy while the AFUE efficiency takes cycling into account. Figure 11.3 shows the problem of thermal lag associated with cycling, specifically that the boiler must heat up before it can start producing steam. At the very start of the "on" cycle, the nominal efficiency is zero because all of the heat going in to the boiler is being used just to heat the combustion chamber and to warm the liquid water up the boiling point. Only after the boiler is hot enough will it begin to produce steam.

fastfacts

Heating oil has an energy content of about 140,000 Btu per gallon, while natural gas has an energy content of about 1000 Btu per cubic foot.

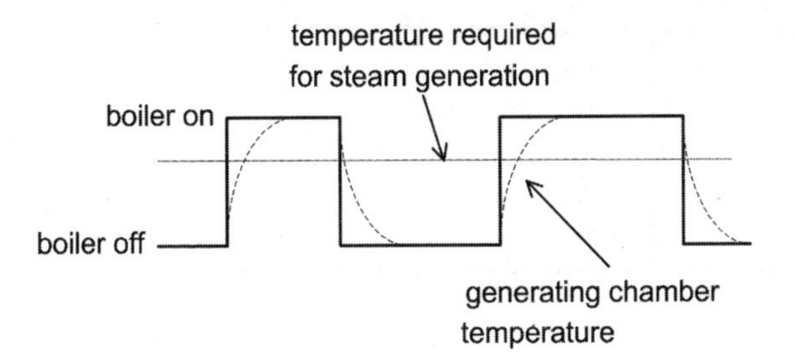

FIGURE 11.3 Effect of boiler cycling.

Ignition

There are a number of ways to light gas or oil in a boiler. A *standing pilot* is a constant flame much like that on a residential furnace or water heater. The pilot is always lit and allows for quick lighting of a gas manifold. An *intermittent pilot* is lit only when the boiler is about to come on. The intermittent pilot is lit by spark ignition. *Spark ignition* forces a voltage differential across a gap that produces a high temperature spark, much like the spark plug in a car. The spark then ignites the gas or oil in the vicinity.

CONTROLS

There are two goals for the boiler control algorithms: (1) maintain steam pressure setpoint or hot water temperature setpoint, and (2) ensure high operating efficiency. The second goal becomes more important as the size of the boiler increases. The fuel costs alone on a large boiler can be hundreds of dollars per hour, so it makes sense to ensure that the fuel is being used correctly.

If using a hot water boiler, it is a good idea to use a primary loop configuration with a three-way valve and a nulling loop (Figure 11.4). This allows you to keep the boiler temperature constant and keeps condensation out of the boiler under low load. If you have a cast iron section boiler, it is a particularly good idea to keep water temperature constant to help prevent leaking boiler gaskets that can occur from constant expansion and contraction. The water temperature setpoint in the primary loop is usually between 150° and 180°F.

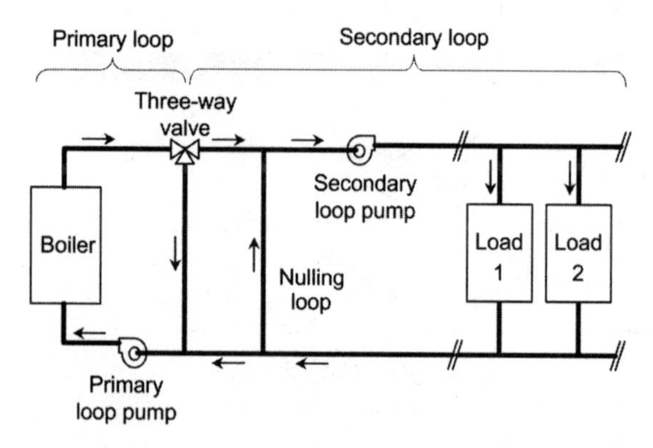

FIGURE 11.4 Primary/secondary loop configuration for hot water boiler.

The most common method of control in small boilers is a simple on/off control as illustrated in Figure 11.3. As the pressure goes below the setpoint, the boiler turns on and heats the water. As the pressure goes above the setpoint, it turns off. A deadband is incorporated into the controller so that the boiler is not constantly turning on and off right at the setpoint. On/off control may lead to short cycling of the boiler, since the pressure oscillates between the upper and lower limits of the dead band.

High/low/off control provides three different levels of heat in a firebox. This is similar to on/off control but with an intermediate step for low loads. The low-fire level keeps the boiler from shutting off altogether, helps maintain a higher overall boiler efficiency, and allows for quick capacity increases as necessary. The low fire level is generally about 20 percent of the full capacity of the boiler.

Larger boilers have more sophisticated control and safety systems for a couple of reasons: (1) the cost of the controller is cheap relative to the cost of the rest of the boiler, and (2) the need for better control is critical for human and equipment safety. Continuous, modulating fire control is used in large boilers where the cost of a gas flow regulator is relatively low and can be easily incorporated into the rest of the control panel (Figure 11.5). Modulating control allows for full, variable control from the low-fire level all the way up

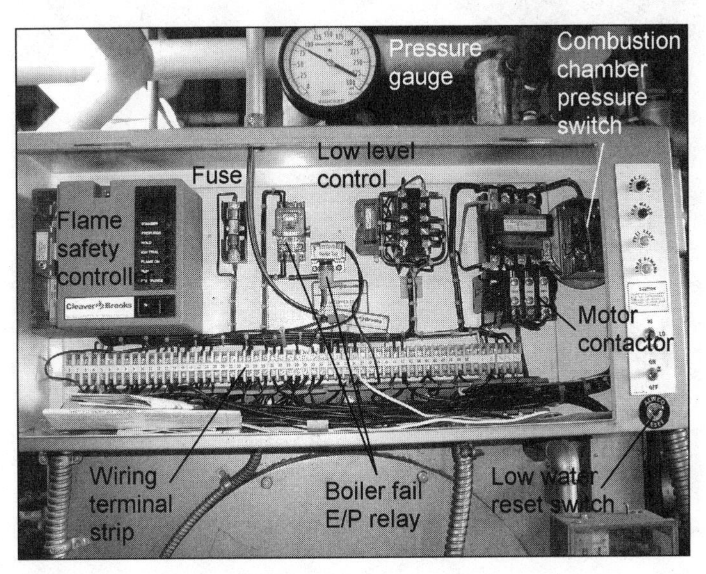

FIGURE 11.5 Boiler control panel.

to full capacity. Figure 11.6 shows an electric actuator that is connected to both the gas valve and the flue damper to modulate and maintain the boiler capacity.

The blower controls on large boilers attempt to match the supply of combustion air and fuel to maintain the highest efficiency and pollution control of the flue gas. The operation of a combustion-based

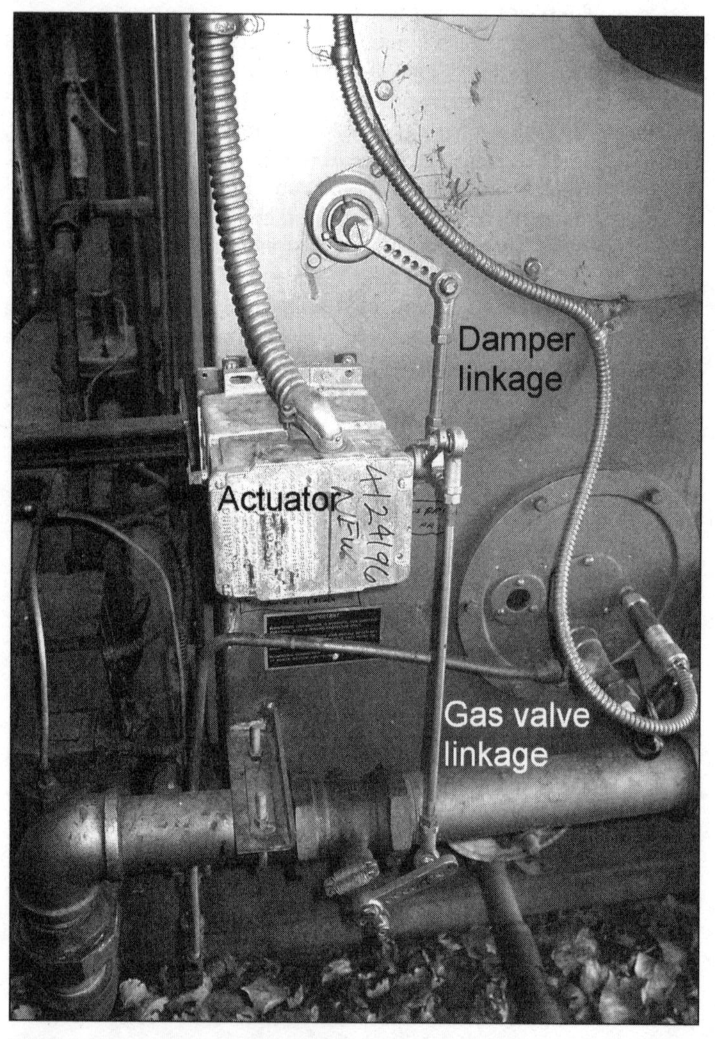

FIGURE 11.6 Electric actuator for gas and damper adjustment.

boiler must meet several criteria. The air intake should be controlled so that enough oxygen is supplied to provide complete combustion without producing too much carbon monoxide (CO) and other pollutants. But if too much air is provided, then the efficiency of the process decreases because energy is used just to heat up the excess air without going to heating the water. Ideally, the flue gas contains about 3 to 5 percent oxygen (O_2) and 10 to 12 percent carbon dioxide (CO_2). In large boilers the flue gas conditions are usually monitored continuously to make sure that the combustion process is operating as efficiently as possible. Also, it is very important to keep the flue gas temperature above the dewpoint temperature of water so that there is no condensation in the flue or the firebox. Not only does condensation lower the efficiency, but it can also cause corrosion of the iron components. The only exception to this is in boilers made especially for condensing the flue gas. In these kinds of boilers, a secondary heat exchanger is used to extract more heat from the hot gases, thereby increasing the overall efficiency of the boiler. It is necessary, however, to use a stainless steel or other non-corroding metal for the heat exchanger so these can get pretty expensive.

As with all large HVAC equipment, boilers have safety controls that shut off the boiler (or prevent ignition) if a condition exists which is hazardous to the equipment or facilities personnel. The *flame-failure safety shutoff* or *flame relay* checks for the constant presence of flame whenever the boiler is operating. These sensors fall into the following categories:

- A *thermocouple flame detector* is a high temperature thermocouple (see Chapter 5) that produces a small voltage when heated. The thermocouple sits in the pilot flame and provides a constant electrical signal so long as there is flame.
- An *ultraviolet flame detector* is basically a light sensor that is keyed to respond to the specific wavelengths of light in a flame (that's 0.185 to 0.260 microns, for those who care). They can be configured to provide a switch closure or electric signal in response to the presence of flame.
- A *flame rod* consists of a central electrode surrounded by a metal sheath. The electrode sticks out of the end of the sheath and sits in the flame. A voltage is applied across the electrode, and the flame provides an ionized "path" between the two terminals of the electrode, allowing current to flow.

The function of the flame relay sensor is to cut the main burner gas flow if for some reason the flame is extinguished. This prevents gas from building up inside the combustion chamber. Since thermocouple flame sensors have some weight to them, they stay hot even when the flame is cut off and may not shut down the boiler for many seconds. UV sensors and flame rods, on the other hand, respond immediately to loss of flame. In fact, in most large boilers with UV flame detectors, the main gas valve shuts off instantly on flame-relay cut-out.

Combustion boilers typically have a *prepurge* cycle in which the blower turns on and the flue opens before the gas valve is opened. This purges the combustion chamber of any unburned gases before the ignition circuit is activated. Once the combustion chamber has been purged, the damper closes.

Hydronic boilers have a water flow interlock to prevent boiler operation when there is no water flow. The water flow interlock should be tied directly into the boiler start circuit. A lack of water flow should kill the flame and shut off the boiler immediately. This may not be enough, however, since even the residual heat in a boiler can flash standing water.

Boiler Safety

Depending on local code and the size of the boiler, you may have to install a double block and bleed valve assembly (Figure 11.7) to ensure that the gas line does not stay charged when the boiler is off. A double block and bleed consists of two gas valves with a solenoid valve in the middle that vents gas to the atmosphere. The two valves operate in series. If one valve fails, the other still blocks the gas flow and it bleeds off through the solenoid valve. On a call for gas, both gas valves open and the middle bleed valve closes. The opposite occurs when the valve is shut off.

Boilers also shut off on high or low gas pressure. If the pressure is not within the specified range for that boiler, the gas valve may not be able to properly control the amount of gas going to the burners. This can lead to an overheating of the boiler. The high pressure cut-out (Figure 11.8) is located upstream of the gas valve.

Boilers have a number of water level switches. The level control in a hot water boiler is a special float switch used as a safety that shuts off the boiler if the water level gets too low. In a steam boiler there is a level control and a low-water control that each work with their own float switch or with two switches on a single float (Figure 11.9). The level switch turns on the feedwater pump that takes water from the condensate return to fill and maintain the proper

FIGURE 11.7 Double block and bleed valve gas actuator.

FIGURE 11.8 Boiler gas valve and high pressure cut-out.

water level in the boiler. The low-water control shuts the boiler off if the water level gets too low. This is to keep the tubes in the boiler from uncovering. If they do get uncovered they can become red hot and melt. In theory, this condition should never exist because the boiler should always be full of water. If it does run low, any new water that does enter the boiler can immediately and explosively turn to steam. For this reason, boilers always have a pop-off pressure safety valve. These operate on temperature only, temperature and pressure, or pressure only. Most steam boilers use pressure-only pop-off valves. Sizing of the pop-off valves depends on the size of the boiler. It is very important to make sure the pop-off valves operate correctly—you should replace them every two or three years just in case. Always use the correct pressure and pound-per-hour ratings when replacing the pop-off valves. If you lift a pop-off valve to test, it usually never seats properly again.

FIGURE 11.9 Fill sensor and low-limit cutout on steam boiler.

fastfacts

It is generally cheaper to replace a pop-off valve than to let it leak. A leaky valve allows oxygen into the boiler through the increased make-up water. This can cause all kinds of rusting problems in hot water boilers and scale problems in steam boilers.

Hydronic boilers have high temperature cut-outs, also called high limit controls. These turn off the boilers when the temperature gets too high. For hydronic boilers, the temperature setpoint is typically 140° to 180° unless the boiler is used for a specific high temperature application (such as a hot water absorption unit).

Hot Water Reset

To conserve boiler energy, the hot water temperature setpoint can be adjusted based on the outdoor temperature. When it's cold outside, the setpoint is adjusted upward. Note that this is, strictly speaking, not a control process but rather a dynamic setpoint modification. Figure 11.10 shows a simple version of how the temperature reset control would work.

SAFETY

Among the various components of HVAC equipment, boilers should command the most respect. They operate at lethal temperatures and pressures. Boilers are dangerous to be around and dangerous to work with. General and frequent inspections can be a big help in identifying problems before they occur.

- Make sure there is enough make-up air. Combustion boilers use a lot of air and there should be a sufficient supply of outside air directly into the boiler room. Never block the make-up air intake, even briefly.
- Make sure the main gas valve shuts off if the pilot goes out.

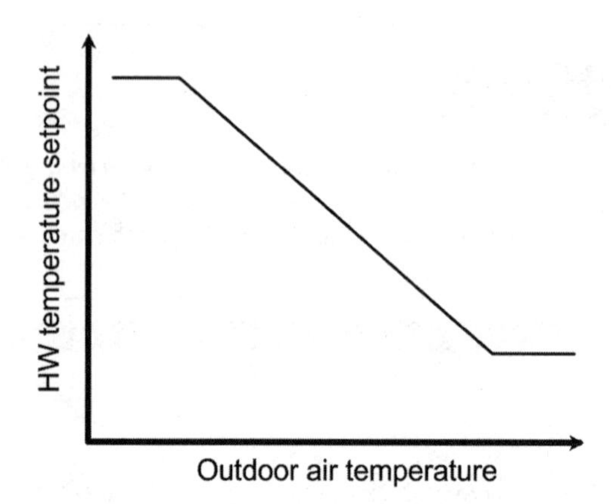

FIGURE 11.10 Hot water temperature reset.

- Make sure the flame relay is working. One way to check this is to start the boiler and then physically disconnect or remove the flame detector. The boiler should shut off immediately. UV detectors should be replaced every three or four years—if you go longer than that you will probably have operational problems with them. It only takes a few minutes and it's a good idea.
- If your boiler uses a flame rod, make sure the flame rod is in the middle of the pilot. If not, the boiler may not start or may cycle unnecessarily.
- Large intermittent pilot boilers have a timer so that if the flame relay sensor does not detect a pilot once the boiler starts it locks out the boiler. This *proving time* is typically three to five seconds, although on big hot water boilers it may be much shorter. Check your local code as it usually dictates the timing.
- Make sure the pilot has enough spread so that when the gas valve opens, it lights the full manifold. You can check this by watching the manifold as the gas valve opens. The entire manifold should light more or less at once. If not, you may need to increase the spread, or *fan*, of the pilot.
- Never try to light a gas or oil boiler manually.
- As silly as it sounds, make sure the pilot assembly has not fallen off. This sometimes happens, usually after maintenance when someone discovered that the pilot assembly mounting screws are hard to put in and then just shoved it into place. While

mildly amusing, this can be inconvenient (the pilot will not light the main burners) and dangerous (uncontrolled gas flow will cause an explosion in the boiler).

- Make sure everything is clean. Dirt on the gas manifold (or in an oil line or injector) can prevent the boiler from lighting evenly if at all. Dirt in the pilot can prevent the boiler from starting.
- In bigger boilers, the gas valve is separate from the pilot valve. Usually these are electro-hydraulic valves and they can fail shut if they run out of oil. You will have to replace the valve if the oil leaks out.
- On hydronic boilers, make sure the high limit cut-out works (don't set it over 200°F). On steam boilers, make sure the high-pressure cut-out works. One way of testing the high pressure switch is to turn it down while the boiler is operating. When the cut-out setting reaches the actual steam pressure, the boiler should turn off. Remember to restore the high pressure setpoint when you're done with the test.
- Make sure the flow switch works on hydronic boilers. If you turn off the pump, the boiler should shut off. If the boiler continues to operate with no water flow it can get hot enough to blow up. Alternatively, the pop-off valve will blow off and all the water will leak out of the system. Now you not only have a big mess but you also have no heating. Check the sight glass and make sure the water levels are always within the proper range.
- If the boiler has a draft damper, make sure that the linkage is sound and that the damper is working properly. Failure to maintain the proper draft though the boiler can lead to boiler inefficiencies and can also feed fire back into the mechanical room.
- The pressure relief valve on a steam boiler should be checked every year. Check by pressurizing the system and see if the valve pops off at the proper pressure. You can also bring the valve to an authorized service station for reseating and testing.

fastfacts

Check the low-water cutoff switch once a month. Replacing a defective switch is much cheaper than replacing a melted heat exchanger.

fastfacts

Never put an air-handling unit in the same room as a natural draft boiler. The air-handling unit can create a negative pressure in the room. If there is not enough make-up air, the boiler fire can shoot out across the floor.

Proper air flow is essential to the operation and safety of combustion boilers. Make sure that nothing is obstructing the stack, both in the mechanical room as well as up on the roof. On the intake side, ensure that make-up air is always available to the boiler. If the make-up air on a natural draft boiler is restricted, the flue flow can reverse and you can have fire shoot out the front of the boiler. Most people get a little disturbed to find open flame spreading across the floor. Not only that, but it also means that you are not heating the water with the flame.

Keep an eye on the boiler gas valves and make sure they are working properly. They should not be corroded or making any unusual sounds. If you have even the slightest suspicion that a gas valve is faulty, turn off the boiler (lock and tag out if necessary) and check the valve. It is far cheaper to replace a leaking or broken gas valve than it is to replace everything else in the mechanical room.

Boilers and vapor compression refrigeration equipment should never be in the same room. This goes for refrigerant storage canisters as well. If the refrigerant should ever leak, it will turn into phosgene gas as soon as it hits the flame in the boiler. Local codes usually dictate that you cannot have boilers and chillers together, but don't even think about getting away with it otherwise. Always remember that boilers can explode from gas build-up. If you smell gas, shut off the boiler and try to ventilate the mechanical room.

Boiler Controllers

The boiler controller is the heart of the boiler. Forced draft boilers must have a controller and you should periodically check them by watching them through a few cycles. Make sure the timing is correct. All pertinent information is fed into the controller, such as high and low gas pressure cut-outs, high-temperature and low water relays, etc. These controllers are replaceable in case they break. A good rule of thumb is: when in doubt, throw it out. If the flame controller does not work, replace it. True, they are not cheap (around $500) but it is better to replace them than to allow the boiler to run with improper control. The controller is cheap compared with trying to explain to the insurance company why the boiler blew up. Most controllers will last a good long time, but they are designed to fail safe, that is, with the gas valve shut. Obviously, if they fail safe they don't work anymore. Sometimes they don't even fail safe.

Do not try to repair a broken boiler controller. It is not worth it. Do not let broken ones sit on the shelf in the shop. Smash them and throw them away. Not only will this keep others from trying to use them, but it will probably make you feel better as well.

When replacing a boiler controller, make sure you match the exact part number with the one you are replacing. Many of them look similar but they are not the same. To make matters worse, some manufacturers use the same sub-base for connecting the controllers to the boiler, but they make different controllers for different boilers. When replacing a controller, be very careful that the wires are connected correctly. If not, you can have a gas valve opening when it's not supposed to.

With proper water treatment and frequent blow downs, a boiler will not build up scale during its lifetime. However, this requires constant maintenance and many boilers experience scale and lime build-up. You should blow the minerals out the bottom of the boiler every year, even in glass-lined tanks. Check the water quality every month for the presence of minerals, since this can give you an indication that you need to blow down the boiler more frequently. Keep in mind that well water is usually worse than city water.

TROUBLESHOOTING

Boilers are broken down into two basic categories: hot water (hydronic) boilers and steam boilers. Figure 11.11 gives a troubleshooting flowchart for hot water boilers, but note that steam boilers are really nothing more than hot water boilers half full of water, and some boilers can be used for both steam or HW. The troubleshooting is basically the same for natural draft or forced draft. The primary difference is that the operational control in a steam boiler is the steam pressure, while on the water boiler it is the water temperature.

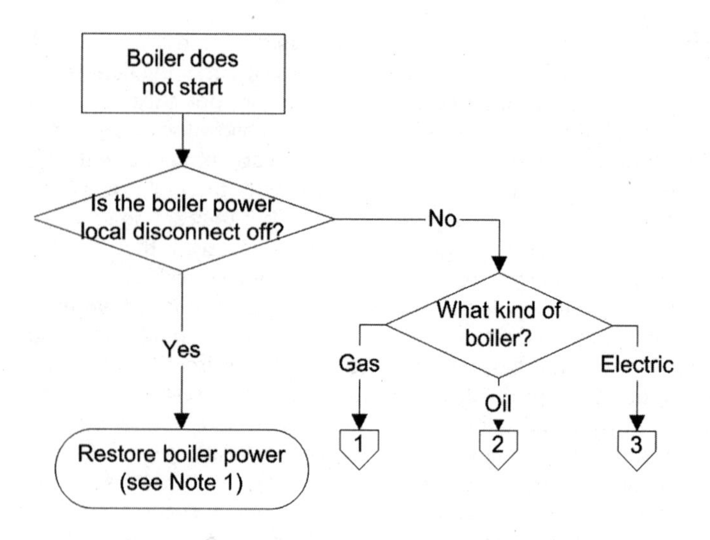

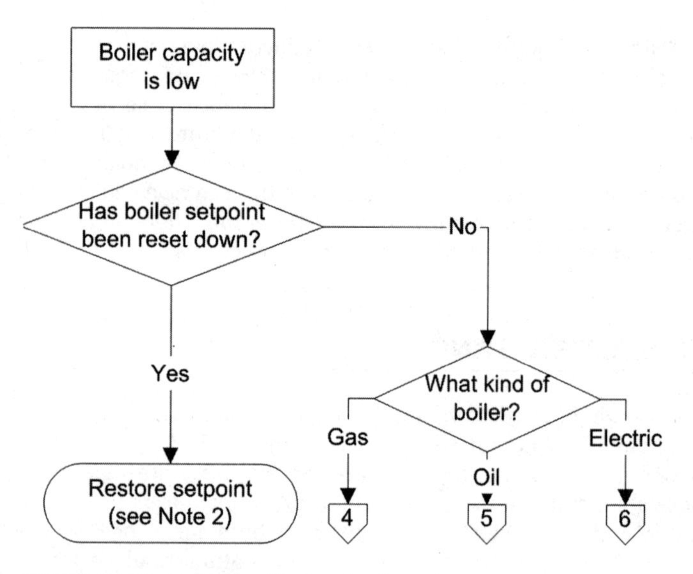

FIGURE 11.11 Boiler troubleshooting flow chart.

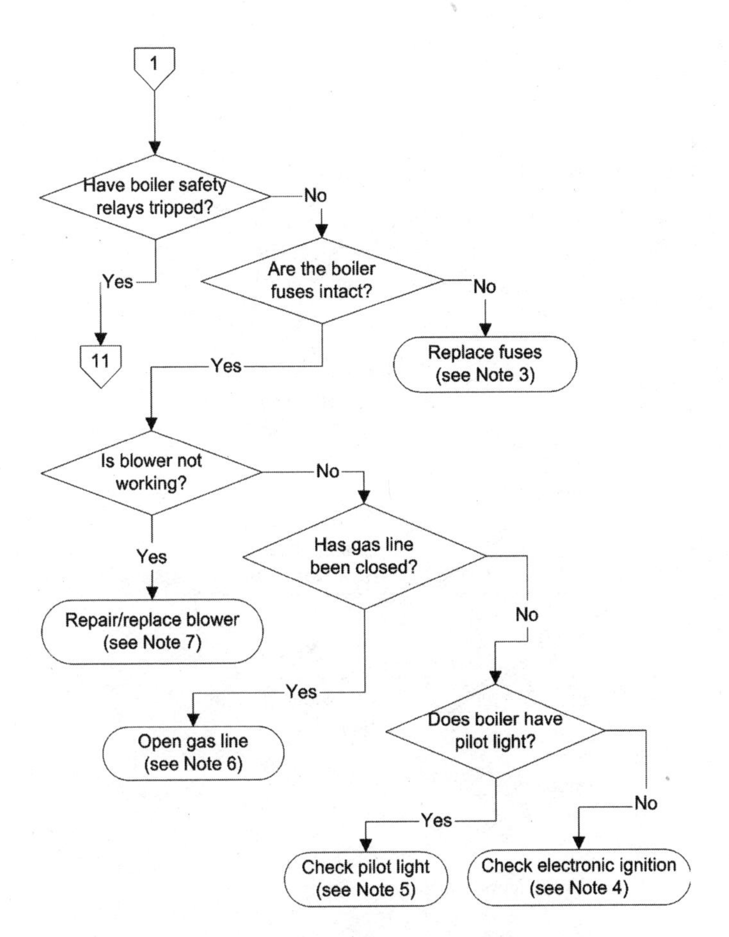

FIGURE 11.11 *(continued)* Boiler troubleshooting flow chart.

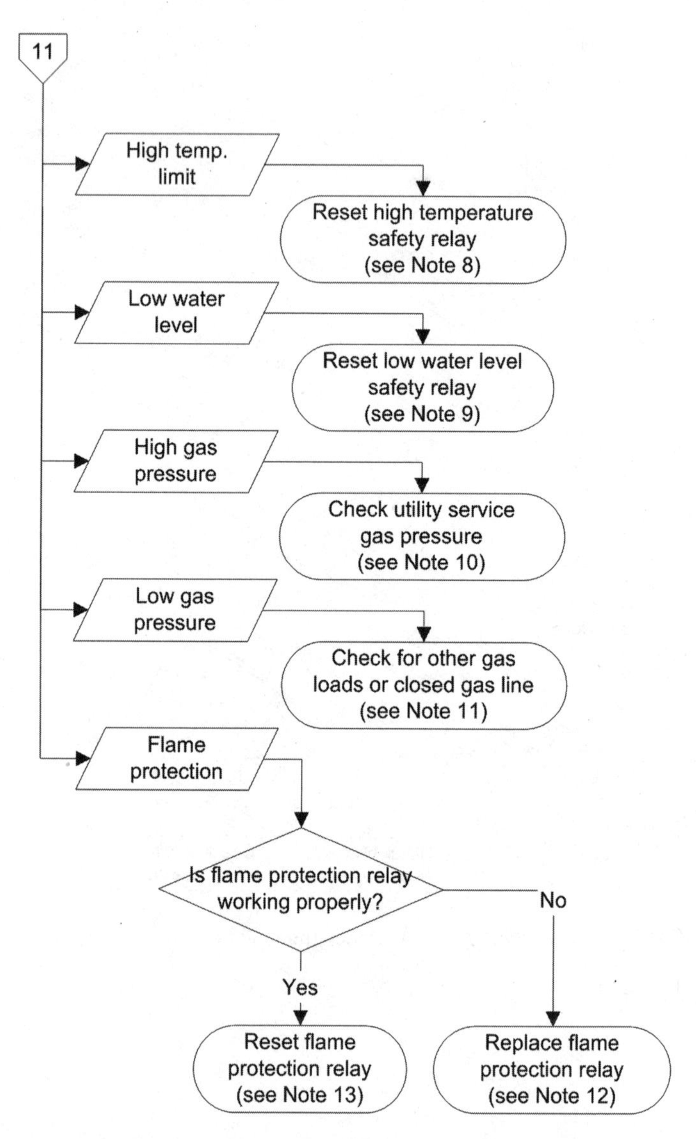

FIGURE 11.11 *(continued)* Boiler troubleshooting flow chart.

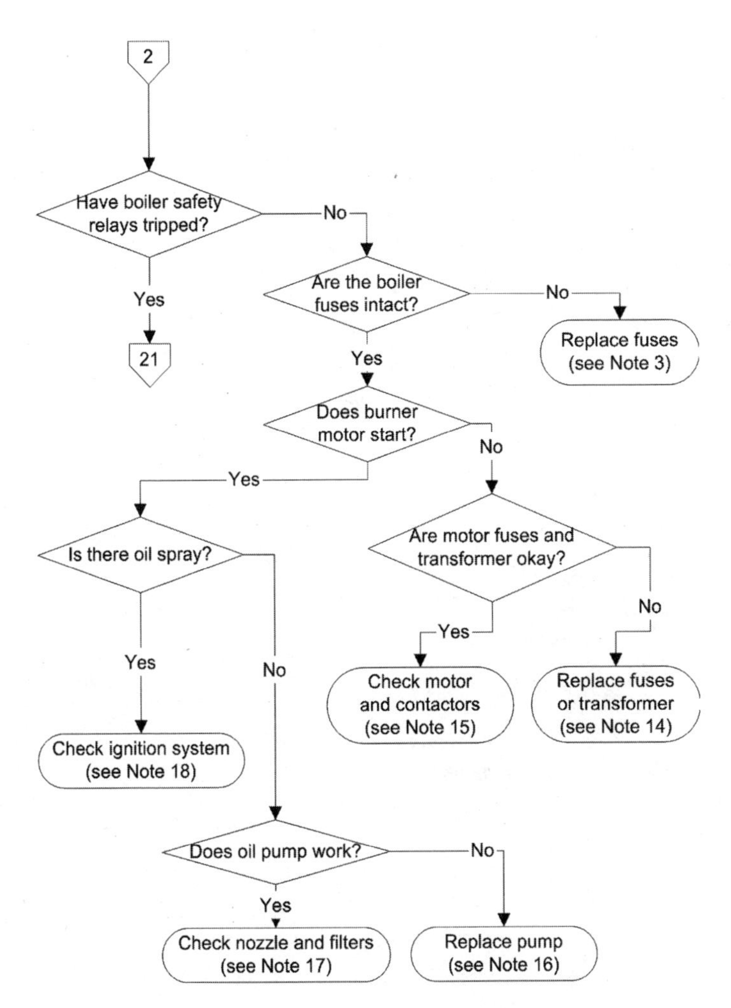

FIGURE 11.11 *(continued)* Boiler troubleshooting flow chart.

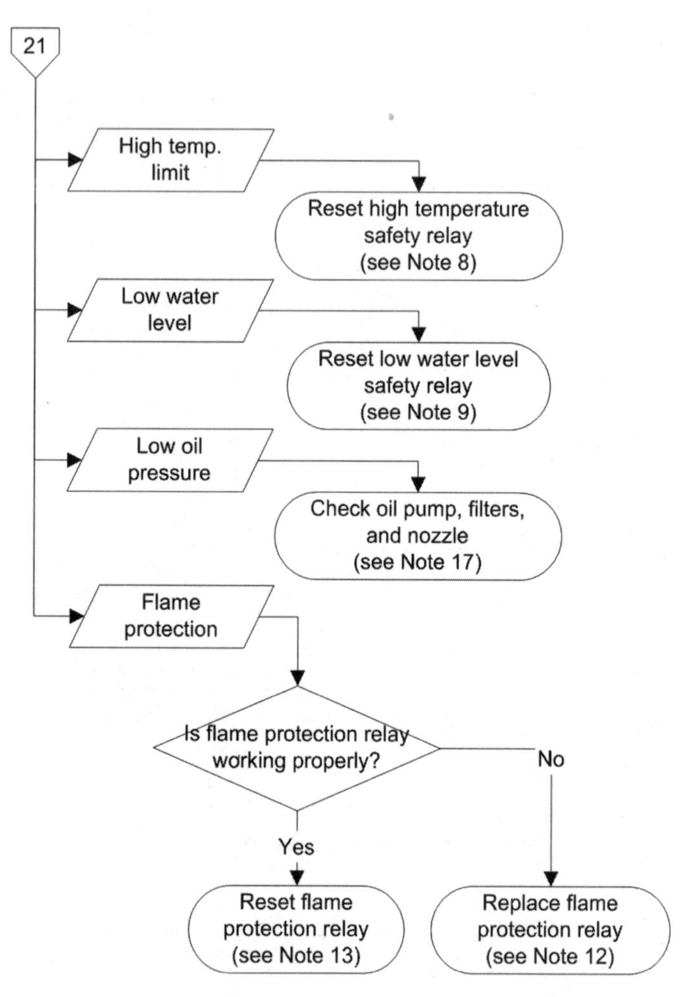

FIGURE 11.11 *(continued)* Boiler troubleshooting flow chart.

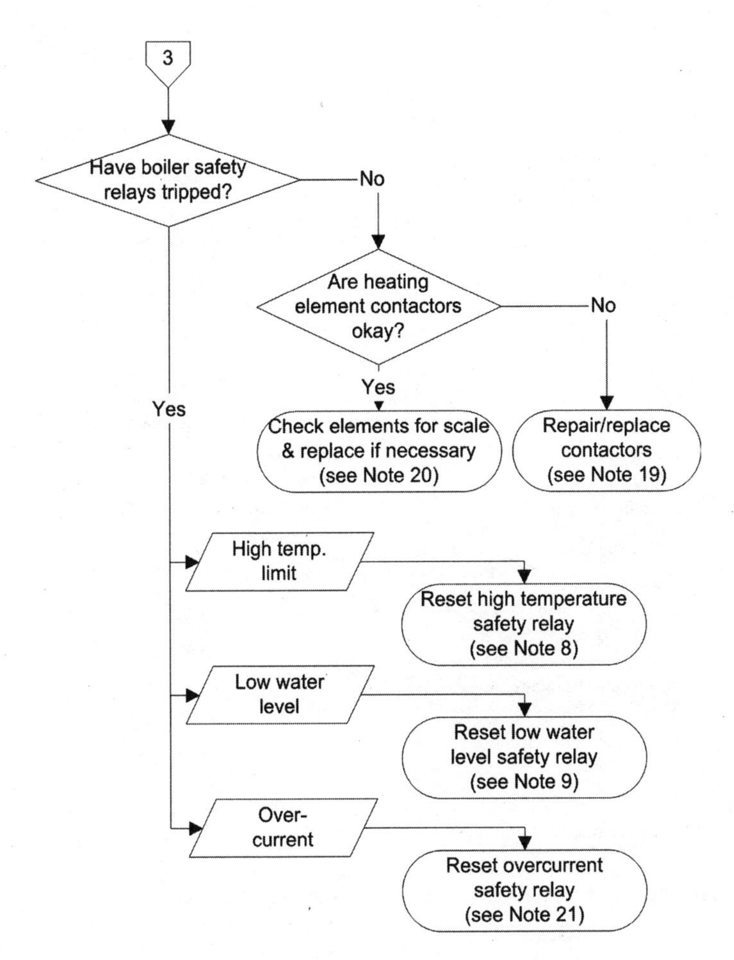

FIGURE 11.11 *(continued)* Boiler troubleshooting flow chart.

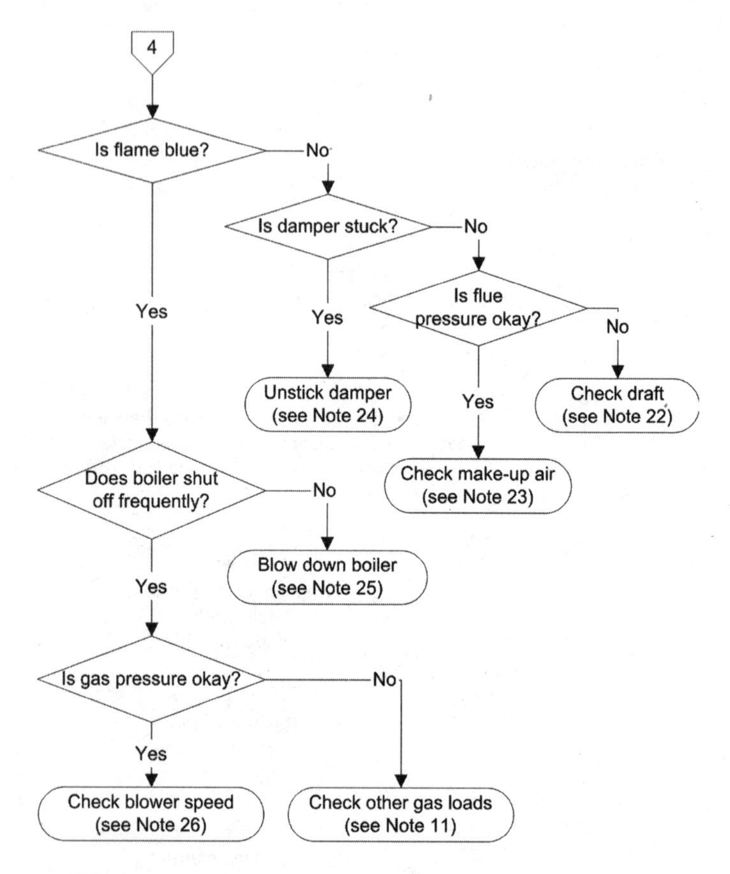

FIGURE 11.11 *(continued)* Boiler troubleshooting flow chart.

Note 1: Never turn on a local disconnect until you have identified why it was shut off in the first place. If there is maintenance or other servicing of the equipment, turning on the boiler can lead to injury to personnel and/or damage to the equipment.

If the electrical connections are all valid between the motor control center (MCC) and the boiler, then the problem may be a blown fuse or that the power is shut off at the MCC. You can check for fuse continuity using a standard ohmmeter. Make sure the MCC power is turned off before accessing the fuses. Do not attempt to override the safety latch on the MCC access.

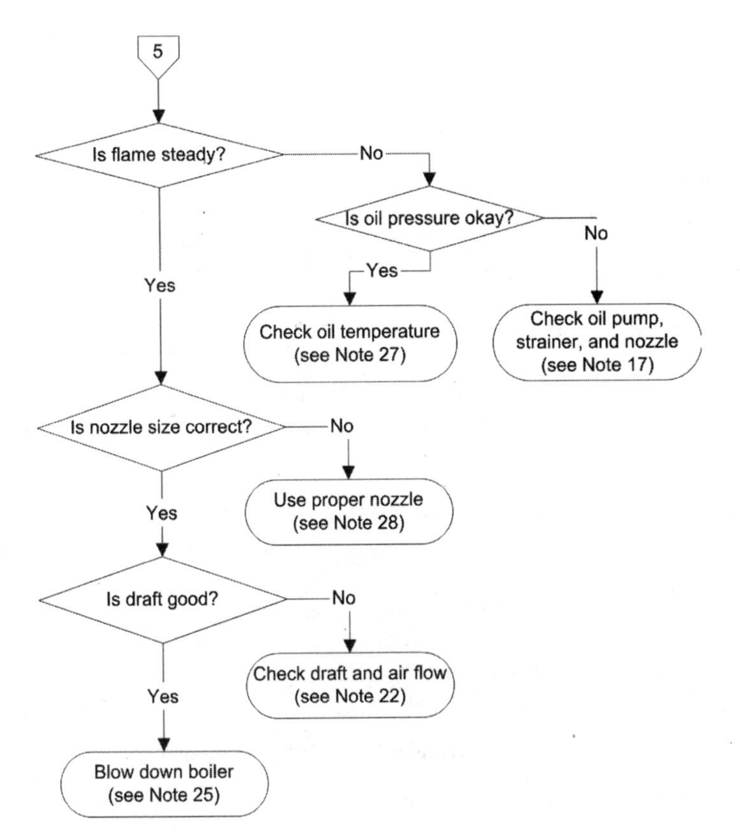

FIGURE 11.11 *(continued)* Boiler troubleshooting flow chart.

Note 2: In some cases, there may be seasonal adjustments of the hot water boiler setpoint. If the water temperature is not hot enough or the steam pressure not high enough, check the internal setpoint of the boiler and adjust upward as necessary. Alternatively, the boiler may be subject to temperature or pressure reset control. If this is the case, check the compensating temperature sensors in the reset circuit to make sure they are operating properly.

Note 3: If the fuses continue to blow after installing new ones, then the problem is probably a faulty blower motor.

Note 4: Check for the presence of spark from the electronic ignition system. You should be able to hear a characteristic snapping noise if

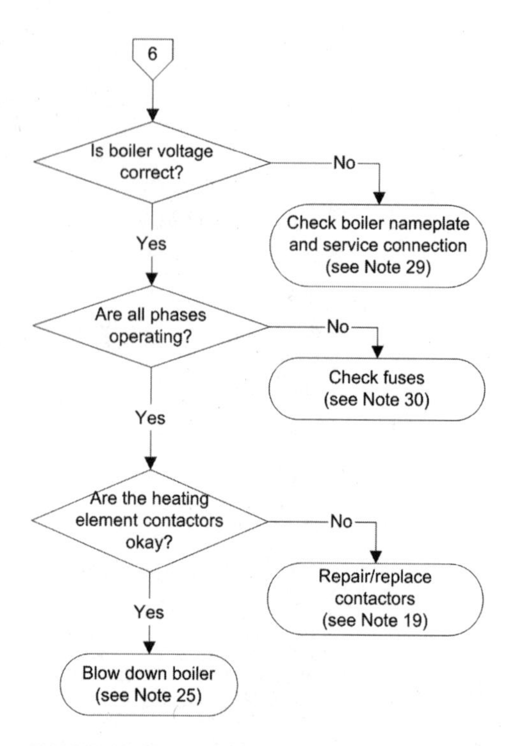

FIGURE 11.11 *(continued)* Boiler troubleshooting flow chart.

the system is successfully achieving a spark. If not, check the ignition plugs and power supply. If the gas solenoid opens and you don't see a spark, then the problem could be a bad ignition transformer or the ignition rod has shorted out. Sometimes the porcelain insulation breaks off on the ignition rod. Even a small crack in the insulator can short to ground and kill the spark. This can be very hard to find and it might be easier just to replace the part.

If the spark is okay, then another problem may be that the gas valve is not opening on call for ignition. If possible, check the voltage on the gas valves to see if it is correct. Electro-hydraulic gas valves tend to last longer than solenoid valves, but they can run out of oil if not maintained. If a solenoid gas valve stops working, it may be difficult to see the problem without taking the solenoid apart. With a faulty electro-hydraulic valve you can easily see oil on the floor and know that the valve has a problem.

fastfacts

Never reapply power at the MCC if you do not know why the power was turned off. Normally the MCC would be tagged and locked for maintenance or other service reasons, but this step is sometimes overlooked. Never energize a circuit until you have verified that there is no threat to personnel or equipment.

Note 5: Make sure that the pilot light is in good proximity to the gas manifold and that any kind of thermocouple or flame rod detector is sitting in the middle of the pilot flame. If the pilot is lit manually (and let's hope it's not), then you will have to light it. Always make sure that the combustion chamber is free from gas fumes before attempting to ignite a pilot light. Allow time for the pilot detection system to acknowledge the presence of the pilot before attempting to start the boiler. In the case of a thermocouple-type pilot sensor, this may take a minute or so. If the pilot continues to go out after ignition, check for downdrafts from the flue or replace the pilot sensor. On natural draft furnaces, watch for backfires. Don't look under the cover while lighting, particularly when it's cold outside and there's no draft taking the flame up the flue. You can easily burn your eyelashes off if the boiler pops and sends a ball of flame into your face.

Note 6: Check the gas line going to the boiler (or the fuel oil line, if appropriate) to see if it has, for some reason, been shut down. This could occur either in the line or at the gas regulator. Sometimes the fuel may be shut off at the regulator to avoid burning the pilot light during the cooling season. However, if there is any reason to believe that the line may have been shut off for maintenance, do not open the line until you have verified that it is okay to do so.

Note 7: It may be necessary to operate the blower motor independently from the boiler itself to check to see if the motor is in fact working. Most large boilers have the ability to test the blower motor. If the boiler has a pre-purge or post-purge cycle, then you can also check for blower operation during this time.

Note 8: If the high temperature relay continues to fail after reset, it could indicate a problem with either the boiler temperature control, the boiler setpoint, or the high temperature relay. To check the boiler temperature control, start the boiler and decrease the setpoint. You should see a decrease in the flame. If this does not occur, you may need to replace the setpoint thermostat or the gas regulator. If the temperature control appears to work correctly and the boiler setpoint is correct, then replace the high temperature relay.

Note 9: If the low water relay continues to fail after reset, it could indicate a problem with either the feed water supply, the condensate return pump/receiver tank switch, or the low water relay. Do not ignore continued trips of this relay, as low water levels in the boiler can lead to a flashing of the water and potential personnel injury and equipment damage.

Note 10: A high gas pressure cut-out may indicate a problem with the gas pressure regulator at the service connection to the building, or it could be in the pressure regulator leading to the boiler. In either case, the higher gas pressures could mean that other gas-driven devices in the building are overpressurized. This is a potential problem. If you find high gas pressure at the boiler, it is a good idea to check for similar high pressures at space heaters, furnaces, water heaters, etc.

Note 11: As with high gas pressure, a low gas pressure cut-out can point to problems with the gas pressure regulator at the service connection to the building or in the pressure regulator leading to the boiler. However, a more likely cause is that there are other pieces of equipment in the building that also use gas and that they are all operating at the same time. This can prevent the boiler from starting, and can also lead to intermittent operation of the boiler as it turns on and off as the gas pressure fluctuates. If you have this problem frequently, you may want to check with the utility about higher-pressure service to the building.

One way to check for low gas pressure with a forced draft boiler is to see if the pilot is on when the fan is running. Usually you can hear the pilot solenoid click in, and when it does, check to see if there is a spark from the electronic ignition. If you do see a spark and the solenoid valve is open and there is still no flame, then the gas pressure is probably low (or there may be no gas at all).

Note 12: The flame protection relay can be based in temperature (thermocouple), electrical (flame rod), or optical (UV sensor). Depending on what part is broken, it may be possible to replace only the bad section of the sensor.

Note 13: If the flame protection relay continues to fail after reset, it could indicate a problem with either the gas/oil supply, the blower, or the flame protection relay circuit.

> **fast**facts
>
> *If the flame will not light in a hot water boiler, check to see if there are any leaks in the water piping. If the boiler springs a leak internally, water can come down and knock out the fire.*

Note 14: The burner motor operates in a pretty harsh environment. If the motor shaft seizes or if the motor itself is not properly maintained, you can have poor air flow and lowered boiler output. This can also cause the motor to draw excess current. If the fuses continue to fail, consider replacing the burner motor. Also check the transformer that powers the burner motor relay. If this is not working then the blower will not come on.

Note 15: Check to make sure that the burner motor contactors are not dirty and that the motor is not stuck or burned out. If it is necessary to replace the burner motor, make sure that the replacement motor is the same size as the existing one. The wrong size could lead to the incorrect air flow through the boiler and over- or under-heating problems.

Note 16: Before you replace the pump, it might be a good idea to check to see if you actually have oil in the storage tank. If the oil level is low enough (or if someone forgot to pay for the oil delivery) the oil line may have gone dry. Also check the coupling on the oil pump, since a broken or slipping coupling can lead to a pump that seems to be working but is not actually pumping any oil. If you do replace the oil pump motor, make sure to use the same size motor. A motor of different power will probably cause problems unless the nozzle size is also changed.

Note 17: Over time, the oil nozzle can get clogged with bits of dirt and debris that come with the oil. There should be a filter upstream of the oil pump that removes this dirt, but usually these filters are not cleaned very often. Another thing to check for is an air leak in the oil line, since this can create air lock and prevent oil from reaching the nozzle (you will have to bleed the line to clear the air).

Note 18: If there is oil spray but no ignition, then the problem is probably electrical. Check the ignition electrodes for dirt, gap width, and loose connections. If possible, remove the ignition element and check the porcelain to see if it is cracked. If everything appears okay, then check the upstream circuitry (such as the transformer) to make sure the proper voltages are present.

Note 19: Each element or set of elements in the boiler is activated by a separate contactor. The contactors can either fail open so that the element cannot provide heating or can fail closed (through "welding" of the contacts) so that the boiler is always heating. When trying to determine if the contactors are opening and closing correctly, be aware that on some electric boilers the contactors dispatch in more or less random order each time they start. This is to prevent any single element from being used more that the others and burning out faster.

Note 20: The heating elements can accumulate scale over time, effectively insulating them from the water. You may be able to clear them by blowing down the boiler. If not, remove and replace them with the same wattage elements.

Note 21: An overcurrent safety is usually a sign of another problem in the boiler. This could be caused by old electrodes, an undersized boiler, or a problem in the control circuitry. If you experience overcurrent cut-outs frequently, do not keep resetting and ignoring.

Note 22: You should have a steady blue flame when looking through the sight glass into the fire box. If the flame is yellow or "floats" above the gas manifold, there is not enough combustion air reaching the combustion chamber. Check to make sure the vents are not blocked and that the flue is not blocked either at the firebox or at the roof exhaust. The flue pressure above the firebox should be between 0.02 and 0.05 of an inch W.G. negative. If it is positive then you have flow down the flue. If you are not already unconscious because of carbon monoxide inhalation, then adjust the flue damper controller to open a little bit more.

If you live in a particularly windy area, or if the flue roof vent is simply badly located, you may have frequent downdrafts when the outside conditions are right. You can either install a forced air boiler (if you have the money) or install a wind block on the top flue vent.

Note 23: Almost all codes now require dedicated venting for make-up air into the boiler room. Never block these vents or try to steal the air for another combustion process. Keep in mind that air handlers located in the boiler room may depressurize the space and draw make-up air away from the boiler. Insufficient make-up air can lead to a low flame and low boiler capacity.

Note 24: The damper linkage is often connected to the gas valve. Keep the linkage in good shape so the damper can move freely. If the damper is preventing the proper draft, the flame will not be hot enough to provide enough heating.

Note 25: Over time, scale and other solids build up inside a steam boiler. You should blow down the boiler at least once a month, but if the water is particularly hard you may need to do so more frequently. As the solids build up inside the boiler, they act as insulating material on both the heating elements and the sensors. At best, this decreases the capacity and efficiency of the boiler. At worst, it makes the boiler work much more than it should, shortening the life of the equipment and creating a fire hazard. When you blow down a boiler, the water should come back in pretty quickly. If not, the piping may not be clear.

When performing a blow down on a steam boiler, whether auto or manual, the law requires you mix the blow down with cold water and keep the mixture below 120°F when it discharges into the sanitary sewer. This is done to protect the sewer and any personnel who may be working in the sewer.

Note 26: Boiler blowers sometimes have several windings to operate at different speeds, depending on the load. If the speed is incorrect (due, perhaps, to the motor getting stuck on one set of windings), the heating capacity may be decreased because the

fastfacts

A technician was investigating a no-heat complaint and found that the boiler stack was red-hot. The boiler had operated for many years without proper water treatment and had never been cleaned out. Over time, the solids had built up so deep that the heating surface was completely covered and the boiler couldn't heat the water. The solids were actually covering up the operating and high-limit controls as well. In other words, the boiler didn't see the temperature at all and took a long time to react because the thermowell with the temperature bulb was completely covered with solids. It was fortunate that the building did not burn down.

flame is not high enough. One way to identify this is by rapid fan cycling or frequent boiler cycling.

Note 27: If the oil storage tank is located outside, or the oil runs along a pipe fixed to an exterior wall, the oil can get cold and viscous. You may need to pre-warm the oil or increase the pump capacity.

Note 28: Oil nozzles are sized for a specific capacity and pressure. If the nozzle has been replaced recently, check to make sure that the proper size was used. Also check to make sure that the nozzle is angled correctly according to the manufacturer's recommendations.

Note 29: Stranger things have happened. It is possible that the boiler was connected to the wrong electrical service. Make sure the line voltage is the same as the boiler nameplate voltage.

Note 30: It is possible to lose a phase and still have the boiler appear to operate normally, albeit at a decreased capacity. Use a clip-on ammeter to make sure that all legs of the boiler are drawing about the same amount of current. You don't need an exact number—you are just trying to make sure that no fuses have blown.

Boiler Is Too Hot

Usually the problems with boilers is that they do not start or they are not producing enough hot water or steam. Fortunately, most safeties in boilers fail in such a way as to turn the boiler off. However, there can be cases where the boiler is putting out too much heat. This is not only inconvenient, but dangerous.

If the boiler is producing too much heat, the first thing to do is check the flame size. If it is too large, the regulator is set too high. Use a manometer to determine if the gas pressure is correct and adjust as necessary. Another thing to check is the burner orifice. If this is too large you may need to replace it with the correct size. If you cannot adjust the flame and the orifice is the correct size, the regulator is probably defective and should be replaced.

The most dangerous situation occurs when the boiler is on and does not turn off automatically. At this point you sincerely hope that one or more of the safety switches will trip off before you need to call the fire department. If the burner will not turn off, check the obvious things first: Is the thermostat set too high or is it too cold? Is the high limit switch set too high? Is the fan set to manual on? If these all appear normal, then there may be a short circuit somewhere in the boiler control circuit.

fastfacts

Never jumper out any control on a boiler, even temporarily. The controls are there for a reason, and defeating them significantly increases the risk of explosion, property damage, and personal injury.

PRACTICAL CONSIDERATIONS

The common theme throughout this chapter has been to respect a boiler and its power. To do so you must keep the boiler in good shape. There is another reason to do this: in most municipalities, annual boiler inspections are required on large (over 1 million Btu/hr) boilers. You should keep boilers maintained so that the inspector does not red-tag it. There are steep fines for operating boilers that are "condemned." If the inspector tells you to do something (no matter what you think of the inspector), the best strategy is to do it.

Water Treatment

Water treatment is more important in a steam boiler than in a hot water boiler. This is because the hot water boiler is a closed system, but in the steam boiler, oxygen can enter the system through the make-up water. The constant supply of make-up water also means that the water must be treated, particularly if the steam is used for humidification.

If too much air gets into the condensate lines, that implies that carbon dioxide (CO_2) is also making it into the system. The CO_2 can react with the condensate water to form carbonic acid (H_2CO_3). This lowers the pH of the water, requiring additional water treatment. It is best to identify all places in the system where air can penetrate and to take the proper measures to prevent this. Some places where air commonly gets into steam systems include any valves or flanges in the piping, make-up water, and any condensate pumps.

In addition to oxygen, you also have minerals and solids to contend with. In fact, small steam boilers are about as labor intensive as big steam boilers because of the solids. Try to keep the solids in suspension until you can blow down the boiler. Watch the water treatment and blow down the boiler on a schedule based on periodic water tests. Blow downs are automated on large boilers but require human intervention on small boilers.

An occasional flue gas analysis is a good idea to make sure that the fuel/air mixture is correct. On larger boilers the flue gases may be automatically monitored, but it is typically not done on smaller boilers. Testing the gases several times a year under different loads is a good way to find out if you are operating your boiler as efficiently as possible. The cost of the tests is quickly recovered in fuel savings.

There are a lot of hot water boilers out there that are over 40 years old. Many of these still work, although it is tough to find replacement parts. Generally, these older boilers are also not very efficient. The newer ones are very efficient, but the replacement parts are proprietary. Parts must be ordered from the manufacturer and the sophisticated controller electronics can make these parts very expensive.

Safety

There are two ways you can successfully blow yourself up with boilers. The first is a gas explosion if the fire control doesn't work properly. This can happen if the gas valve sticks open (fortunately, this doesn't happen often unless someone has modified the valve). The other way is if the water level gets too low in a steam boiler. When cold water rushes in it turns into steam faster than the pop-off valve can handle and blows the ends off the boiler. There is quite a bit of power involved.

More than any other piece of HVAC equipment, boilers must be treated carefully and with respect. Do not go off and leave a boiler that

fastfacts

Despite being made of metal, boilers can and do burn down. In one case, both the feed water pump and low-water cut-out stopped functioning at the same time. Unfortunately, the boiler did not. It continued to operate until it got so hot that the control wiring melted.

fastfacts

A novice technician once propped open the door on an air-handler in an effort to use warm air from the boiler room to keep the air handler freeze protect from shutting off the fans. Not only did hot air go into the air-handling unit, but also gas and carbon monoxide. Fortunately, the flame was pulled out from under the natural draft boiler and burned all the insulation off the boiler control wiring, shutting down the boiler controls before the building burned to the ground.

you are not confident about. If you are not satisfied with any repairs or maintenance, do not turn on the boiler. It's as simple as that.

It is also good practice to keep a register of all boilers in a building or on a campus. Replace the flame detectors every few years and blow the boilers down at least every month. It is important that the sight glass works properly on each boiler. Check the glass to make sure you can see water and make sure it's at the right level. When you blow down the boiler, also blow down the sight glass and level control regularly to keep sediment out. Make sure the pressure gauges work properly and the high pressure cutout works as it's supposed to.

Be careful about the placement of the make-up air vents. In freezing weather, the make-up air can flow across water piping and freeze it on the way to the boiler. Plant design should consider the placement of pipes and make-up air vents to keep cold air from drafting across pipes in boiler rooms.

Troubleshooting

It is often easier if a boiler just flat-out quits and won't start. It is more difficult to track down intermittent problems. Sometimes a boiler runs for a while and then stops working. If this happens and you can't easily identify the problem, you may have to backtrack from the effect to find the cause. For example, if a boiler keeps shutting off, you should check the flame safety circuit first. If the flame detector sensor is working, there might be other problems (like short circuits, improper flame sensor location, etc.) that cause the boiler to quit. Likewise, if the boiler keeps tripping out on high temperature, it may be that the hot water pump shuts off at the same time as the boiler, leading to a rapid rise in the temperature of the hot water sitting in the boiler even if the boiler is off.

If the boiler starts at some times and not at others, check the wiring, especially on forced draft units. The vibration of the blower motor can vibrate wires loose—it is not unusual to see screws backed out on terminal strips. The vibrations can even short out wires in conduit and mess up relays. In the worst case, the vibrations knock fire brick loose so that it falls down on top of the main burner and knocks the pilot assembly out.

If there are two or three boilers on one meter, the activation of more than one at the same time can pull the gas pressure down far enough to trip out the low gas pressure safety. This can also happen if there is a domestic hot water heater on the same gas line. Even if there is no low-pressure safety, the insufficient gas pressure can prevent the flame rod or UV detector from seeing the pilot flame so the main gas valve won't open.

Backfires in a boiler can blow the pilot out. This happens on natural draft boilers if the pilot is unable to maintain a slight draft before the main burner comes on. It can also happen on forced draft boilers at start-up if the pre-purge cycle did not work correctly or was improperly timed. Backfires are dangerous if the explosion is large enough and you are standing nearby.

If the stack temperature is too high, the boiler tubes may need to be cleaned out. This condition implies that too much heat is going up the stack instead of into the water. Possible causes include scale build-up on the water side of the boiler or soot build-up on the burner side. As with the flue gas analysis, the cost of cleaning the boiler will be quickly recovered in the extra efficiency.

chapter 12

STEAM DISTRIBUTION SYSTEMS

Many commercial buildings have a steam distribution system for providing space heating and domestic hot water. Pressurized steam travels around a building without a need for pumping. While convenient, these systems are often wrought with problems. This chapter describes steam distribution systems.

THEORY OF OPERATION

A typical steam system (Figure 12.1) uses a boiler to produce steam that is distributed by a series of pipes throughout a building or facility. As the steam travels through the piping, some of it condenses before it reaches the load. The resulting condensate should be collected and returned to the boiler.

The steam loads consist of heating coils, freeze-protection coils, humidification devices, steam-to-hot-water converters, etc. Some of these loads remove steam from the system (humidification) while others convert the steam from its gaseous phase back to the water phase, releasing heat. In all cases, it is desirable to remove any liquid water from the steam distribution system. Liquid water left in the piping can cause steam pressure loss and pipe corrosion. Too much water in a steam coil prevents the steam from flowing effectively through the coil and reduces the heating capacity. Steam traps are used to remove the condensate.

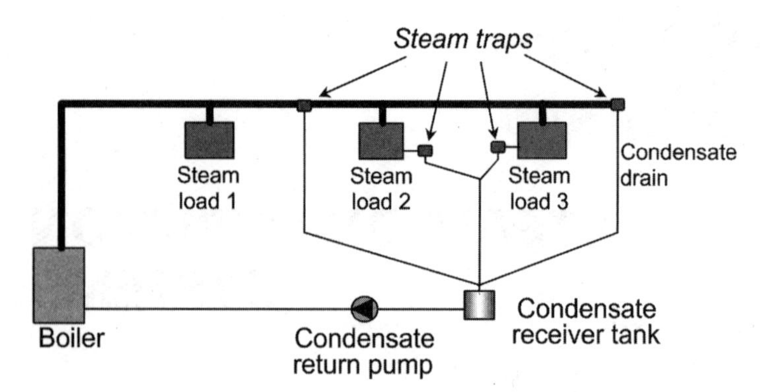

FIGURE 12.1 Example layout of steam distribution system.

Condensate removed from a steam system can either be dumped down the drain or returned to the boiler. The second option is preferred since boiler water is usually treated with any number of chemicals such as corrosion inhibitors. Also, it is much more energy efficient to re-use warm or hot water than to heat up cold water from the mains. Condensate is returned either directly to the boiler or to one or more receiver tanks. As these tanks fill up, a level sensor is used to run a pump that returns the condensate back to the boiler water feed line. Larger steam systems may contain a *de-aerator tank* or *DA tank* that are used to keep the condensate piping from rusting out. These work by heating up the feed water to drive out the oxygen.

Some older steam systems operate as *vacuum systems* where the condensate pump actually pulls a vacuum on the condensate line. This makes for a fairly efficient system, particularly when using steam radiators. These are generally not being installed anymore since hot water heating systems are far more common. If you are working with a vacuum system, make sure the vacuum pumps are maintained.

Another type of steam distribution system is the single-pipe steam heating system. These also exist mostly in older buildings. In a single pipe system, the steam and condensate come down the same line. The pipes are generally larger diameter lines to allow the condensate to return without interfering with the steam. These systems are mostly obsolete, but some are still out there. The biggest problem with a single-pipe system is not understanding of how the system operates. It's all done by gravity, with steam flowing in one direction up the pipe and condensate running back down in the other direction. If this is not the easiest image to conjure up, maybe you should convert to a hot water heating system.

In some campus situations, large boilers are used to produce high pressure steam (120 to 150 psi) that is then sent to a number of buildings. At the buildings, the high pressure steam passes through a pressure reducing station (Figure 12.2) and comes out at 25 psi or even lower. Steam stations are high maintenance items and must be sized correctly for the building steam loads. If undersized, there can be noise in addition to the capacity problem. There should always be some form of steam pressure relief valve (Figure 12.3) after the steam station that is set to blow at about 10 to 15 psi higher than the desired building pressure. This is done so that if the steam station "runs away" then the coils and converters in the building do not experience the high campus distribution pressures. It is a good idea to vent the relief valve to a safe location outdoors to keep the steam from entering the mechanical room.

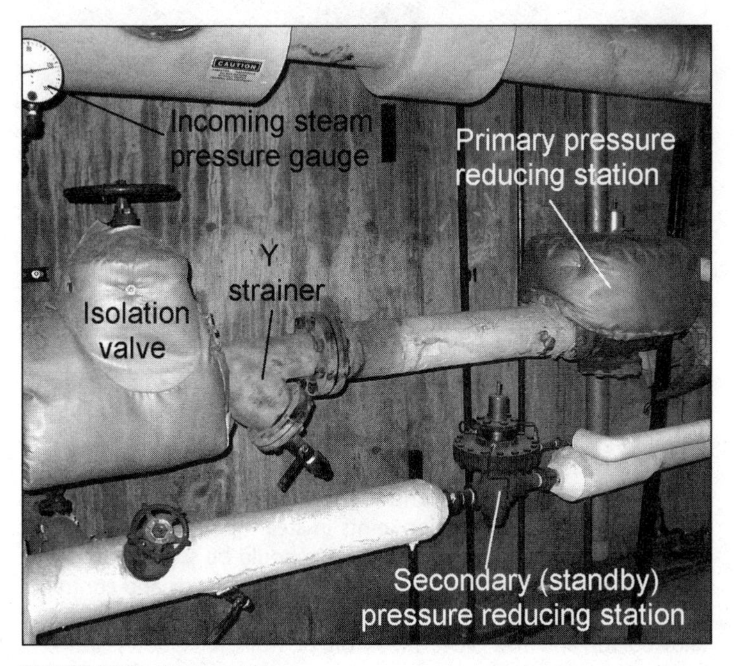

FIGURE 12.2 Steam station.

FIGURE 12.3 Steam pop-off valve (rated at 30 psi).

Steam Piping and Valves

The piping in a steam distribution system should be sloped so that the condensate gathers in a central spot where it can be removed by a steam trap. In addition to the steam traps, there is usually a vacuum-breaker valve at the highest location in the piping. The function of this valve is to prevent low pressure in the steam piping (as might occur with condensation) from drawing condensate back up through the condensate return lines and steam traps.

Steam piping expands considerably once it has become hot. It is necessary to install expansion joints in steam piping to account for the change of pipe length. The piping should be supported so that no binding or kinking occurs when the pipe expands and contracts. It is not unusual to see an installation where the expansion isn't allowed for and the pipes break off of walls, or elbows and tees break off the main runs.

fastfacts

Steam pipe expansion joints usually have some kind of packing. The joint needs to be tightened often or the packing will start leaking. If not kept tight, the packing can blow out and need to be replaced completely. Design of a steam system should allow for easy access to expansion joints.

It goes without saying that steam lines should be insulated. This is done not only to limit the amount of condensate build-up but also to protect personnel working around the piping.

Steam valves are basically the same as water valves and are discussed in detail in Chapter 6. In general, there should be a strainer located ahead of the control valve (Figure 12.4) to prevent large objects in the steam system from damaging the valve. If possible, steam valves should be mounted sideways to reduce the amount of convective heat reaching the actuator. There are special silicone diaphragms (Figure 12.5) that can be used in steam valves that seem to last longer in the high temperature applications.

FIGURE 12.4 Steam valve and strainer.

FIGURE 12.5 Pneumatic steam valve actuator with rubber (top) and silicone (bottom) diaphragms.

It is good practice to always install vacuum breakers after steam valves. If a coil is full of steam and the valve closes, the steam eventually condenses. As the steam condenses, the pressure decreases dramatically—by a factor of about a thousand. If the condensate is held up in the system, the pressure difference between the outside of the coil and the inside can cause the coil to implode. This is referred to as a *collapsed coil* and happens in steam-to-air coils and in the tube side of steam-to-fluid heat exchangers. Vacuum breaker valves prevent this from happening.

Collapsed coils can be bad for a number of reasons. The main one is that you lose heating capability in the coil. The other is that a collapsed coil can leak and let contaminants back into the steam system. For example, if there is a collapsed coil in a steam-to-hot-water converter, you may have glycol entering the condensate return system. This can really foul up the boiler and can clog up the traps.

fastfacts

Steam systems should be designed so that the pressure dropped from modulating valves is at least 50 percent of the pressure drop across the entire system.

Steam Traps

Steam traps are required to keep steam lines from filling up with condensate. If steam hits a big pocket of condensate it can flash the condensate and hammer pipes. In theory, steam traps allow water to drain from the steam piping without letting the useful steam escape. In practice, a good deal of steam and heating energy is wasted by leaky steam traps. Steam traps are not all the same and different steam applications require different types of steam traps to operate properly.

There are several different types of steam traps. Disk traps (Figure 12.6) operate by assuming there is a little bit of condensate on top of the disk. When the trap is hot (indicating steam in the pipe next to the trap) the condensate turns into steam and holds the disk closed. When the trap is cold—indicating condensate in the steam piping—the steam above the disk condenses and the disk opens. Disk traps usually have a characteristic "clicking" sound as the disk snaps open and closed.

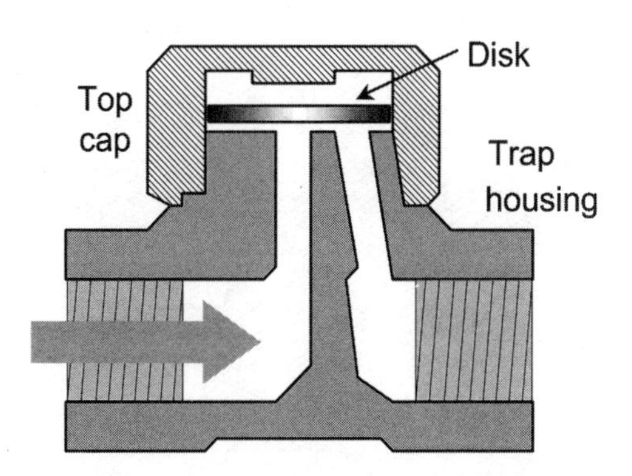

FIGURE 12.6 Schematic of a disk-type steam trap.

Thermostatic steam traps (Figure 12.7) use a temperature-dependent valve to open the trap when there is cool condensate present. The sensing material can be a bi-metal thermostat or some kind of pneumatic bellows. These kinds of traps can leak a lot of steam if the resistance of the valve, the temperature sensitivity, and the line steam pressure are not properly equated.

Mechanical steam traps (see Figure 12.8) generally use some kind of float sensor to open the trap once it has filled with condensate. As the liquid is forced out by the steam pressure, the level of water drops and the valve closes again. Obviously, a trap like this must be oriented in the upright position to work properly. In most low-pressure steam heating applications float and temperature or bellows-only steam traps are used.

Steam traps should be located at the end of steam lines or at low locations. The condensate piping should be pitched downwards from the traps towards the condensate receiver. In some cases, condensate lines span several floors so proper labeling can be helpful.

If possible, try to locate steam traps where they are at least marginally accessible. This helps future technicians make sure that the steam traps work properly. They are notoriously leaky and should be inspected frequently to make sure steam isn't blowing through. If steam ends up in condensate lines then the lines can expand and contract and bang around a lot. If the steam goes backwards through a trap and backwards through a coil, there can be heating occurring when it's not supposed to be. A good rule of thumb is to inspect traps every five years or so to make sure they're working properly.

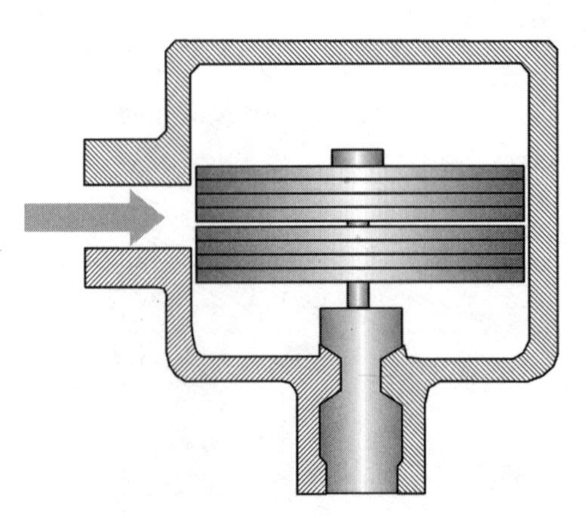

FIGURE 12.7 Schematic of a thermostatic steam trap.

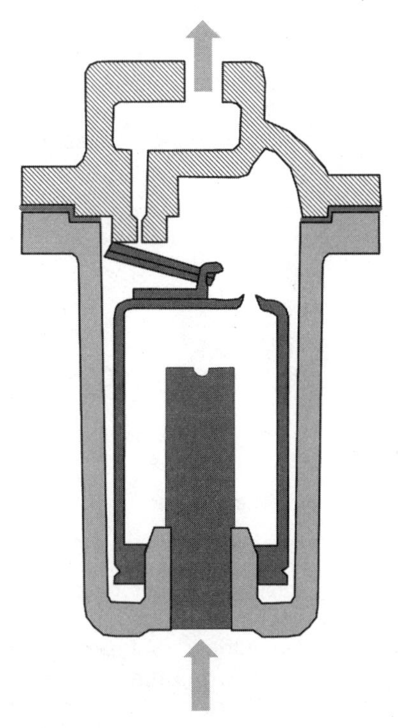

FIGURE 12.8 Schematic of a mechanical, inverted bucket steam trap.

Steam-to-Air Heat Exchangers

Steam-to-air heat exchangers (Figure 12.9) are used in areas where rapid heating of air is required. Examples include preheat coils of air-handlers and reheat coils of terminal boxes. As steam condenses in the coil, it must drain out at the bottom, so always put the down-stream steam trap and condensate drain at the lowest point. Don't try to use steam pressure to lift condensate out of the coil. Under part load conditions the coil will be half full of condensate and half full of steam and the coil will have low heating capacity control and uneven heat output. Cold spots in the coil output develop where the condensate collects, and in freezing weather these cold spots may freeze and split a steam tube. In addition, the coil may hammer due to steam coming in contact with the cooler condensate. In severe

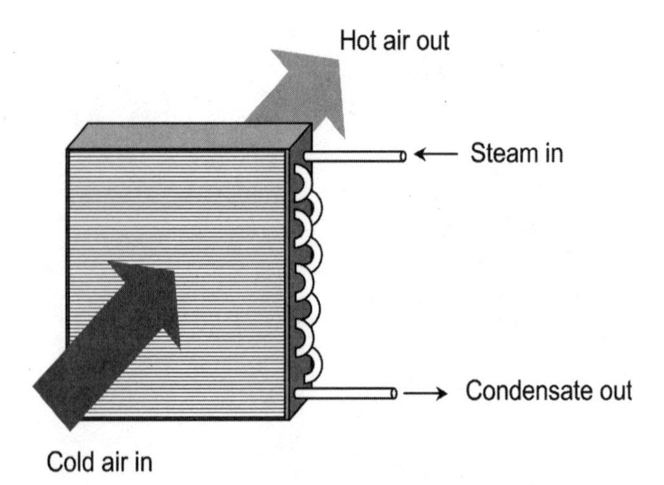

FIGURE 12.9 Steam-to-air heat exchanger.

hammering conditions steam tubes, coil headers, steam traps and connecting piping may become damaged.

If steam coils are used in outdoor air streams, you should use face and bypass dampers as discussed in the next section. This type of arrangement allows full steam pressure to the steam coils in freezing weather to avoid a frozen steam coil while controlling the heat output of the coil with moveable dampers.

Steam-to-Water Heat Exchangers

A *steam-to-water heat exchanger*, often called a *converter*, is like a heating coil except that it is much smaller and heats water, not air (Figure 12.10). The water and steam never mix; they just exchange energy. Converters are used for generating hot water at remote

fastfacts

In order to operate correctly the steam coil must be able to gravity drain condensate under all conditions and coil loading.

points in a building. They are used instead of multiple water heaters. The steam flow through the converter is controlled to maintain the hot water temperature setpoint. Converters are usually shell-and-tube heat exchangers with the steam running through the tubes. They should be heavily insulated, as shown in Figure 12.11, to limit the amount of heat lost to the environment and to protect people working around the converter. Converters can range in size from a few feet long, handling perhaps 50 GPM of water, up to ten feet or more, handling hundreds of gallons per minute.

Converters should include vacuum breakers after the control valve but before the converter body. If you collapse a coil in the converter, you can have city water coming back in through the condensate line and quickly flooding the condensate pumps, creating a bad mess. This can also throw boiler treatment off.

Condensate Return Systems

The condensate return systems are located at the lowest part of the building so that the entire system can be gravity-driven. Often the receiver tank is located in a sunken area in the mechanical room floor so that even steam traps in the mechanical room drain correctly.

The condensate gathers in a receiver tank that usually holds anywhere from tens to hundreds of gallons, depending on the size of the building and associated steam distribution system. The tank and the condensate return lines should be insulated to both preserve the

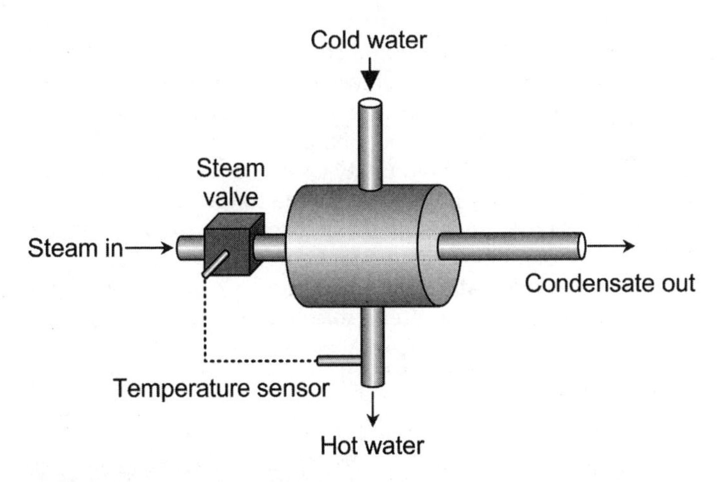

FIGURE 12.10 Steam-to-hot-water converter.

FIGURE 12.11 Insulation on steam-to-hot water converter.

heat of the condensate and to protect any personnel working in the area. The receiver tank has a level switch that empties the tank when it gets full. Some condensate pumps are electric powered, others are steam or air powered. Electric pumps have high maintenance requirements because of a high duty cycle and the hot water. Also, leaking steam traps may put steam into the condensate system, and when steam hits the condensate pump it could cause cavitation and possibly loss of pumping capacity. Steam or air pumps actually work quite well in condensate systems. In such a system (see Figure 12.12) the float switch energizes a solenoid valve to let compressed air into the tank. The air pressure then pushes the water from the receiver. In either case, water from the receiver returns to the boiler water feed line.

You must properly support and insulate the condensate lines to avoid losing too much heat. This can save a lot of energy and operating expense. At the same time, make sure that the condensate

FIGURE 12.12 Condensate receiver tank, air-pump driven.

temperature is not too high, or else there can be cavitation problems in condensate pumps and feed water pumps.

CONTROL

Most steam flow control systems are automated to provide space and water heating or freeze protection. The control loops are usually controlled by a temperature sensor that opens up a steam valve when the temperature goes below a setpoint value.

Steam-to-Air Coils

Steam-to-air coils are used to provide rapid heating of air streams. Typically, such coils are used for reheat systems or in freeze-protection applications.

When controlling the temperature of a steam-to-air coil, an averaging bulb type of sensor or transmitter mounted across the air discharge of the coil should be used (Figure 12.13). On larger coils careful placement of the averaging bulb or sensor is required to insure that the temperature being sensed represents as close as possible the actual discharge temperature of the coil. In some cases, better control of the steam-to-air coil may be obtained by placing the averaging bulb or sensor farther downstream of the coil to get a better mix of the air and reduce the effects of cold spots that may be present in the coil. This technique is appropriate on oversized steam valves or coils or on draw-through applications (with the sensor on the discharge side of the fan).

In some 100 percent outside air applications, face and bypass dampers (Figure 12.14) are used to bypass air around the steam coils to maintain a constant discharge air temperature. These are difficult to control, especially at sunset when the outdoor air tempera-

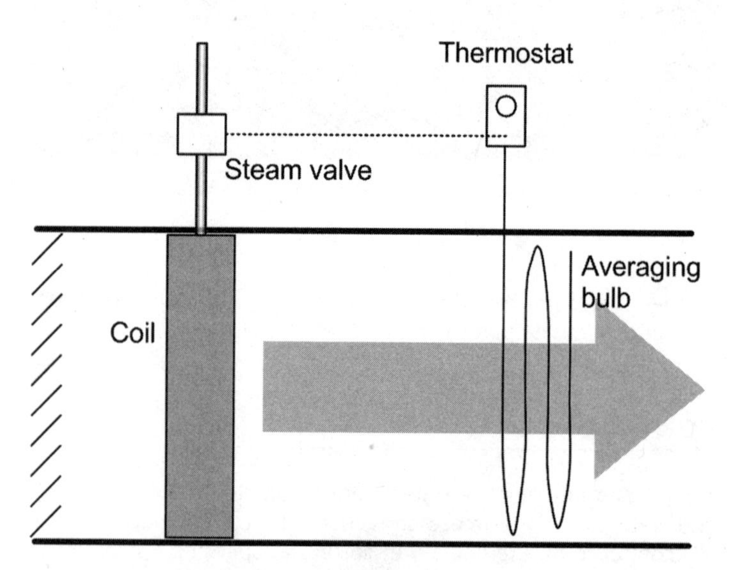

FIGURE 12.13 Schematic of averaging coil outlet temperature sensor.

ture is changing rapidly. The following control sequence allows the steam-to-air coils to operate under all outside air temperatures while providing a stable coil discharge air temperature without incurring low limit fan trips or uncontrollable discharge air temperatures:

- When the outside air is above 35°F, the bypass dampers are closed (that is, full flow through the steam coil) and the steam valve is modulated to maintain the required discharge air temperature.
- When the outside air is below 35°, the steam valve moves to the full open position and the bypass dampers are modulated to maintain the required discharge air temperature.

Low temperature limit controls, also called *freeze stats*, are used to prevent steam-to-air coils from freezing. These are long capillary bulbs connected to a lockout type thermostat that trips out if any section of the capillary senses a temperature below the setpoint. The low temperature limit also fails safe (trips out) if the bulb becomes damaged.

Careful placement of the low temperature limit bulb is required to make sure that the coil is protected from freezing. The bulb should be placed on the leaving airside of the steam-to-air coil and

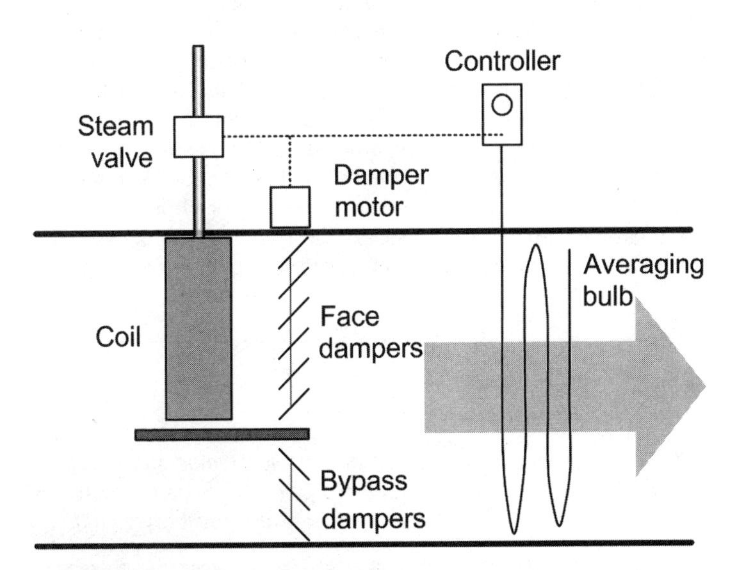

FIGURE 12.14 Steam coil with face and bypass dampers.

cover as much of the coil face area as possible. Larger coils should use two or more low temperature limits wired in series to insure that the coil is protected.

False trips of the low temperature limit may occur if there is a gap around the coil through which cold air may come and encounter the capillary bulb.

Steam-to-air coils may also have a high temperature limit lockout device located in the discharge air. This is wired to lock out the fan controls to prevent a runaway steam coil in the event of a failure of the valve or controller. These kinds of safeties can be useful in 100 percent outside air make-up fans where it is necessary to prevent overheating of a space while still allowing steam to the coil to prevent the coil from freezing.

Steam-to-Water Coils

The steam valve on a steam-to-hot-water converter modulates to maintain the outlet water temperature. The water temperature is measured with a sensor mounted in a thermowell installed in the water pipe. To ensure good control, the thermowell should be located within ten feet of the converter discharge. There is already enough lag time because of the thermal inertia of the whole system without including the water travel time as well.

The complete length of the thermowell must be in contact with the liquid in the pipe or the temperature sensor or transmitter will not function properly. On smaller diameter piping the well should be placed in an elbow or tee as close as possible to the discharge converter. Strap-on type temperature transmitters will work in this application, but close control of the discharge liquid temperature may not be possible; the steam valve will cycle or hunt.

On large steam-to-liquid heat exchangers a one-third/two-thirds steam valve arrangement should be used to provide close control of

fastfacts

NEVER jumper out a low temperature or high temperature limit sensor. The result may be a damaged coil or coils that may be beyond repair. This can cost a lot of money and possibly your job.

fastfacts

You do not want any steam flow through the converter when there is no water flow. A flow switch in the piping or a differential pressure switch across the suction and discharge of the pump are normally employed in conjunction with a steam valve to shut off the steam when there is no flow.

the exchanger temperature output. This consists of one small steam valve in the steam supply that operates up to one third of the total steam flow to the exchanger. A larger steam valve supplies the remaining two thirds of the total steam flow to the exchanger. Both valves are staged to provide the proper steam flow in accordance with the output of the discharge liquid temperature controller.

An improperly sized steam valve or converter or some other misapplication may cause the steam-to-liquid heat exchanger to be uncontrollable even though the controller has been tuned and all components are functioning correctly. A redesign of the application may be required to correct the problem.

Humidification

In particularly dry climates, steam humidification may be desired to increase the amount of water in the air supplied to the occupied zones. The steam is injected through multiple port injectors located within the air duct (as shown in Chapter 15, Figure 15.5). It is important to have dry steam in these applications because water droplets collecting on the bottom of the duct can lead to rusting ducts and microbial growth.

Steam humidifiers are controlled either by outdoor dry bulb temperature or by actual measurement of the duct air humidity. If measuring the duct air humidity, be sure to place the sensor far enough downstream to allow the steam to be fully dispersed. Otherwise you may have local "streams" of high humidity air that can throw off the control.

When sizing a steam humidifier, it may be better to purchase a unit that is slightly *undersized* rather than oversized. The traditional HVAC sizing technique is to calculate the capacity of your equipment based on design conditions and then multiply by some safety

fastfacts

Using steam injection for humidity does not increase the air temperature very much.

factor. The inevitable consequence is that the equipment operates at part load for most of the time. In steam injection systems this can mean more chance for condensation of the steam and standing water in the duct.

SAFETY

Obviously, steam is hot and can cause severe burns if not carefully controlled. Steam piping should be carefully constructed and insulated so that there is low potential for burns from direct contact or leakage. It is a bad idea to jury-rig steam systems, especially since it can end up harming those who work on it in the future. Even 5 psi steam can cause severe burns—that's why steam radiators don't go over 5 psi. Steam at over 10 psi can cause cast iron radiators to explode, and a steam pipe at 150 psi can be as dangerous as a high voltage line. Treat steam with respect, just like electricity.

Most boilers have a number of safety systems to prevent temperatures and pressures from getting too high. High pressure lock-outs on boilers (like in Figure 12.15) are common and practical and should be used on all steam systems. However, you should not rely on these safeties to ensure that the steam system is safe. There may be dangerous pressures and temperatures in the pipes even if the boiler is off.

Be very careful when working around steam lines. Most mechanical rooms and utility corridors are clumsy places to be and it is easy to grab onto a hot steam line without realizing it's hot. If the line is uninsulated, you can get a severe burn almost instantly. Even 5 psi steam is pretty hot. Steam over 15 psi can be hot enough to melt through your shoes, so be very careful when steam lines run across the floor. Do not wear polyester when working around steam lines unless you want it to become a permanent addition to your skin. A cotton shirt is best since this keeps you cool and insulates you at the same time.

FIGURE 12.15 High pressure steam lockout control.

Remember that you can't see steam at higher pressures. It is a colorless vapor, but at high enough pressures it can cut through your fingers. In theory, condensate lines should be at atmospheric pressure, but that's not always the case. The unalterable rule is: *make sure to isolate steam lines and condensate lines you are working on.* You may think a line is dead but then a valve opens somewhere and you get a face full of steam.

Never walk through a steam stream. If you hear steam shooting out of a pipe or flange, stay away. Never work on a steam line that is under pressure. It doesn't matter if it's a simple task like adding a gauge tap or weldlet. Welding can weaken pipe and if it breaks loose, you don't want to be anywhere close to it. Listen, don't look, for steam.

If you are isolating a steam coil, always turn off the steam gate valve ahead of the control valve but leave the condensate valve

fastfacts

The higher the steam pressure, the higher the temperature. The higher the temperature, the greater the expansion of the steam piping.

open, especially if there is no vacuum breaker. If you don't do this, the steam may condense and create a very low pressure in the coil. This can collapse the coil and prevent steam from flowing through it in the future.

Remember that steam lines grow when the steam is turned back on. If the steam is allowed back in too fast, the pipe can grow fast enough to snap loose from the mounting brackets. Always open steam valves slowly so that the expansion is gradual.

It is good practice to install steam pressure gauges at several points along the steam lines. A gauge is cheap and can prevent you from making a big mistake. Periodically check the gauges to make sure they are working. A few extra bucks for a quality steam gauge is a good investment. Whenever a steam pressure gauge is attached to the steam line, make sure there's a loop in the connector to keep from frying the gauge (Figure 12.16). Condensate gathers in the lower turn of the loop and prevents water from getting into the gauge mechanism. This should also be done when connecting an open steam connection to any other kind of sensor (Figure 12.17).

If you are working on steam lines that were installed before 1980, there's a good chance that the insulation contains asbestos. In fact, asbestos was used on a lot on steam lines after 1980 right up until it

fastfacts

When designing a steam distribution system that goes through an access corridor or tunnel, always have pressure relief valves vent outside to a safe location. If the valve vents into the tunnel, the steam can displace the oxygen in the tunnel and suffocate any personnel working there.

FIGURE 12.16 Steam pressure gauge with protective loop.

was banned. So even if the lines are covered with silica blocks, the joints, tees, and elbows may still be coated with an asbestos plaster.

While we're on the subject of hazardous materials, note also that steam lines are often treated with amine-based chemicals. These have been known to accumulate in condensate lines that are not used very much. These show up as fragile blue crystals that can blow out of the pipe. If you are taking apart condensate lines, let the air out of the line slowly. This stuff is not really poisonous, but you don't

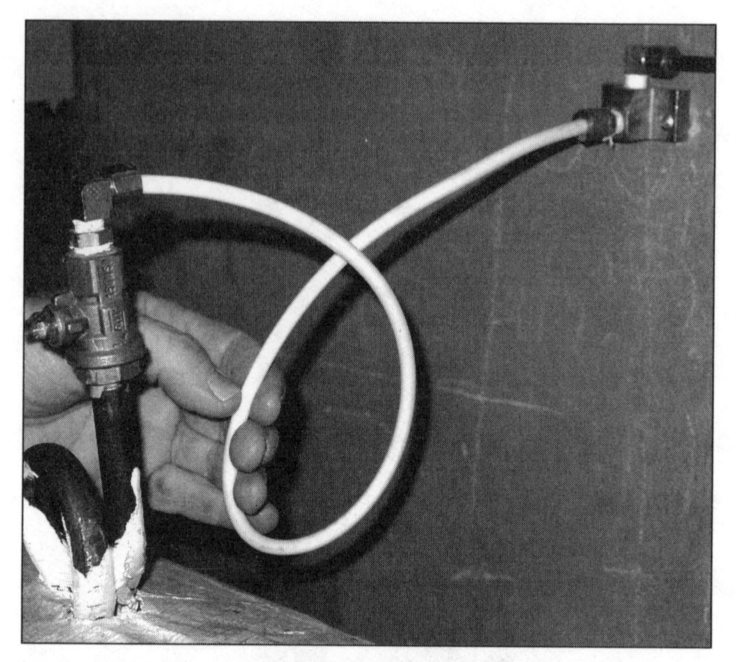

FIGURE 12.17 Transmitting steam pressure signal with Teflon tubing.

want to breathe it or get any in your mouth. The biggest problem with the amines is that the crystals can get into control valves and jam them shut.

Steam Valves and Traps

Small steam valves are typically used in radiation systems such as baseboard heaters and reheat coils. A small steam radiation valve actuator has a short lifetime due to the excessive heat that conducts up the valve from the steam piping. This is true for both pneumatic and electric valves. Pneumatic actuators may last around five years and electric actuators even less than that. In the pneumatic valve actuators the diaphragm has a tendency to disintegrate and leak. In the electric actuator the gears can turn brittle and split. Also, the wiring can get so hot that the insulation burns off and the signal short circuits. In both cases you need to exercise caution to avoid hazards caused by unconstrained high-pressure systems or by the metal radiator components being inadvertently energized. If an electric valve uses 120 VAC to drive the solenoid, it is possible that the baseboard fins will carry this voltage, or that the slightest disturbance to the actuator housing will provide for an unexpected shock. This can have bad consequences, particularly if you are working near hot piping. Similarly, pneumatic tubing that lies on the piping or baseboard fins can melt away, causing pressurized air leaks and loss of valve control.

Another problem with electric actuators in steam valves is that the heat can "cook" the oil right out of the actuator. The oil can leak or even blow out. This can happen unexpectedly, so always use eye protection when working near the actuator.

Large steam valves can be in the range of 1 inch up to 8 inches or larger. Special care needs to be taken with these valves since the pressures are generally higher and the amount of steam that can eject during a leak is larger.

When working with steam valves or other steam system components, make absolutely sure to isolate the valve before taking off the actuator or valve bonnet. There can be a big problem if the valve stem flies out during disassembly. Not only can the flying valve stem cause injury, but there is also a serious risk of scalding from hot steam coming from the hole where the valve stem used to be. In addition, any hot condensate in the pipe or valve can also come gushing out. It is best to isolate the valve and wait for the temperature and pressure to drop. Use steam gauges on the piping if necessary to determine when it is safe to work on the system.

fastfacts

In high-pressure steam systems the steam is usually in a superheated state. You can't see the steam even if it is leaking out of a valve or coupling. This can cause severe burns and the high pressure steam can cut into skin like a knife. Always be careful around steam piping and steam valves, especially if you are working in a confined space.

Always use lock out/tag out procedures when working on steam and always let the steam line cool down before taking it apart.

When working on steam systems, always isolate the section of pipe you are working on. Whenever necessary, perform lock out and tag out procedures that guarantee that no one will heat or pressurize a system that you are working on. Five minutes of prevention at the beginning of a job can save you a trip to the hospital or worse.

Make sure the steam line is clean of condensate. This prevents pipe banging and hammering, but is also important for safety. If steam encounters a big slug of condensate in the line, the condensate can flash and create dangerously high pressures in the pipe. The pressure surge can run down the length of pipe and blow elbows and tees clean off.

Trap maintenance is a critical item and all traps should be inspected every five years, more frequently if the system is experiencing hammering. In a float trap you sometimes find the float lying in the bottom of the trap. Be sure to isolate the trap when working and to put all the parts back together as instructed and then test. Never leave a trap without testing.

TROUBLESHOOTING

Steam systems can have problems in the heating load components, steam traps, and the steam generation equipment. For troubleshooting boilers, see Chapter 11. Some of the more basic problems can be identified in simple ways. For example, a steam system that is usually quiet may start making loud noises or whistling sounds. This usually indicates a problem that could be caused by one or more of the following problems:

- Defective steam traps
- Waterlogged condensate return system
- Defective condensate pump
- Defective boiler feed water pump

This section describes some of the basic components of steam distribution systems and how to solve various problems found within these components.

Steam Valves

Many of the valve troubleshooting techniques discussed in Chapter 6 are valid for both water and steam valves. Steam valves have a unique subset of problems, however, due to the higher pressures and gaseous nature of steam. For example, if a steam valve is not maintained properly, the valve packing can deteriorate to the point where steam can shoot up inside the actuator and destroy it. For electric actuators this happens through melting or breaking of gears and short-circuiting of the electronics, while in pneumatic actuators the problems occur when the spring gets rusted out and does not return the valve to an open position. When these problems happen with water valves there is usually a good indication by the presence of a puddle on the floor or water dripping from the actuator housing. On the other hand, with steam you generally cannot see the leak until the problem manifests itself in occupant complaints or catastrophic failure of the piping, valve or actuator. You can't see steam, but the effect is still there. One effect that you may be able to detect is corrosion around the valve stem (Figure 12.18) that may include crystals that form from the treatment used in the boiler water.

Even a small leak through a steam valve can lead to overheating problems in occupied spaces. One way to check for an internal steam valve leak is to force the valve closed, wait for a little while, and then check the temperature of the pipe on each side of the valve. Ideally, one side is hot and the other side is cold. It may be

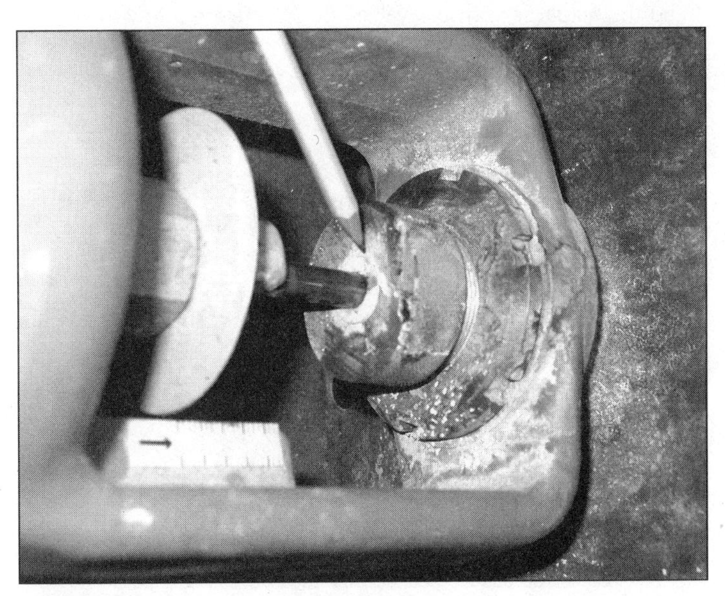

FIGURE 12.18 Corrosion on steam valve from leaking valve seal.

necessary to feel the pipe a little distance from the valve to avoid the effects of heat conduction along the pipe. If the downstream pipe is just warm, then the valve is probably okay. If the downstream pipe is hot, then there is probably a leak in the valve. Once you have a little experience with this, you may even be able to determine the extent of the leak by the distance of hot pipe from the valve. If the downstream pipe is appreciably hot for more than 5 or 6 feet from the valve, then the valve probably needs to be replaced or rebuilt.

One cause of zone overheating is when the valve diaphragm fails and the air pressure signal cannot close the valve. A squeeze bulb can be very useful to determine whether or not a pneumatic valve has a perforated diaphragm. To perform such a test, disconnect the compressed air signal and use the bulb to pressurize the diaphragm (see Figure 12.19). If the diaphragm holds pressure for several minutes, it is probably okay.

If a large steam valve is leaking badly you may have to remove the entire valve assembly from the piping and take it apart to find out what's going on inside. Figure 12.20 shows the relative size of a six-inch valve housing to the actuator. Keep in mind that these are quite heavy. If you are removing the valve from a high location, always use the proper equipment and support to keep the valve from crashing down. Once you have removed the top bonnet and freed the actuator from the valve body, there are a few things you

FIGURE 12.19 Squeeze bulb on pneumatic valve.

should look for. Inspect the seat at the point where it contacts the plug (Figure 12.21) to check for a rough surface, mud or corrosion build-up, and wire draw. The surface should have a bright, polished appearance. Any dirt that is sticking to the seat should be removed and the seat should be refinished if necessary. Wire draw appears as a thin line across the surface of the seat and represents a developing crack in the seat. You may need to replace the seat if the condition is beyond repair. You should also check the valve plug on the valve, especially where it contacts the seat (Figure 12.22). This should be smooth and clean, with a slightly shiny finish.

Larger pneumatic steam valves tend to use positioners to ensure proper control. See Chapter 5 for more information on positioners. If the valve you are working on uses a positioner and there are problems with the valve, check not only the valve but also the positioner and positioner diaphragm.

If there is full pressure on the input of the positioner and the valve doesn't move, then pull off the line between the positioner and the actuator. Try to blow up the positioner diaphragm with a squeeze bulb and see if it holds pressure. Alternatively, put a gauge on the

FIGURE 12.20 Six inch steam valve disassembled for rebuild.

FIGURE 12.21 Valve seat on six inch steam valve.

FIGURE 12.22 Valve plug on six inch steam valve.

output of the positioner; with full pressure (~15 psi) on the positioner input, you should have full pressure on the positioner output. Note that this applies to only low-pressure spring-loaded valves.

Sometimes a large steam valve is not working simply because it is oversized. The valve may have been sized for maximum operation, but because it rarely works in this range, it is difficult to control. In fact, many large steam valves typically operate in just the first third to half of their potential travel range. If you suspect this is the case, you may want to use a one-third/two-thirds configuration with a split-range positioner (or slaved electric actuators with the slave potentiometer set to kick in about halfway up.)

Receiver Tanks and Condensate Return Systems

The receiver tanks collect condensate from the steam traps and hold it until the level in the tank is high enough. At that point, the condensate return pump activates and returns the hot condensate back to the boiler. There is sometimes a check valve between the feed water pump and the boiler that prevents steam from flowing back into the feed water receiver tank. An overheated feed water receiver tank or cavitation in the feed water pump may be due to a leak in this check valve.

An unusual amount of steam vented from a condensate pump or boiler feed water receiver tank indicates trouble in the steam system and should be investigated and repaired to prevent damage to the steam system.

Steam-to-Air Heat Exchangers

Steam-to-air heat exchangers are used in reheat coils in single-deck VAV systems and as a crucial part of a freeze protection system. In the reheat situation, the steam coils may be the only way of heating the space while in the freeze protection mode the coil prevents costly damage to downstream air system components. If you are not getting enough heat from a steam-to-air heat exchanger, refer to Figure 12.23 for a troubleshooting flow chart.

Note 1: It doesn't work unless it's turned on: if there's no steam the coil can't work. Check to make sure the valve isn't isolated and that the boiler is operating.

Note 2: Check for dirt, bent fins, clogged filters, or anything else that may be obstructing the flow of air through the coil.

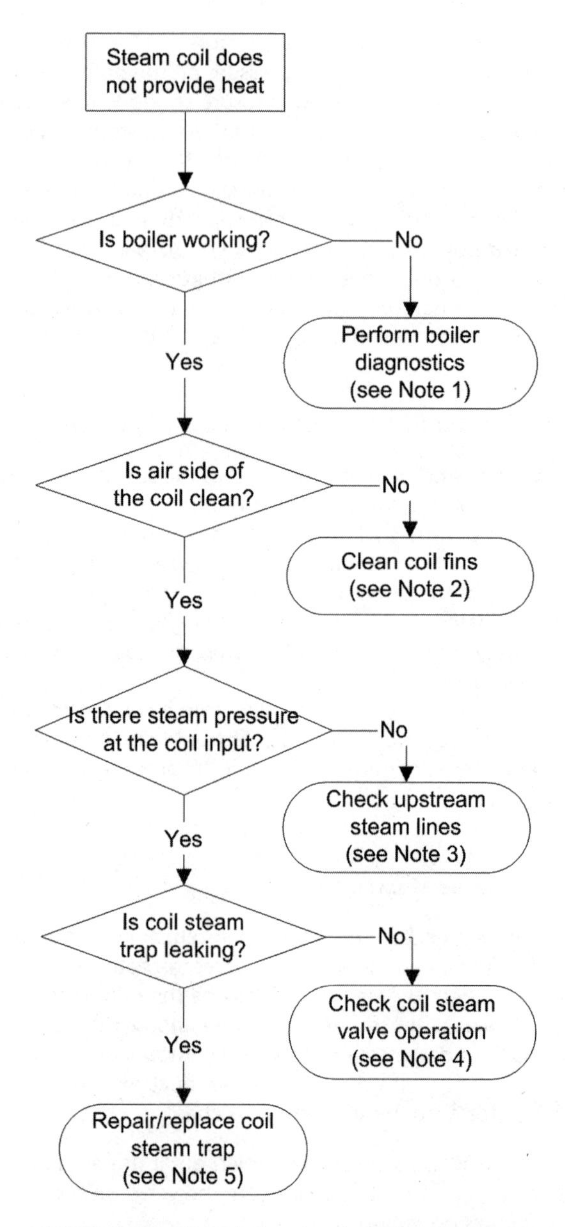

FIGURE 12.23 Troubleshooting flow chart for a steam-to-air heat exchanger.

Note 3: Trace back upstream to see if you can find a closed isolation valve or malfunctioning steam trap. If the HVAC plant undergoes seasonal operating changes, see if perhaps a steam valve was closed when it wasn't supposed to be. You may be able to identify the location of a leaky steam trap by checking the steam pressure at the inlet to other coils. Of course, if that much steam is escaping through the trap, you probably also have a problem at the condensate receiver tank.

Note 4: If you have steam pressure and the coil steam trap is working, your steam valve may not be opening. If it's a pneumatic valve, check to make sure the pneumatic tubing is on good and tight and that there is a signal. Otherwise, check the valve diagnostics in Chapter 6.

Note 5: If the coil steam trap is not working then the coil can fill up with condensate and not allow steam through. This is a double problem: no heating is possible and the water in the coil can freeze and burst the coil. The way to check for a malfunctioning steam trap is to first shut off the steam to the coil using an isolation valve (if it's below freezing then also shut off the fan). Next, open the condensate line after the steam trap and then slowly open the isolation valve to supply steam to the coil. You may need to turn the fan back on to create enough condensate. If only condensate comes out of the trap then the trap is okay. If a lot of steam comes out with only a small amount of condensate, or if no steam or condensate comes out, then you need to repair the trap.

Steam-to-Hot Water Heat Exchangers

Steam-to-hot water heat exchangers are used to increase the water temperature in space heating and domestic hot water systems. If you are having problems with a steam-to-hot-water heat exchanger,

fastfacts

A steam-to-air coil that is clogged with dirt on the air side has a reduced heat output even though the discharge temperature may remain the same.

fastfacts

A steam trap that is "blowing through" or allowing steam to pass without trapping reduces the heat output of the coil. A steam trap that is "stopped up" (and does not allow condensate to pass) may cause the coil to hammer and during freezing weather may trip the fan off on low temperature safety. In the worst case you can split a tube.

first follow procedures similar to those for the steam-to-air heat exchanger, and specifically check to make sure the steam trap is operating correctly and has not allowed the steam side of the converter to flood with condensate. If this seems to be okay, you should check for a ruptured tube in the heat exchanger. This is done in the same test for checking the steam trap, but now there is no need to apply steam to the coil. Shut the steam off to the exchanger and open the condensate line *ahead* of the steam trap. If there is a steady flow of liquid from the condensate line, then you should pull and inspect the tube bundle.

If the tube bundle is okay, it may be necessary to run a condensate check for total dissolved solids using a TDS meter. This lets you check for crud build-up on the inside of the tubes as well as for a small leak from a ruptured tube. Obtain a sample of condensate from the condensate line after the steam trap. Place a sample of condensate in the sample cup of the TDS meter and record the readings of the meter. For best results repeat several times until a consistent reading is obtained. If the reading is above normal condensate conductivity then pull and inspect the tube bundle. If the reading is below the make-up water conductivity then the tube bundle is probably okay.

Due to steam hammering or through normal use, the tube sheet gaskets on a tube and shell heat exchanger may break internally and cause an internal short circuit of the steam or fluid in the tube bundle. If this occurs, the heat output of the exchanger may decrease to the point that it will not keep up under heavy loads or may stop altogether.

Troubleshooting Steam Baseboard Heaters

Baseboard heaters are often used in single-duct VAV systems where there is a need for some form of reheat in the space itself. This reheat method places the heating elements along the walls and windows opposite the diffusers in order to counteract drafts that may accompany cold primary air descending from the ceiling. There are several potential problem areas with baseboard heaters, not all of them located in the heater itself.

As a general rule, if there is a report of overheating in a particular zone with baseboard radiation heating, the first thing to do is look at the thermostat. If the steam valve fails to an open position, a lack of signal from the thermostat can lead to continuous, uncontrolled heating of the space. If the thermostat is working correctly, then go ahead and check the rest of the system. See Figure 12.24 for a flow chart used to troubleshoot steam baseboard heaters.

Note 1: Most electric thermostats use a 24 VDC or 24 VAC power supply. Check to make sure that there is sufficient input voltage at the thermostat itself.

Note 2: Run the thermostat setting up and down and see if this generates any output signal. Check to see if the thermostat is located in an area where it may experience a cold draft (for example, near a window or directly under a diffuser). If the thermostat is located on an exterior wall, the wall itself may be cold and "fooling" the thermostat into thinking the zone is colder than it really is. It is also possible that infiltrating air is passing through the wall and into the area behind the thermostat. If this occurs, you may be able to cure the problem by sealing the space behind the thermostat with a gasket or by spraying in some kind of expanding foam. Finally, make sure the thermostat is mounted in the correct orientation. Mercury bulb thermometers will not work if they are not mounted properly.

It is also possible that the thermostat is located away from the zone that is too hot. This can happen in circumstances where remodeling has occurred and new walls or partitions have been built. It is not unheard of to have a thermostat remounted on the opposite side of a new wall from the zone it is supposed to be controlling!

Note 3: If the thermostat checks out okay, then the problem is probably in the steam valve. Most electric baseboard actuators are 24 V, two-position valves, although some use a 4 to 20 mA or 0 to 10 V signal for modulation. Run the thermostat up and down and see if the actuator moves. If it doesn't, then take the actuator off

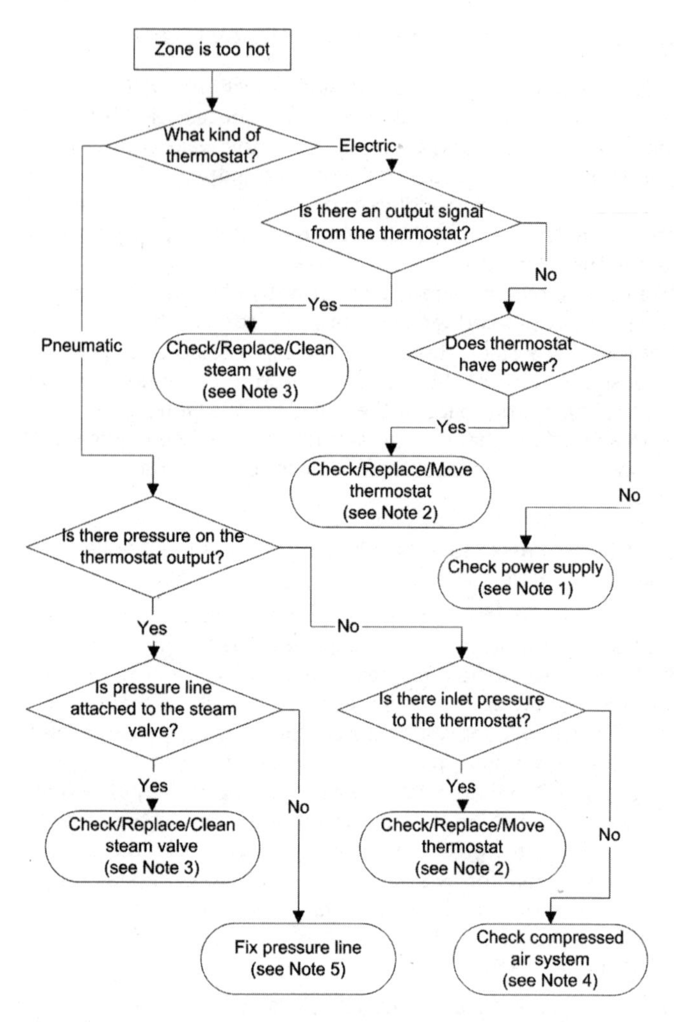

FIGURE 12.24 Troubleshooting flow chart for a steam baseboard heater.

and try to repeat this test. If the actuator still doesn't move, make sure there is voltage at the actuator input terminal and check for loose wires. If it is a tri-state valve (one terminal for open, one for close, and one common) then switch the leads and see if the actuator moves. If there is still no motion, then the actuator probably needs replacing.

If the actuator is pneumatic, check to make sure the compressed air line is firmly attached to the actuator input. Then check for any leaks coming from the actuator diaphragm. You can usually feel or hear any leaks. If the diaphragm is blown, remove the air line from the actuator and the actuator from the valve stem. The diaphragm usually blows out around the side. Replace the diaphragm, reinstall the actuator on the stem and then reconnect the air line to the actuator.

Note 4: There should be inlet pressure on the thermostat in the 15 to 20 psi range. If this is not the case, check the compressed air system and tubing to make sure it is working and there are no leaks.

Note 5: If the actuator is pneumatic, the constant cycling of the steam valve over time can lead to a loosening of the compressed air piping connected to the barbed fitting of the actuator. Sometimes the pressure line is slightly loose and leaking, or it may have popped off altogether. If the line is not attached tightly, pull it off, cut a new area for the barb to grip on, and reattach.

The polyethylene tubing on pneumatic actuators tends to swell when it gets hot. If the pipe is attached to a barbed fitting on the actuator, the pipe can get hot enough and increase in size to where it blows off the fitting. A useful cure for this is to use an extension of ¼ inch copper tubing three to four inches long. Connect one end to the actuator and another to the tubing as shown in Figure 12.25. The copper dissipates enough heat to help keep the compressed air tubing cool.

If the zone is too cold, then follow the steps outlined above to make sure that the thermostat and the actuator are working correctly. However, since steam valves generally fail open, a zone that is too cold usually implies one of three things:

- The thermostat is in a bad location: check to make sure that there are no heat sources near the thermostat that could make the thermostat think the zone is hotter than it really is. For example, if there is a copying machine, coffee maker, or computer monitor next to or under the thermostat, the thermostat will be at the setpoint temperature while the rest of the room is too cold.

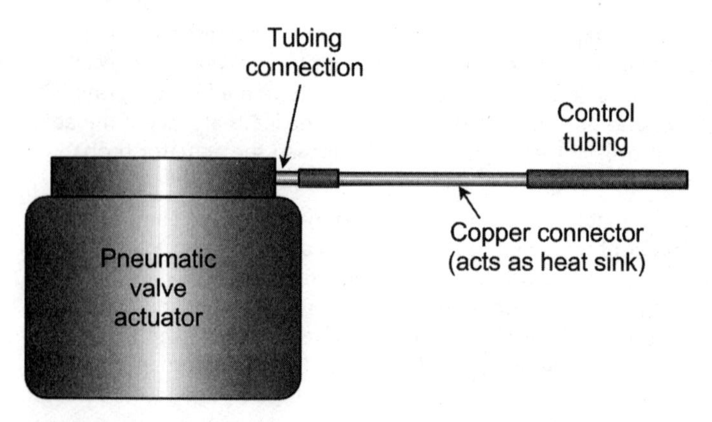

FIGURE 12.25 Steam valve extension to prevent overheating of compressed air tubing.

- A leaking steam valve can send moisture into the valve actuator. In an electric actuator, this can allow condensate to form on the circuits or can melt the gears, forcing the valve to remain shut. On a pneumatic actuator, a steam leak can rust the spring on the valve so that it never returns to the open position. Since the leaks are almost always on the seals of the valve, the steam (which may not be visible) will be directed straight at the actuator.
- If there is still a problem after checking the thermostat and the actuator, take the actuator off and manually check to see that you can move the valve stem up and down. In particular, make sure it hits the valve seat on the way down and then comes all the way back up. If you detect a restricted valve stem travel, the valve could very well be filled with dirt. If you suspect this is the case, isolate the valve with an upstream shutoff, pull the valve bonnet, take the valve top off and inspect the seats, plug, and stem. You may need to clean the inside of the valve body and possibly flush the valve out.

PRACTICAL CONSIDERATIONS

Steam Traps

Many problems in steam systems arise from faulty or broken steam traps. Make sure the steam traps are oriented correctly. Unlike most

fastfacts

As long as you have the valve apart, go ahead and repack the valve to save yourself a trip later on. If convenient and you have the time, replace the plug and stem as well. You will have to do this eventually and it's best to do this all at once. Remember that the labor cost is usually higher than the cost of the valve parts.

other piping, the traps depend on gravity to drain the condensate and, in the case of float valves, to work properly. A sideways or inverted steam trap can be a continuous vent of steam to the condensate return. Also make sure the steam lines are clean. Disk traps can catch dirt between the disk and the body, allowing steam to escape.

If you notice cavitation occurring in the condensate pumps, check to make sure that the traps are operating correctly. Since the condensate is so hot, even slightly higher-than-normal temperatures in the condensate return system can lead to cavitation in the pumps.

A lack-of-heat complaint on a steam system if often caused by a faulty trap. The question is finding the right trap. If you have a steam baseboard heater that does not heat and everything appears okay at the baseboard, it may be that a defective trap at the end of the steam line has caused the whole line to fill up with water.

fastfacts

A steam baseboard unit would not heat, but the automatic control valve and steam trap checked out fine. A close inspection revealed that the steam trap at the end of the steam line was blowing through, heating up the condensate return line. The heat from the condensate return line was conducted to the baseboard trap, keeping it closed. The baseboard soon filled up with water and was unable to heat the space.

The bottom line is that steam leaks are expensive. On top of coil failures and heating complaints, there is the cost of sending steam out of the system without using it. Each leaky steam trap can cost you the equivalent of hundreds of dollars per year in wasted boiler fuel.

Steam Valves

If a steam valve won't shut off and everything else looks okay, chances are the valve seat or valve disk may be defective, such as wire draw or chunks knocked out. Wire draw is when the valve seat cuts down the middle. For plug valves under 2 inches, it is recommended that you use a brass seat with an EPDM rubber disc. EPDM is a special high temperature rubber that gives good shutoff characteristics for the price. For steam systems up to about 10 psi, brass seats are usually okay in conjunction with composition or stainless steel discs. If the pressure is higher than 10 psi, then you probably should use stainless steel for the seats and composition disk, or stainless steel for both. If you use brass in higher pressure systems you run the risk of having wire draw on the seat. Of course, it's a trade-off since brass seated valves are typically much cheaper than stainless steel. If your system has very high pressures (greater than 20 to 30 psi), you should probably use an industrial quality valve. Even at low pressures, industrial quality valves generally last much longer. The high capital cost pays itself off in lower maintenance costs.

When designing a steam system, it is a good idea to double the close-off rating when specifying the steam valves where you want tight shut-off (for example, in an exterior zone where the heating and cooling loads vary greatly over the course of the year). Some valves simply don't have tight shut-offs, despite the manufacturer's claim otherwise. If the tight shutoff valves are too expensive, you may be able to get by using two valves in series (like a steam butterfly valve and then the letdown valve for pressure control). Make sure to place the shutoff valve ahead of the control valve.

fastfacts

A positive positioner is recommended for pneumatic steam valves of 1 inch and larger. These allow for more accurate control than adjusting spring ranges.

fastfacts

A typical no-heat call is due to old valve packing that has allowed steam to escape and rusted to the return spring in the valve. The valve plug can be forced down by the actuator but the spring cannot open the valve back up again.

Due to the excessive heat of the working environment, steam valve actuators generally do not last very long. The diaphragms on pneumatic actuators tend to crumble and disintegrate over time. If replacing a diaphragm you may wish to consider using silicone rather than the standard rubber. The silicone diaphragms tend to last longer.

Another victim of the hot environment is the valve stem seal. Experience shows that Teflon seals tend to last longer than rubber seals. Sometimes the seal won't leak steam but it can fill up with condensate and then it can start leaking hot water. This can destroy the valve packing. Unfortunately, the leak is not apparent to the casual observer and is only identified when the associated steam traps fail and the subsequent water leak becomes apparent. Graphite/Teflon rope packing tends to last longer than conventional packing. In any case, you should figure on repacking steam valves every three or four years.

Heat conducts between the steam pipe, valve body, and valve actuator. In baseboard applications, the distances are relatively short and electric motors tend to cook within a few years. Typical replacement costs are around $50 for an electric valve motor replacement and $10 for a pneumatic diaphragm replacement, plus labor, so this should be figured into your maintenance budget.

Complete closure of steam valves is critical, particularly in steam baseboards, if you wish to avoid occupant complaints. In the half to three-quarters inch ranges, the plug and seat combination lasts about five years with constant use. Usually the seat is a composition disk or elastomeric compound that becomes brittle and splits or just simply disintegrates. Sometimes the plugs can be replaced, but care must be taken when using brass-body valves that have raised seats. These can develop roughness or wire draw that prevents them from sealing correctly. Some valve manufacturers have resurfacing tools that can be used to reseat valves. However, some cage valves require that you replace the whole valve and plug assembly.

In bigger steam valves there is more distance between the valve body and actuator. Some valve brackets even have extensions to increase the distance. The idea is to keep the actuator motor or diaphragm as far away from the heat as possible.

Another way to prevent the heat from rising up to the valve actuator is to simply orient the valve on its side. This way the hot air rising from convection will not encounter the actuator. If you lay the steam valve on its side, however, be sure to make sure that the motor is capable of operating in such a position. In some electric motors, the oil reservoir or other components must be in a specific position for the actuator to work properly. If it is not possible to reorient the valve position, try to blow some supply air across the actuator to keep it cool.

In one case that the authors are aware of, a side-mounted pneumatic valve actuator has been in continuous service for more than 40 years, compared with about five years for a typical upright orientation. It must be noted that sideways-mounted actuators can have the problem of excessive valve stem wear on one side of the stem. The bigger the valve, the greater the stem wear. This occurs because the disk holder can be quite heavy (~10 pounds) and in the horizontal orientation all this weight bears on just one side of the stem. Thus there is a trade-off on the valve orientation between diaphragm replacement and possible stem and packing replacements.

Steam-to-Water Heat Exchangers

Do not try to lift condensate out of a coil or converter with steam pressure. This causes operating problems and leads to flooded coils. The converter condensate should gravity drain through a steam trap

fastfacts

Under most situations it is easier to replace the whole valve on a baseboard. However, if the valve has odd dimensions (such as with older valves) or if it is in a very awkward location, it may be easier to try to fix the valve stem seal or plug in place.

to the condensate line or condensate pump. If you do try to lift the condensate out of the converter, the result is a converter that is half full of condensate at low loads. This makes it very difficult to control the outlet hot water temperature, and if you use float or thermostatic steam traps the trap will never close.

A steam-to-water heat exchanger that is half full of condensate will cause hammering when incoming steam hits the condensate and suddenly condenses. Over time, this hammering damages the tubes and will require the tube bundle to be repaired or replaced when the fluid being heated in the tubes (usually at a higher pressure than the pressure in the shell) leaks into the condensate. In addition the hammering may break off the float in the steam trap.

Severe hammering of heat exchangers has been known to cause condensate and feed water pump failures when tubes have broken off of the tube sheet section of the tube bundle and flooded the condensate system with the fluid being heated in the exchanger. In addition, piping connected to the heat exchanger may fail due to mechanical stresses placed on them as the result of a hammering heat exchanger.

Steam-to-Air Coils

The face and bypass damper arrangement in Figure 12.14 should be used on a steam preheat coil that experiences 100 percent outdoor air. Without this kind of system, you will not be able to supply full steam flow to the coil when the air is below freezing and you run the risk of freezing the condensate in the coil. This can split a tube, or cause overheating due to the high discharge air temperatures needed in order to keep the coil from freezing, or it can trip out the fan on low temperature safety control (if you modulate the steam valve to control discharge air temperature). So-called freeze-proof coils WILL freeze under certain operating conditions.

If you do freeze and split a tube in a steam-to-air coil, it may be better to cut the tube and solder the ends close as possible to each header. If the repair of the tube is attempted at the middle of the coil without steam flow the tube will fill with condensate and freeze and split again.

FANS AND DAMPERS

Fans and dampers are used to move and control airflow in buildings. The fans force air into ducts that then route the air to where it is needed. Dampers are used to modulate the amount of air entering the building, flowing across coils, and flowing into occupied zones.

THEORY OF OPERATION

Fans

Fans fall into two categories. *Centrifugal* fans (Figure 13.1a) use a rotating drum to "throw" air into a duct, while *axial* fans (Figure 13.1b) are propellers that "sweep" air into a duct. The blades on centrifugal fans can be forward or backward-inclined. Large centrifugal fans usually use backward-inclined blades as these tend to be quieter. The blades on axial fans tend to force the air to the side as well as downstream, giving a high flow rate but at a low pressure rise. They are also relatively noisy. Most axial fans are tube-axial as shown in the figure, while some are vane-axial, which uses a series of straightening vanes to increase the fan pressure rise and decrease the fan noise.

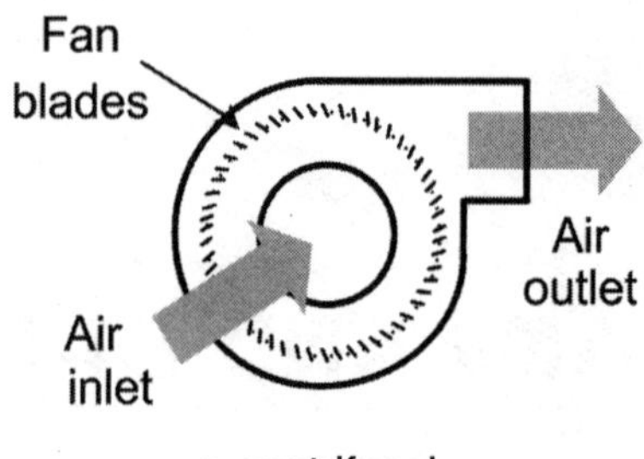

a. centrifugal

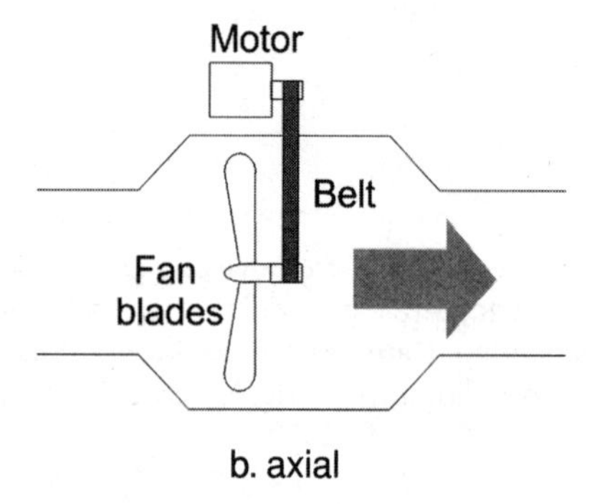

b. axial

FIGURE 13.1 Fan schematics.

The performance of a fan depends on the power of the motor, the shape and orientation of the fan blades, and the inlet and outlet duct configurations. Air passing through a fan is subject to an increase in pressure and a slight increase (2° or 3°F) in temperature. The temperature change results from energy added to the air through the mechanical agitation of the airstream. The power added to the air is found by multiplying the flow rate and the total pressure rise,

$$\text{Fan Horsepower} = \frac{\text{CFM} \times \text{DP}}{6356 \times \eta}$$

where CFM is the fan flow rate in cubic feet per minute, ΔP is the pressure rise across the fan in inches W.G., and η is the fan efficiency, usually around 70 to 80 percent.

Fans are governed by certain laws that relate the rotation speed, volumetric flow rate, pressure rise, and power consumption to each other. The flow rate is a direct relation to the fan rotation speed so that as the fan changes speed the flow rate changes proportionally,

$$\frac{CFM_2}{CFM_1} = \frac{RPM_2}{RPM_1}$$

The pressure rise across the fan changes as a function of the square of the fan rotation speed,

$$\frac{\Delta P_2}{\Delta P_1} = \left(\frac{RPM_2}{RPM_1}\right)^2$$

and the power consumption of the fan changes as a function of the cube of the rotation speed,

$$\frac{kW_2}{kW_1} = \left(\frac{RPM_2}{RPM_1}\right)^3$$

The *fan curve* is a graphical representation of the pressure rise across the fan as a function of the flow rate. Figure 13.2 shows a typical fan curve for a fan that can change speed. Note that this is similar to the pump curves in Chapter 6. The principles are similar except one pumps air and the other pumps water. The left side of this graph represents the case where the fan flow is blocked (think of closing dampers on the fan outlet) so that there is a high pressure rise and low flow. On the right side of the graph the fan has unimpeded flow (think of a window fan) so there is not much pressure rise across the fan but there is a lot of flow.

Dampers

Dampers are broken down into two different types, opposed blade and parallel blade (see Figure 13.3). The pressure drop across a damper is given by the loss coefficient C_d (a function of the blade angle):

$$C_d = \Delta P_t \div P_v$$

where ΔP_t is the total pressure loss across the damper and P_v is the velocity pressure. The loss coefficient depends on the damper type. The fraction of full flow through a damper is given by

$$f_{full} = \sqrt{\frac{C_{d,0}}{f \cdot C_d + (1 - f) \cdot C_{d,0}}}$$

where f is the open damper resistance as a function of the total system resistance and $C_{d,0}$ is the loss coefficient for the damper when it is full open. Figure 13.4 shows how the damper loss coefficient changes for different damper types.

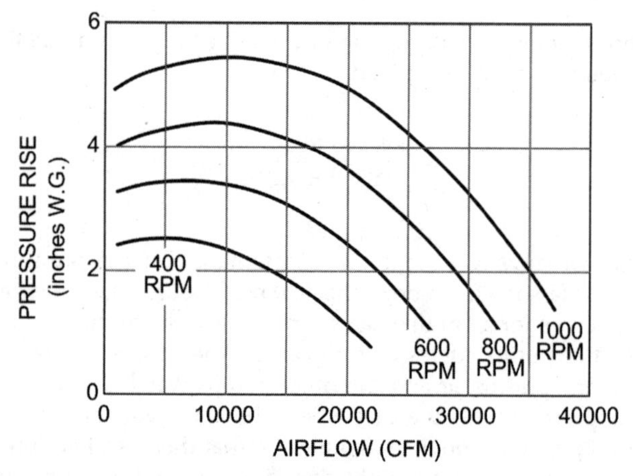

FIGURE 13.2 Hypothetical system and fan curves.

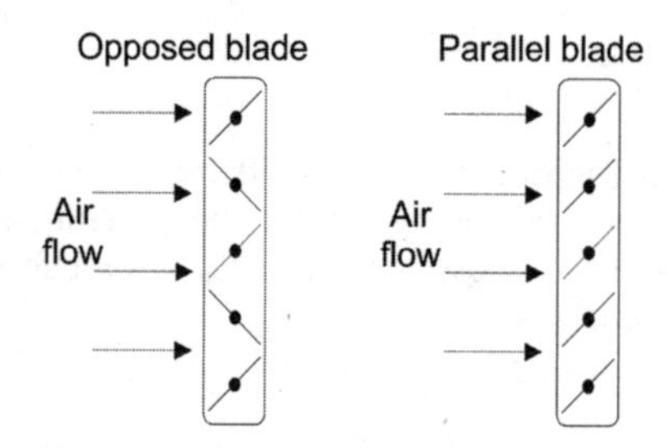

FIGURE 13.3 Schematic of opposed and parallel damper blades.

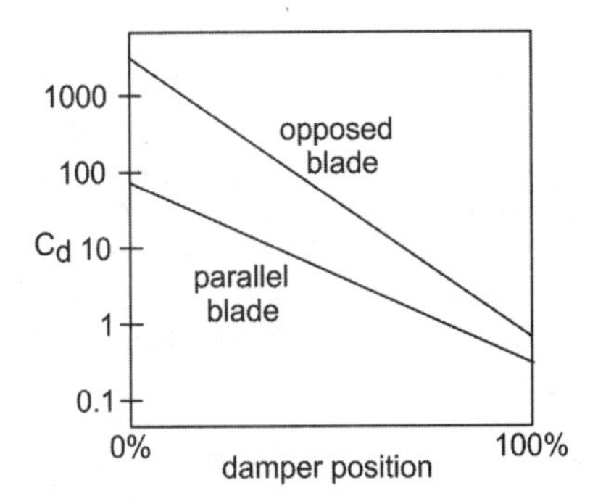

FIGURE 13.4 Damper loss coefficient as a function of damper position.

Opposed blades are preferred for modulating service from noise and energy-loss standpoints. Typical leakage amounts are about 50 CFM/ft^2 at a pressure drop across the damper of ΔP = 1.5 inch W.G. Outside air dampers, however, need to be much tighter to prevent excessive infiltration and coil freezing. This is accomplished by adding a rubber strip across the edge of the dampers to create a seal when the damper blades come together (see Figure 13.5). Tight dampers can have leakage rates as low as 10 CFM/ft^2 at a 4 inch W.G. pressure drop.

FIGURE 13.5 Pneumatic actuator on tight-seal outdoor air dampers.

In smaller round ducts, such as those leading to terminal boxes, the damper may consist of a single, movable plate that rotates around a central pivot. In appearance, these resemble butterfly water valves and are called butterfly dampers (Figure 13.6). These dampers rarely achieve a tight seal when closed, but since most zones require some amount of minimum air flow the leakage is not a problem.

In some cases you may encounter a need to install blast gates. These are usually used to help balance the air distribution system but can also be used to isolate a section of ductwork during maintenance. A blast gate is a simple, movable plate that is held in place with set screws as shown in Figure 13.7. They are usually located at the junction of the take-off and the main duct so that the entire take-off ductwork can be isolated if necessary (Figure 13.8).

Fire Dampers

Ductwork is supposed to allow air free movement around the building. Unfortunately, it also acts as a convenient conduit for heat and smoke, allowing the effects of fire in one section of a building to spread to another section. Fire dampers are used to prevent this from happening and should be located where sections of duct pass

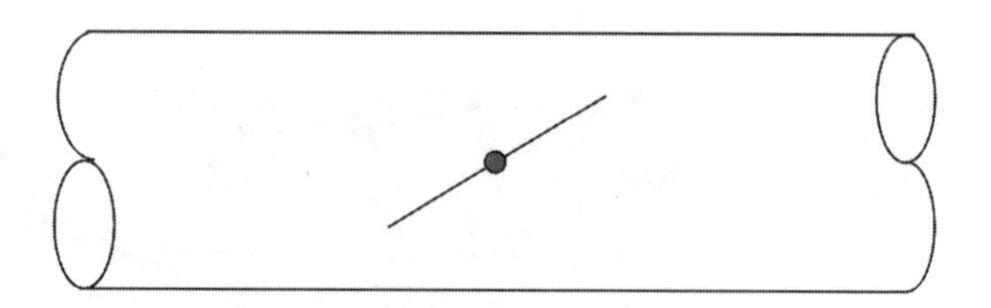

FIGURE 13.6 Butterfly damper in round duct.

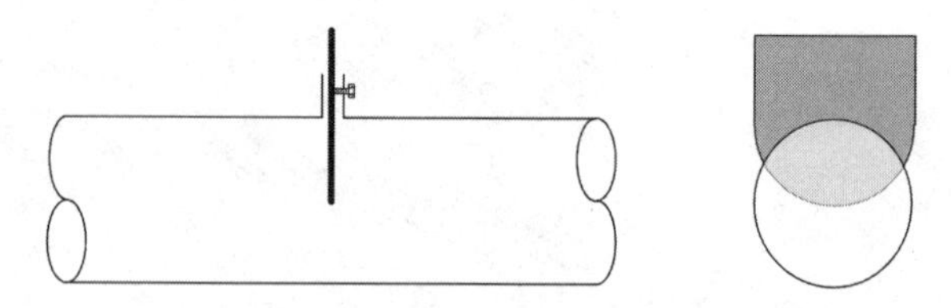

FIGURE 13.7 Gate damper or blast damper in round duct.

FIGURE 13.8 Blast dampers on duct take-offs.

through a partition wall or between floors. Local codes should be consulted to determine when and where fire dampers are required.

Fire dampers are self-contained, spring-loaded dampers that are kept open by a small link of metal. The metal chosen has a relatively low weakening temperature and allows the damper to close when the temperature goes above a certain level. The links can be specified to yield at a certain temperature with a standard rating being around 165°F.

Face and Bypass Dampers

Face and bypass dampers are often used on preheat coil arrangements (Figure 13.9) where a steam or electric coil may be used.

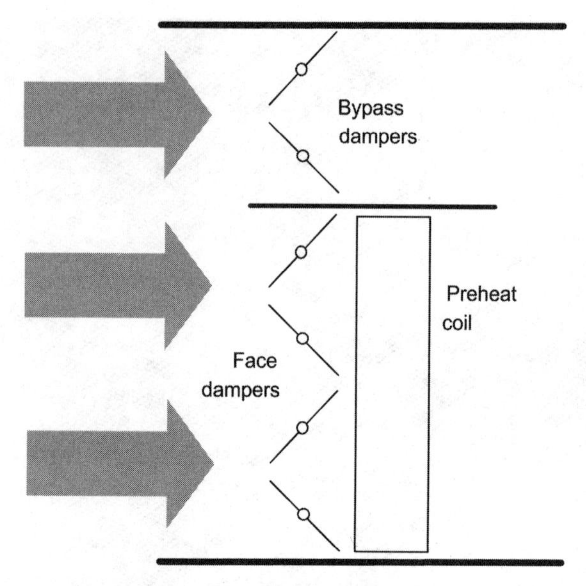

FIGURE 13.9 Face and bypass dampers.

These coils do not offer good modulating control, so the face and bypass dampers are used to provide some kind of air mixing to allow for air temperature control downstream of the coil. These can be difficult to control, however, especially at sunset when the outdoor air temperature is changing rapidly. To ensure some degree of stable control, the bypass dampers should be closed when the outside air is above 35°F so that there is full flow through the steam coil. The steam valve is modulated to maintain the required discharge air temperature. When the outside air is below 35°, the steam valve moves to the full open position and the bypass dampers modulate to maintain the required discharge air temperature.

CONTROL

The control of airflow and duct pressure is critical to keeping indoor air quality in the proper range. Details about duct air pressure control are discussed in the section on air distribution systems (Chapter 15). To control the amount of air flow through a constant speed fan, you can use *outlet dampers* that impede the air leaving the fan (Figure 13.10) or *inlet vanes* that restrict the air entering the fan (Figure 13.11). Inlet vanes are preferred from a noise standpoint. They are also slightly more efficient because they can be used to "pre-swirl" the air. Figure 13.12 shows a typical fan curve for fans with inlet vanes.

Another way of controlling the air flow through a fan is to simply lower the fan rotating speed. Beginning in the late 1980s the cost of variable speed drives (see Chapter 4) dropped enough to allow their widespread use in HVAC systems. They are particularly convenient for varying the airflow on fans. Even though these drives can still cost thousands of dollars, the advantage of such a system becomes clear when looking at the fan curves. Figure 13.13 shows similar

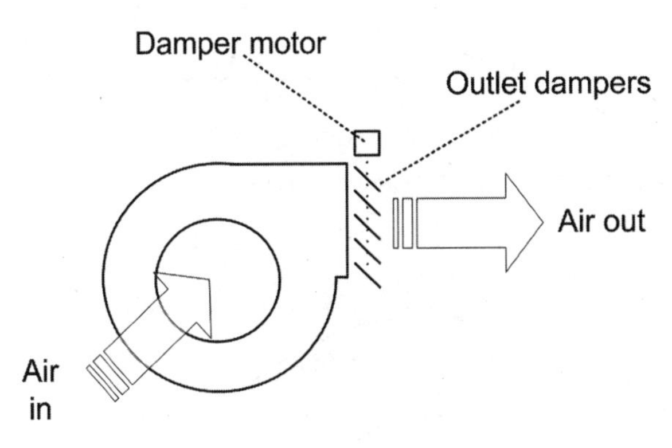

FIGURE 13.10 Centrifugal fan with outlet dampers.

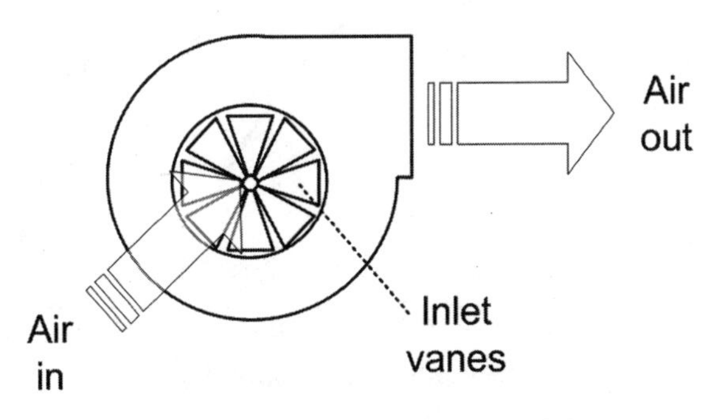

FIGURE 13.11 Centrifugal fan with inlet vanes.

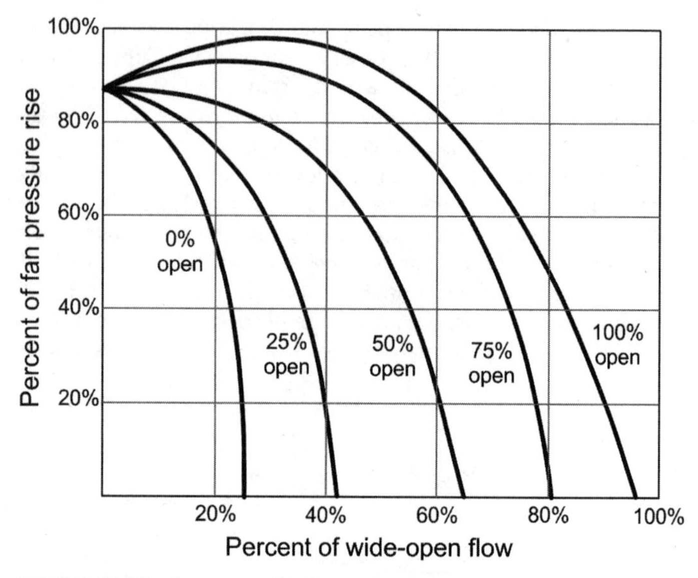

FIGURE 13.12 Fan curves for fans with inlet vanes.

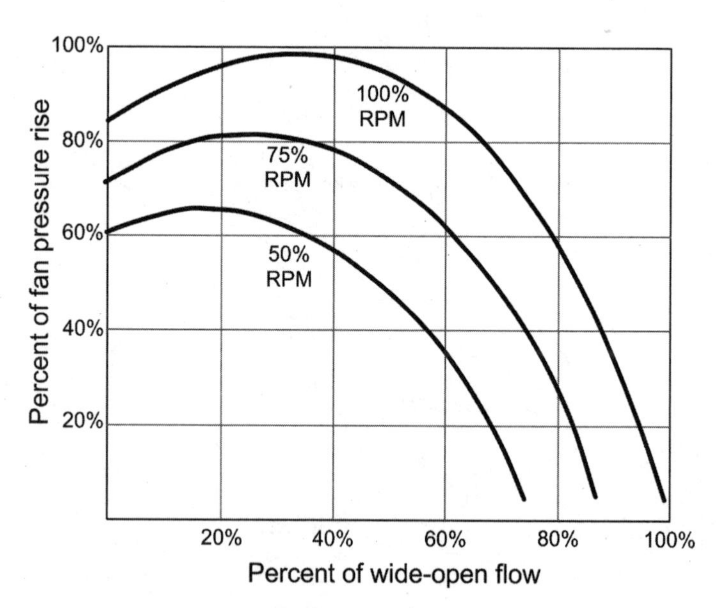

FIGURE 13.13 Fan curves for variable speed fan.

curves for a fan with a variable speed drive. The entire fan curve shifts down and to the left as the fan slows down, meaning that both the flow and the pressure decrease. This can help save a great deal of energy. The fan affinity laws described on page 351 show that the energy consumption changes as the third power of the fan rotation speed, so, for example, reducing the fan rotation speed by one-half cuts the power consumption by $(\frac{1}{2})^3 = \frac{1}{8}$.

Dampers are controlled by pneumatic or electronic actuators and modulate in order to control airflow or resulting mixed air temperature. When used on zone terminal boxes, dampers control the amount of primary air entering an occupied space and the resulting temperature of that space (more details on terminal box damper control are in Chapter 16). In mixed air situations (Figure 13.14) the dampers are modulated to maintain a fixed mixed air temperature, hopefully close to the supply air temperature so that the coils do not have to work as hard.

When the outdoor air temperature is close to the desired supply air temperature, the dampers should be controlled in an economizer mode, where the dampers open to allow 100 percent outside air into the building as shown in Figure 13.15. This reduces the amount of energy required to condition the building supply air.

SAFETY

When working around fans, you can find yourself inches away from hundreds of pounds of rapidly rotating metal that keeps trying to

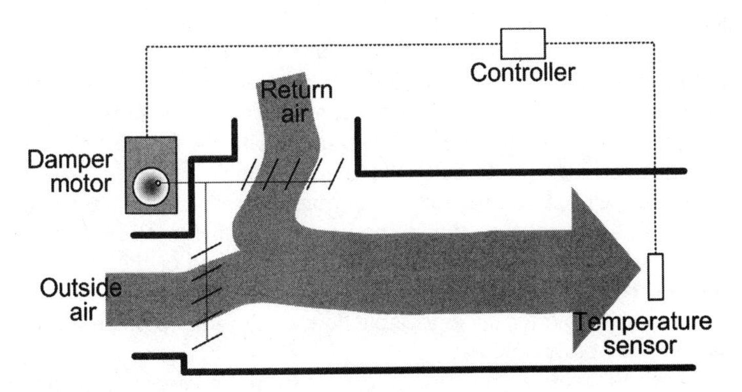

FIGURE 13.14 Mixed air dampers controlling on temperature.

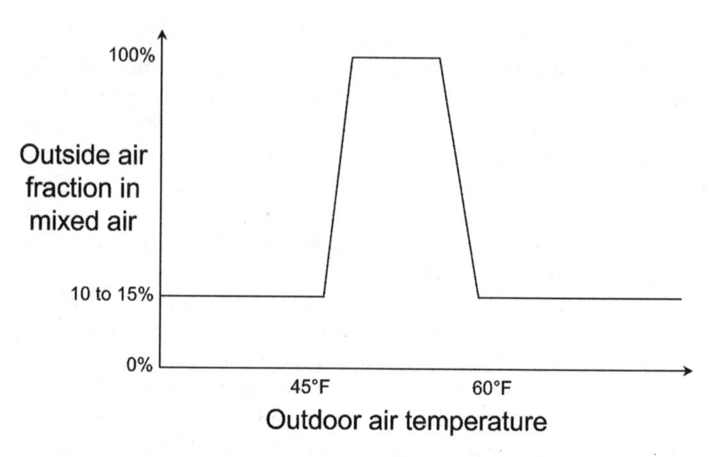

FIGURE 13.15 Outside air damper control for economizer mode.

suck you into it. That alone should be enough to make you cautious. Fan cabinets should be protected from any kind of accidental entry.

As with all other HVAC equipment, fans operate on high voltages and high currents. The electrical connections on motors should be safely protected in grounded enclosures. Never work on a fan motor without shutting off power at a disconnect within line of sight.

Any obstructions in the duct can prevent the air from flowing and create dangerously high pressures in the duct at the fan outlet. Remember that each inch of water pressure is equivalent to a little over five pounds per square foot, and that the pressure rise across a large fan may be rated at somewhere in the vicinity of ten inches W.G. at low flow rates. This would be the equivalent of over 50 pounds per square foot on the ducts, enough to bend or break some duct sections and fittings. On large fans it is a good idea to always include a sail switch (Figure 13.16) in series with the fan start circuit. A timing relay is used to delay the effect of the switch and allow air flow to develop, but if the sail switch does not record flow after, say, four or five seconds then the fan power is shut off. This also prevents the fan motor from running if the fan shaft is seized or the belts are broken. The sail size on these kinds of switches is adjustable so that you can use a bigger sail in variable flow air systems to prevent nuisance trips at low air flows.

Fans can create a tremendous force at both the inlet and the outlet. Of course, it's much better to be blown away from a fan at the

FIGURE 13.16 In-duct sail switch

outlet than sucked into one at the inlet. Be very careful when working close to a fan inlet. Air velocity around the fan inlet can change considerably in the last few feet or inches as you get near the inlet. As the air speed increases from a few feet per minute to hundreds of feet per minute, you can suddenly encounter unexpected forces that draw your hand or hair into the blades. This can be fatal. Also watch for loose pens, eyeglasses, or anything else that could become airborne easily. These objects can hit the fan blades and explode with incredible force.

All fan pulleys, motor pulleys, and belts should be covered with cowlings to prevent accidental contact (Figure 13.17). If you find a fan belt has come off the sheaves and is laying loose around the pulleys with the fan drive motor running, leave the area immediately and

FIGURE 13.17 View of double pulley safely encased in sheet metal cowling.

remove power to the motor at the disconnect or motor control center. Do not enter the fan cabinet until the motor has come to a complete stop and is locked out at the motor disconnect. As the motor slows after being shut off, the belt may become entangled in the motor pulley and whip around with enough force to take off a finger.

Always follow the manufacturer's instructions when replacing belts. Do not force belts over the sheaves to install them as you can internally damage the belts. This shortens the belt life and can pinch you pretty well. Always replace the belt guard before returning a fan to service. Sometimes in tight situations you find that the installing contractor could not fit the cowling on (Figure 13.18). Exposed belts and pulleys are incredibly dangerous and should always be covered.

fastfacts

Never exceed the maximum design speed of the fan. Fan blades can come loose from propeller fans, and squirrel cage fans can explode when operated in overspeed conditions.

FIGURE 13.18 An ideal situation for losing hands or feet.

If you must pull a pulley off a fan shaft, always lock-out and tag-out the fan motor. If the fan starts spinning once you have a good grip on the pulley with the puller, you can end up with a wheel of spinning clubs. If you are unfortunate enough to be holding onto the puller at the time, the motor can throw you a good distance. If you happen to be on a rooftop at the time, this can have some particularly unpleasant consequences.

Dampers

Damper actuators usually generate significant forces to close and open dampers—sometimes on the order of several hundred pounds of pressure. This can be the case on large outside air dampers that need to close tight. Needless to say, you should be very wary of such large forces. If you are working next to a damper and get caught, there may be no way for you to deactivate the damper. If you are inside a fan cabinet or air-handling unit, there may be no way for you to call for help, either. Remember that modulating electronic damper actuators can have a travel time on the order of tens of seconds, but pneumatic and spring-loaded dampers respond instantly to changes in the control signal. You may have no warning that a damper is going to close until your hand gets squashed.

TROUBLESHOOTING

Fans are relatively low on the list of troublemakers. Most problems with fans that don't start or that have low flow rates can be traced to the motor, the belts, and/or the bearings. If any of the fan bearings fail then the shaft can seize and stop turning. This can ruin the belts and burn out the motor. Fortunately, the bearings usually announce their impending doom by starting to chirp. This can give you enough time to replace the bearings.

Fan Doesn't Start or Has Low Flow

If the fan won't start and the electrical connections are all valid between the motor control center and the fan motor, there is a good chance the problem is a blown fuse or that the power is shut off. You can check for fuse continuity using a standard ohmmeter. Make sure the motor control center power is turned off before accessing the fuses and do not attempt to override the safety latch on the access door.

If a fan fuse keeps blowing, there may be a short somewhere in the line between the motor control center and the fan motor. Check the connections in the local disconnect as well as the wire nut connections within the motor housing that connect the motor to the line voltage. If these connections are mechanically and electrically secure, then a constantly blowing fuse may indicate that the motor is nearing failure. If the motor control center uses thermal breakers, check for the presence of variable speed drives anywhere on the system (not necessarily the fan motor). In some cases, VSDs can introduce harmonics into the main line voltage that can cause thermal breakers to trip repeatedly. This usually happens at certain VSD frequencies. If possible, identify those frequencies and set the VSD controller to avoid them.

fastfacts

The fan blade assembly can be extremely heavy. Be sure to properly jack up the shaft and follow all manufacturer's instructions when replacing the bearings. Better yet, avoid the problem by keeping the bearings lubricated correctly.

fastfacts

Never reapply power at the motor control center if you do not know why the power was turned off. Normally the motor control center would be tagged and locked for maintenance or other service reasons, but this step is sometimes overlooked. Never energize a circuit until you have verified that there is no threat to personnel or equipment.

If you have a low-flow or no-flow condition even with the motor running and the belts intact, the problem could be a slipping pulley on the fan shaft. Shut off the fan motor power and check to make sure the set screw holding the pulley has not come loose. If it has, and if the fan has been running for a long time like this, you probably have mechanical damage to the fan shaft.

If the pulley is okay, check to make sure that the motor and fan are in fact turning in the right direction. It is not uncommon after a motor replacement or any electrical work to find that the fan has not been connected correctly. Switching any two phases on a three-phase motor causes the motor to turn backwards. If the wiring appears correct and the motor is still turning backwards, check for other fans in the area that may be running backwards as well—the problem may be upstream of the motor control center.

Damper Does Not Operate

Damper motors fail more often than fan motors. If you have a damper that is not working or is not responding correctly, refer to the troubleshooting flow chart in Figure 13.19.

Note 1: Most electronic damper actuators are powered by 24 or 120 VAC power supplies. If the power is not available to the damper motor, it may not move (some actuators are spring-loaded to return to a NC or NO position upon loss of power). If there is a transformer used to provide the 24 VAC power, make sure that the proper connections are intact on both the primary and secondary sides of the transformer. Also, many transformers have multiple winding configurations, allowing for a wide variety

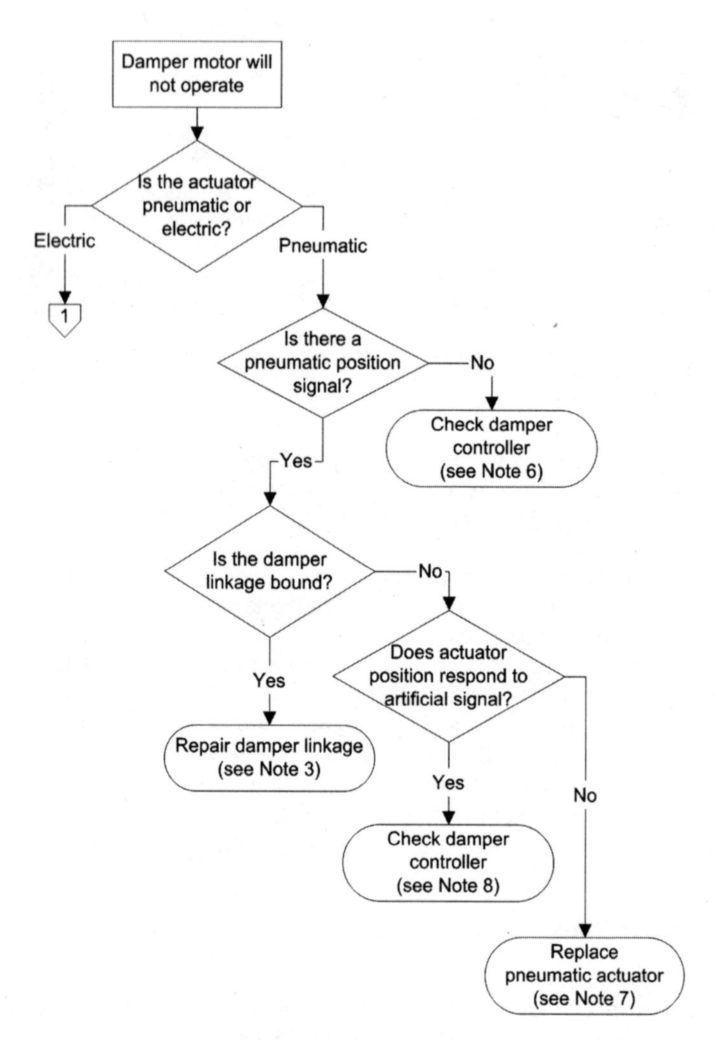

FIGURE 13.19 Fan damper troubleshooting flow chart.

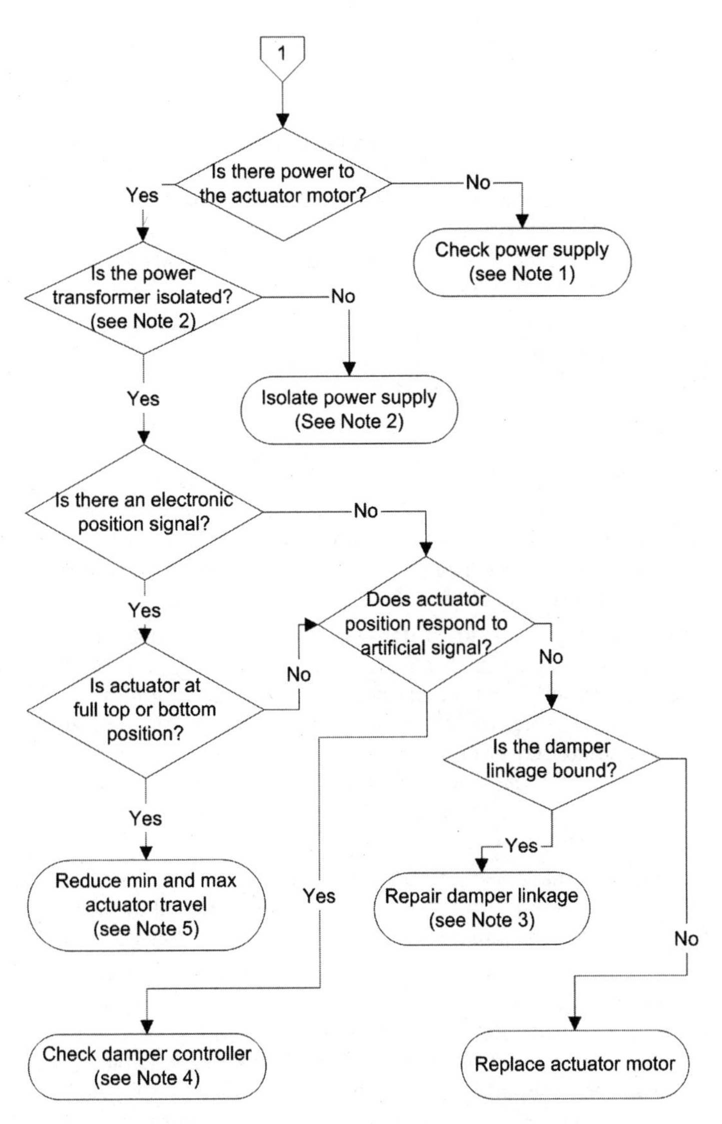

FIGURE 13.19 *(continued)* Fan damper troubleshooting flow chart.

of output voltages. If the transformer is not providing the correct output voltage, you may need to re-wire the transformer depending on the input voltage.

Note 2: Many electronic actuators rely on isolated power supplies in order to work properly (see Chapter 4 for additional information on isolated power supplies). If the actuator is powered by a transformer that also is connected to other actuators or is referenced to earth ground, then it is probably not isolated. A good rule of thumb is one transformer per actuator.

Note 3: The damper linkage can bind up if the linkage is rusty, bent, or at a minimum or maximum extreme. Examine the linkage for binding and repair or replace as necessary.

Note 4: If the actuator has a valid power source connected, you can check to see if the actuator responds to a control signal by connecting a 1.5 or 9 VDC battery to the electronic signal inputs. Disconnect the existing signal wiring first and make sure that you connect the positive side of the battery to the positive side of the input signal terminals. You should be able to drive the actuator from one end to the other if the signal is in milliamps or volts.

Note 5: Under some circumstances, an electronic actuator motor can run the actuator shaft off the threads of the electronic motor. This can happen when the actuator shaft is at its minimum or maximum extreme. You can usually inspect this visually by taking off the actuator cover or by gently forcing the shaft up or down until the gears engage again. If this is the problem, you may need to reposition the motor on the actuator shaft or (if available) by setting the extreme limits.

Note 6: Pneumatic dampers do not operate unless there is a control signal of sufficient pressure to drive the actuator. Check the signal pressure and make sure it is somewhere in the 3 to 20 psig range.

Note 7: You may be able to see if the actuator can respond by forcing a signal using a hand pump or equivalent (note: you generally cannot generate the necessary pressure with your lungs). If the actuator does not respond to a signal, you may be able to open up the cover and check for holes in the diaphragm or blockages in the ports. If you cannot find any obvious problems, you should replace the actuator.

Note 8: If the actuator responds to an artificial signal (as discussed in Note 7) then chances are there is no problem with the pneumatic actuator. You should check the controller and associated sensors to make sure they are operating correctly.

PRACTICAL CONSIDERATIONS

Most fans and dampers are contained within other pieces of equipment like air-handling units (Chapter 14) or zone terminal boxes (Chapter 16). Information specific to this type of equipment is presented in those chapters.

Measuring Air Flow

If it is necessary to measure the airflow rate downstream from a fan, be sure to use the techniques discussed in Chapter 5 on air flow sensors. The standard advice is to measure the air velocity using a duct-averaging technique in a straight section of the duct many feet downstream and upstream of duct elbows and take-offs. Unfortunately, in most buildings such sections of ductwork are purely mythical. You usually end up taking the duct flow measurements wherever you can get access to the duct and in a location where you can fit a pitot tube or pitot tube array. There are definitely locations you should avoid, however, like anywhere near the outlet of a centrifugal fan. The flow profile can be so non-uniform just downstream of the fan (see Figure 13.20) that the readings will be erroneous even with an averaging array.

Energy Conservation

The modulation of air flow in variable air volume systems is essential to maintaining the duct pressure and the ability of the mixing boxes to operate correctly. While the cost of a variable speed drive may seem like a lot, it will be cheaper in the long run. Consider the

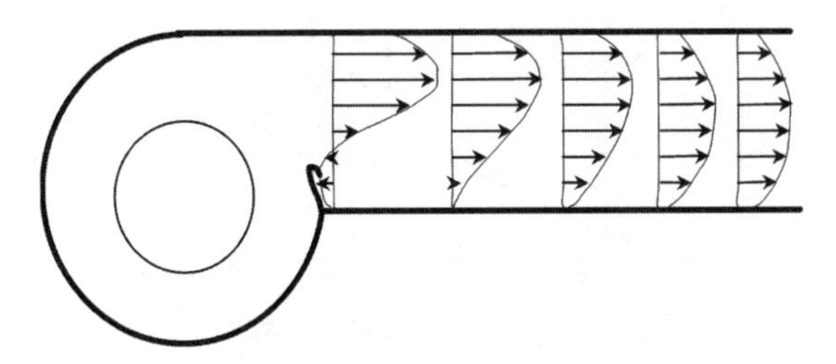

FIGURE 13.20 Flow profile downstream from centrifugal fan.

comparisons shown in Figure 13.21. The graph shows how the power consumption of a fan changes according to the air flow through the fan and the flow control technique. Obviously, if the fan is running at 100 percent flow then it doesn't matter what kind of flow control is being used, but when the fan is running at about half the rated flow, a fan with outlet dampers uses about 75 percent of the rated power while a fan with a variable speed drive uses only about 30 percent of the rated power.

To make this comparison more useful, suppose you installed a 40 HP fan (that's about 30 kW) that runs at half speed for 1000 hours per year. Using a variable speed drive instead of outlet dampers would save

$$1000 \text{ hours} \times 30 \text{ kW} \times (75\% - 30\%) = 13500 \text{ kWh per year}$$

A typical cost for commercial electricity is around five cents per kWh, so the savings would be about $675 per year. Even if the variable speed drive cost $3,000, the payback time is less than five years.

Keep in mind that installing a VSD on a system that runs close to or at 100 percent all of the time will actually cost you more money to operate. Before installing a VSD, make sure that the fan isn't already undersized for the load.

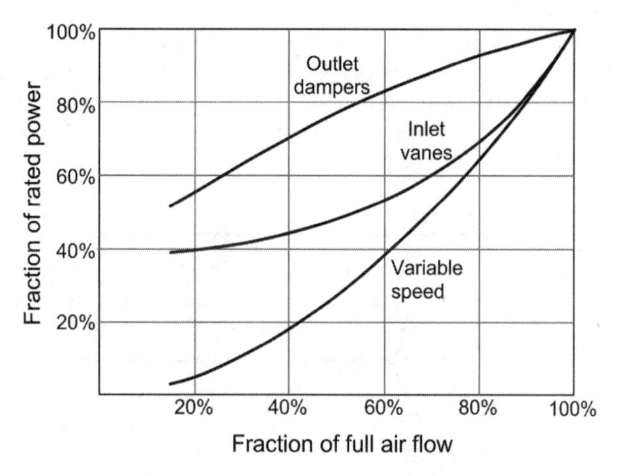

FIGURE 13.21 Energy use versus flow for various fan flow control methods.

chapter 14

AIR-HANDLING UNITS

Air-handling units (AHUs) contain fans, coils, filters, and other components that are used to condition one or more airstreams. AHUs are either built at the factory and delivered intact, or are built up at the job site. For more information about the fan and damper components, see Chapter 13. For more information on entire air distribution systems, see Chapter 15.

THEORY OF OPERATION

There are different types of air-handling units used for different types of air distribution systems. Most units draw in some outside air and mix it with return air from the building so that the outside air does not need extensive conditioning. Some air-handlers, however, use 100 percent outside air. These *make-up air units* are usually used for controlling the building static pressure. *Constant air volume* (CAV) systems provide air at a constant rate into the ductwork. The fan or fans run at full output at all times. *Variable air volume* (VAV) systems vary the airflow rate to provide only as much air as is necessary to the building. The airflow is modified using inlet vanes, outlet dampers, or variable speed drives on the fan.

A *single-duct* unit (Figure 14.1) has a single outlet duct that supplies cool air at around 55°F. The zone terminal box provides any reheating of the air if the zone is too cold. A *dual-duct* unit (Figure

14.2) supplies warm and cool air to two different ducts. The terminal boxes at the zone then selectively mix the warm and cool airstreams to achieve the desired zone temperature. The figure shows each duct with its own fan. Dual-duct systems also can use a single fan to provide the airflow in both ducts. In this case, a set of dampers is required at the inlet to each duct to control the duct pressure and airflow.

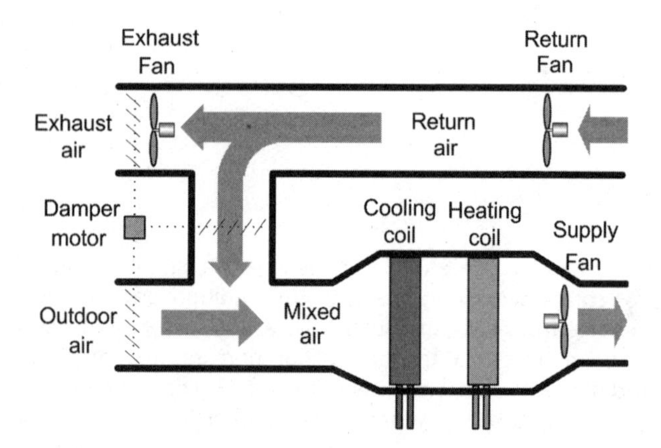

FIGURE 14.1 Single-duct air-handling unit.

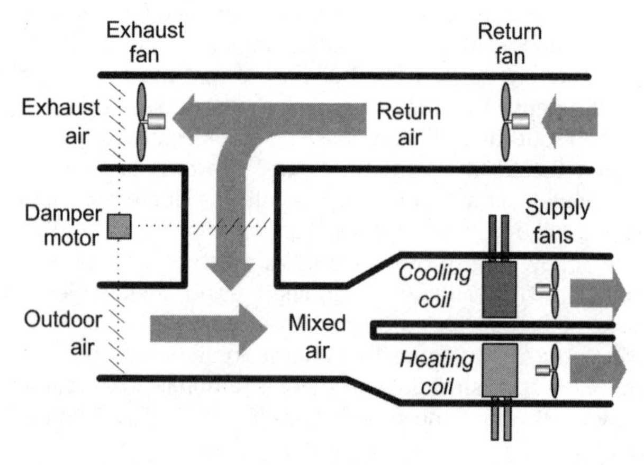

FIGURE 14.2 Dual-duct air-handling unit.

A *multizone* system (Figure 14.3) involves many different outlet ducts, one for each zone satisfied by the AHU. The mixing of hot and cold air occurs at the air-handling unit instead of at the zones. The dampers on each duct are controlled by the thermostat in the zone served by that duct. This kind of system is used for applications such as large, reconfigurable zones such as ballrooms in hotels. The practical upper limit is about a dozen zones per multizone AHU. In practice, multizone units can be quite complex. For example, some multizone units may have a separate heating or cooling coil in each of the ducts leading to individual zones. The configuration depends on the type of zones being controlled, the amount of space available for HVAC equipment, the initial equipment budget, etc.

A special class of air-handling unit is the *rooftop unit* or *packaged unit*. This kind of equipment is very common on large retail buildings, warehouses, and small commercial buildings. The rooftop unit is a self-contained AHU that uses direct expansion cooling and gas heating and usually provides air directly to the zone being conditioned, although it is not uncommon to also use some kind of terminal box. Figure 14.4 shows a schematic of a typical rooftop unit showing the different sections devoted to moving the air and to providing cooling. "Large box stores" and supermarkets may use two dozen such units to provide air conditioning to different sections of the building.

A variation of the rooftop unit is the split system as shown in Figure 14.5. Here the condenser coil of the refrigeration equipment

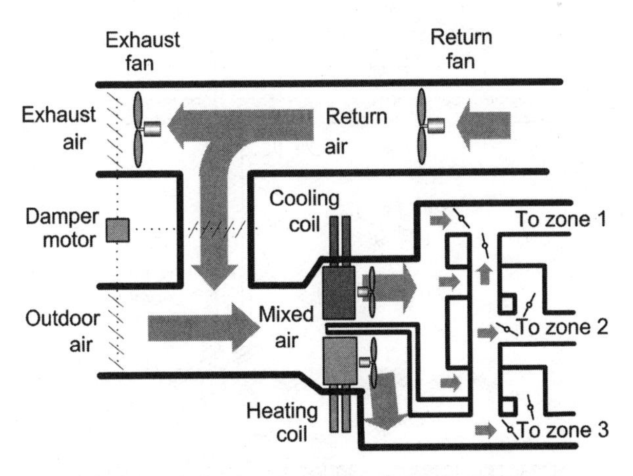

FIGURE 14.3 Multizone air-handling unit (*unit shown serves three zones*).

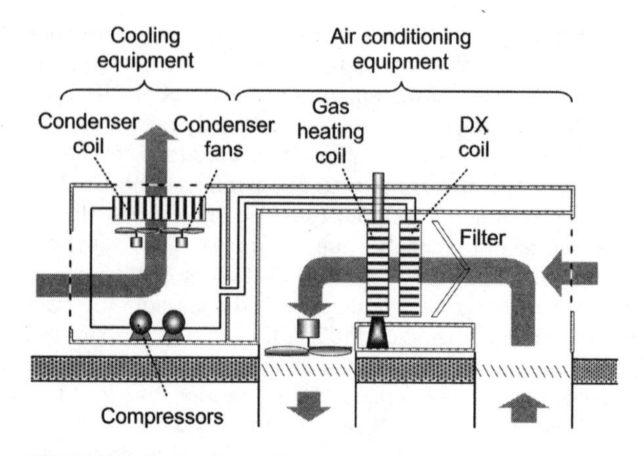

FIGURE 14.4 Rooftop unit.

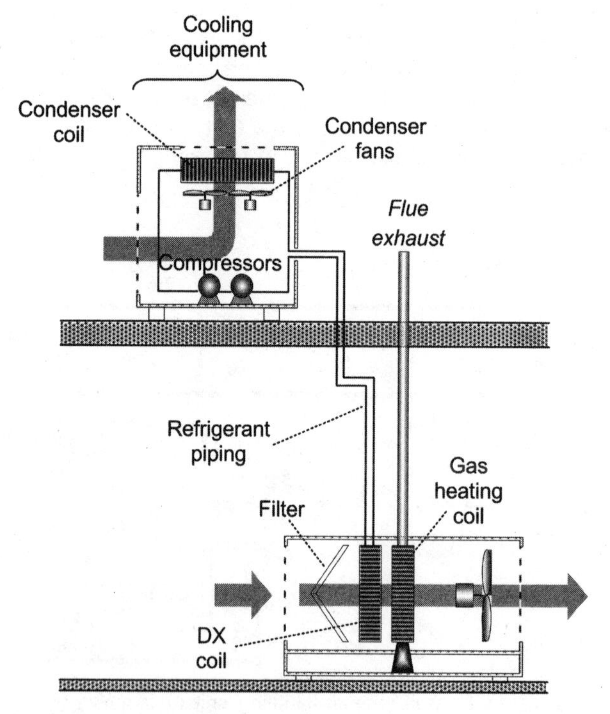

FIGURE 14.5 Split system.

stays on the roof but the air moving section is inside the building. As with other types of AHUs, the final configuration of the equipment depends on the amount of space available, the allowable roof loading, and the distance between the potential rooftop locations and the conditioned zone.

The size of an air-handling unit varies greatly depending on the amount of air passing through the unit. Those that handle less than 5000 CFM are no more than large fan-coil units, often with barely enough room to change the filters. Medium-size units (up to about 20,000 CFM) are the size of cars. Both small- and medium-size units should be mounted on concrete pads with vibration-absorbing springs (Figure 14.6) to minimize the amount of noise they can transfer to the building. Some units are quite large (Figure 14.7) with full-size doors for access during maintenance.

Filters

It is important to filter the air before it enters the coils on the air-handling unit. It is much easier and cheaper to periodically replace filters than it is to clean or replace fouled coils. There are three broad categories of air filters:

- *Fibrous media unit filters* are modular filter panels that are replaced or cleaned when full.
- *Renewable media filters* continuously introduce new filter media into the air stream and remove the old, dirty media.
- *Electronic air cleaners* or *electrostatic precipitators* apply a high voltage between two parallel metal screens that sweep dust out of the air stream.

The efficiency of a filter indicates the percentage of the particles that it removes from an air stream and depends on the type and size of the particles to be filtered. Ideally, this should be 100 percent, but in normal HVAC applications is usually 60 to 80 percent. The exception is in clean rooms and medical facilities where special filter systems are used that have efficiencies greater than 95 percent. The performance and pressure drop across fibrous media unit filters increases as they fill up with dust while the other two filter types perform with essentially constant pressure drop and efficiency.

There is a wide variety of filter types. *Viscous impingement filters* are usually flat panels made of glass fibers or metallic wool that are coated with a viscous substance to catch the dust particles. This type of filter has a low pressure drop, low cost, and a high efficiency for

FIGURE 14.6 AHU vibration mounting.

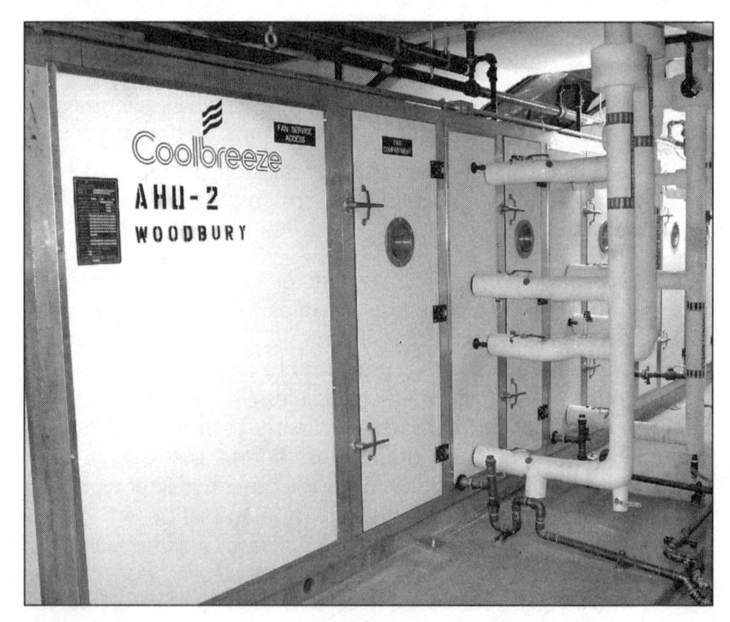

FIGURE 14.7 Large air-handling unit showing access doors.

fastfacts

Sometimes it's useful to combine filters of different types, for example, an inexpensive low-efficiency filter upstream of a more costly high-efficiency filter. This decreases maintenance and replacement costs on the high-efficiency filter.

lint but a low efficiency for dust. Another type is the *dry media filter,* where fibers are much more closely spaced to form a dense mat, without adhesive on the fibers. Their efficiency can be significantly higher than that of viscous impingement filters. These are the types of filters most often used in air-handling units. If especially clean air is required, you should use a high efficiency particulate air (HEPA) or ultra low penetration air (ULPA) filter. These are made of very small glass fibers woven into a paper-like material.

Filters have a specific direction that they must face in order to work properly. Each filter usually has an arrow indicating the direction of air flow.

Most large air-handling units have entire rooms devoted to the filters to make access easy (as shown in Figure 14.8). This is essential when replacing the filters. To avoid the overloading of unit filters they should be inspected periodically or even better, a pressure gauge should be installed (Figure 14.9) to indicate when they need replacing. Most dry media filters have a pressure drop anywhere from 0.05 to 0.25 in. W.G. when they are clean. As the filters capture more and more dirt, the pressure drop increases. Once they reach about 0.5 to 0.8 in W.G. they should be replaced. If you allow the filters to continue to accumulate dirt, the pressure drop starts affecting the fan efficiency, and you may even start collapsing the filters. Once this happens, not only is no filtration occurring but you may also start adding dirt to the air stream from the collapsed filters.

Also, local codes may dictate that you must have fire sprinklers on one or both sides of the filter bank.

Dampers

Most air-handling units incorporate linked return air and outdoor air dampers. The idea is to control the amount of outdoor air that enters

FIGURE 14.8 Filter bank.

fastfacts

A large hotel relied on district heating and cooling for hot and cold water used in the hotel's air-handling units. They also hired the central plant personnel to maintain the air-handling units. Then the hotel owners decided to save money by hiring their own mechanics for the maintenance. This worked for a few years until the hotel lost all its cooling right in the middle of a 3000-person conference taking place at the hotel. Apparently, the hotel's mechanics had removed all the filters from the air-handling units (to save money, of course) and now all of the air-handling unit coils were hopelessly clogged with dirt.

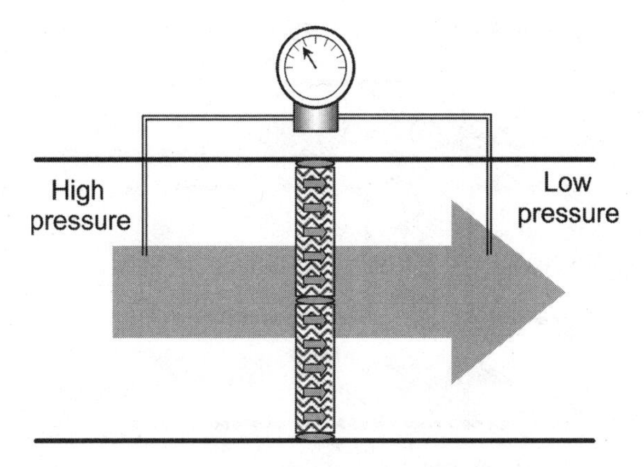

FIGURE 14.9 Measuring pressure drop across the filter bank.

the unit and to make up the difference with building return air. Figure 14.10 shows the typical dampers and upstream fans on air-handling units. Note that the exhaust fan and return fan are not necessarily components in the air-handling unit (nor, for that matter, will you usually have both—just one or the other).

There is a fine balance on the air pressures required to make sure the air flows just as it is supposed to. If the pressure in the return duct is too high, you may end up over-pressurizing the mixed air plenum and preventing outdoor air from entering the building. There is an easy way to determine the amount of outdoor air flow into the building if you can measure the return air temperature, the outdoor air temperature, and the mixed air temperature. The relationship is

$$\text{Fraction of outdoor air} = \frac{T_{RA} - T_{MA}}{T_{RA} - T_{OA}}$$

where T_{RA}, T_{MA}, and T_{OA} are the temperatures of the return air, mixed air, and outside air, respectively.

Coils

Air-handling units have any number of coils, depending on the application. There may be freeze protection coils—usually steam but

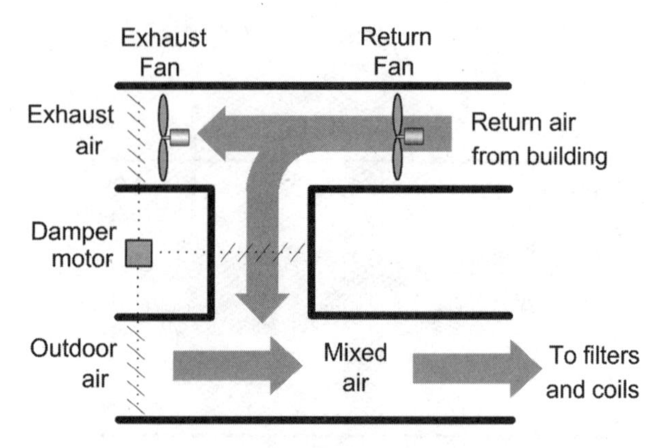

FIGURE 14.10 Air-handling unit dampers.

sometimes electric—on the outdoor air inlet, cooling coils, and heating coils. In a single-deck system, the cooling coil comes first to provide dehumidification and then the heating coil can provide reheat. It is good practice to use a three-way valve as illustrated in Figure 7.10 to allow for constant flow through the coil loop while still varying the amount of flow through the coil itself. If the coil is very large (greater than about 500,000 Btu per hour) you may want to consider using a pumped coil (see Figure 7.12) that has the advantage of always having water flowing through the coil and better outlet temperature control. The three-way valve is used to change the temperature of the coil water loop. Pumped coils offer better freeze protection since there is always flow through the coil, but the initial cost is higher because of the extra hardware and pump.

fastfacts

If the occupants complain about the indoor air quality, check to make sure that the air direction at the outdoor air dampers is into the building, not out of the building.

Humidifiers and Dehumidifiers

In particularly dry climates you can take advantage of evaporative cooling. This also humidifies the air. Evaporative cooling is discussed in detail in Chapter 17. You can also humidify with air sprayers or steam humidifiers.

Dehumidification is almost always done with the cooling coil in an air-handling unit. There are other methods, such as desiccant adsorption systems (see Chapter 15) but that would not be considered part of the air-handling unit. For example, consider an air-handling unit with 100 percent outside air in the summer. The air entering the coil is typical of summer conditions: 95°F and 40 percent relative humidity. As the air passes over the cooling coil, it is cooled to about 67°F before water starts to condense on the cooling coil fins (Figure 14.11). The coil continues to cool the air but at the same time removes water from the air. Saturated air leaves the coil at about 47°F and is then heated to the desired 55°F and 75 percent humidity. The cooling and dehumidification path is shown on the psychrometric chart in Figure 14.12.

The cooling of just the air without removing water is called *sensible* cooling. That is, you can sense the reduced temperature. The removal of the water is called *latent* cooling. Most building cooling

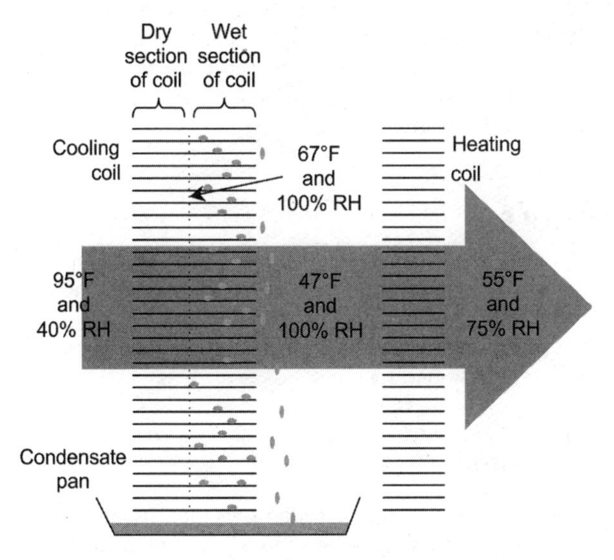

FIGURE 14.11 Cooling and dehumidification with reheat.

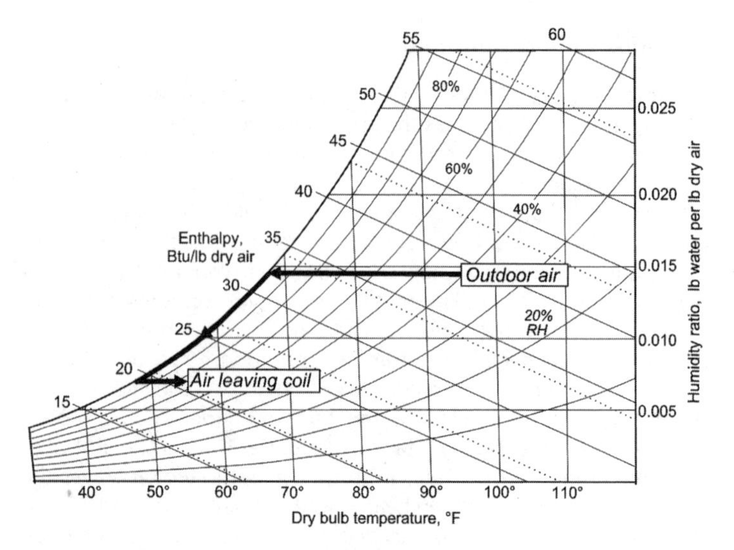

FIGURE 14.12 Cooling and dehumidification path on psychrometric chart.

design calculations report the *sensible cooling ratio,* which is just the ratio of the sensible cooling load to the total cooling load. In very humid climates the sensible cooling ratio can be around 0.5, meaning that half the building cooling load is just removing water from the air stream. On other words, you can't just design the cooling system based on air temperatures alone—you must include the humidity to find the latent load.

CONTROLS

As there are many different types of air-handling units, there are also a large number of different yet appropriate control techniques that can be applied. The control of the AHU depends on the application that they were designed for.

Damper Control

The return air and outside air dampers are usually linked and opposed. The two air streams are mixed to meet a mixed air temperature setpoint. When the building is occupied the outside air damper has a minimum position setting so that at least 20 percent of the total supply air is fresh outside air.

Ideally the mixed air temperature in the air-handling unit is the same as the supply air temperature. This way the coils in the air-handling unit do not have to operate. Of course, this rarely is the case and the dampers are modulated to maintain some mixed air setpoint temperature. As the outdoor air temperature drops or rises, the dampers are modulated to maintain a constant mixed air temperature. If the outdoor air temperature is close to the desired supply air temperature then the exhaust and outdoor air dampers are fully opened and the return air damper is completely closed. Under these conditions, 100% outside air enters the building and the chiller can be shut off because there is no need for cooling the air (Figure 14.13). This is called *economizer mode.*

Strictly speaking, the economizer mode should be used whenever the enthalpy difference between the outdoor air and the supply air is less than the enthalpy difference between the return air and the supply air. However, it is difficult to measure enthalpy so usually just the air temperatures are used. This is okay to do in a dry climate, but in a humid climate you can actually increase the building cooling load. The economizer mode is sometimes used at night to cool off a building mass in preparation for the next day's cooling load. This is called *night purging.*

When the building fans are shut off, the outside air and exhaust air dampers close fully. During the morning warm-up or cool-down

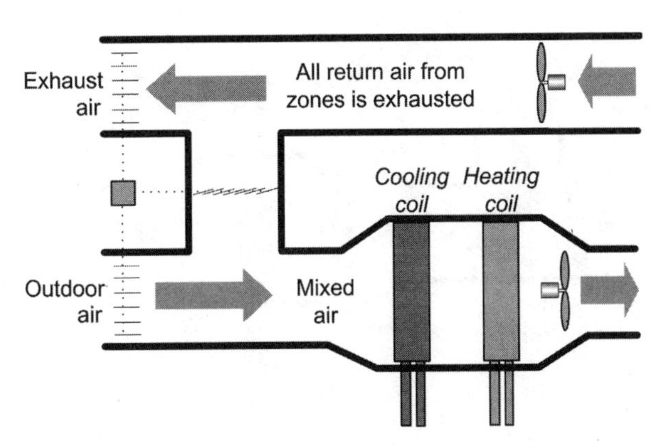

FIGURE 14.13 Air-handling unit showing economizer mode.

period (when the building is unoccupied and there is no need for fresh air) these dampers remain closed to help the building reach the setpoint fast.

Temperature Control

The air-handling unit controls the outlet temperature, humidity, and pressure and airflow rate. In single-duct systems, the outlet air temperature is usually maintained between 50° and 60°F. In dual-duct and multizone systems, the cold air side is between 50° and 60°F and the hot air side is typically between 80°F and 100°F.

In cold climates where the outdoor air temperature may go below freezing, a preheat coil is usually present and would be activated to prevent water in the cooling or heating coils from freezing. Figure 14.14 shows an air-handling unit with a steam preheat coil. In freeze protection mode, for example, when the outdoor air temperature is below 35°F, it may be necessary to turn on all water pumps and force coil valves open. This ensures that there is coil flow and hopefully prevents coil freezing. A combination of electronic and pneumatic sensors and controllers are used to maintain the air-handling unit outlet temperature (Figure 14.15).

In freezing climates, always include a *low limit control* or *freeze stat*. This can be located before the cooling coil (see Figure 5-2) or downstream of the preheat coil. The low limit control should be hard-wired to the fan motor starter so that if the temperature of the

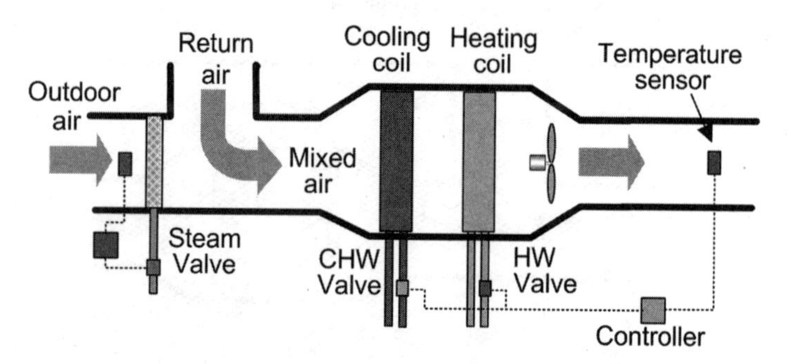

FIGURE 14.14 Typical air temperature controls in air-handling units.

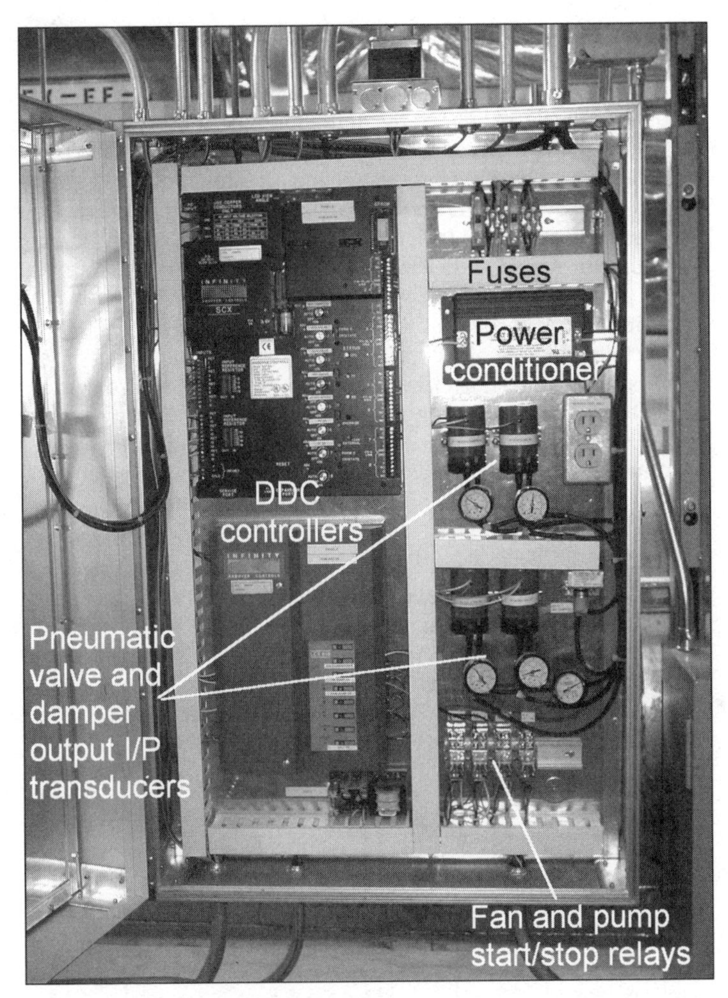

FIGURE 14.15 Fan temperature control unit.

air flowing through the coil falls below 35°F the fan shuts off and the outside and exhaust dampers close. This protects the heating and cooling coils from freezing. On larger coils use two or more low limit controls wired in series to cover as much of the face of the coil as practical. Avoid any small gaps around the coil, as the bulb may sense cold air and cause nuisance trips.

The coil valves on an air-handling unit can be controlled by individual feedback loops, supervisory controllers, parallel controllers, or slave/master valves. For example, Figure 14.16 shows how a single pneumatic signal can be used to control both the cooling and heating coils. By selecting the proper spring range in each actuator (and using a combination of direct and reverse acting actuators) you can have the cooling coil move from minimum to maximum cooling as the air pressure signal increases, and then the heating coil takes over moving from minimum to maximum heating on air pressure decrease. Note that this kind of controller assumes that no reheat is ever required to account for cooling coil subcooling for moisture removal.

Careful placement of the mixed air temperature transmitter or sensor is required to maintain the proper mixed air temperature. Averaging bulbs (Figure 14.17) are normally used for this application in order to read an average of the mixed streams of outside and return air temperatures. One of the problems that can occur is stratification of the air streams in the air-handling unit when cold outside air and warm

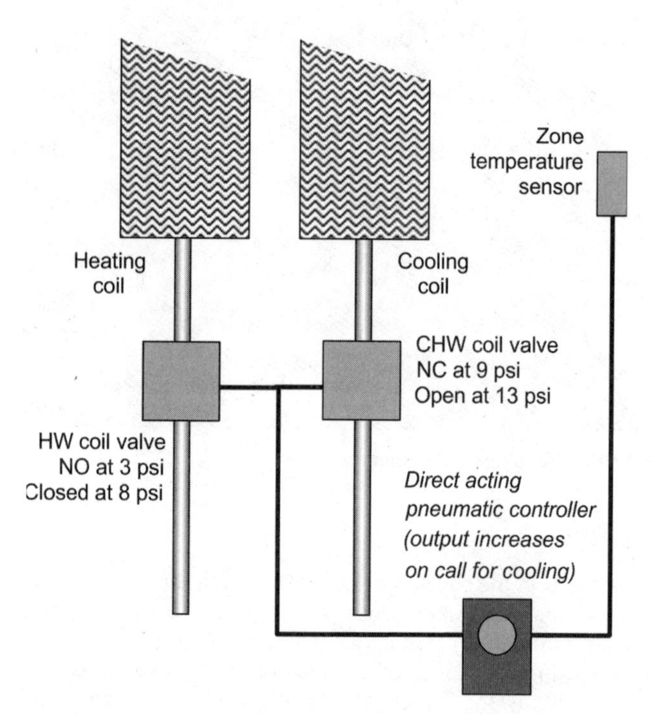

FIGURE 14.16 Using single temperature sensor to operate two valves.

fastfacts

Do not wire a safety through the building automation system. The low limit should be the last line of defense to avoid a frozen coil. Be sure to include some kind of deadband between the limits of the two coil controllers so that you do not have simultaneous heating and cooling.

FIGURE 14.17 Mixed air temperature sensor.

return air do not mix as they should. This is illustrated in Figure 14.18. One way to solve this problem is to use air mixers like those in Figure 14.19 that force turbulent swirling and mixing of the two air streams. The air mixers are usually located upstream of the filters in the mixed air plenum, but can sometime be placed after the filters.

SAFETY

Rotating shafts, motors, belts, and fan blades can be lethal if not paid attention to. All fan belts and pulleys should be protected with a cowling. The cowling generally allows you to inspect the pulleys and belts, but prevents accidental contact and can contain the belt in case it breaks and comes flying off.

Always lock-out and tag-out the fan power when working on motors, belts, and bearings. This ensures that the motor disconnect is locked off and the fan cannot be turned on without removal of a lock that only you have the key to. It is possible that even though lock-out/tag-out procedures are used, the fan and motor may continue to rotate due to the "windmill effect" from air flow through the fan. Other fans that are connected to the ductwork and are still operating usually cause this. When working on a fan beware that this condition may exist, as a fan that was at rest may suddenly start to rotate.

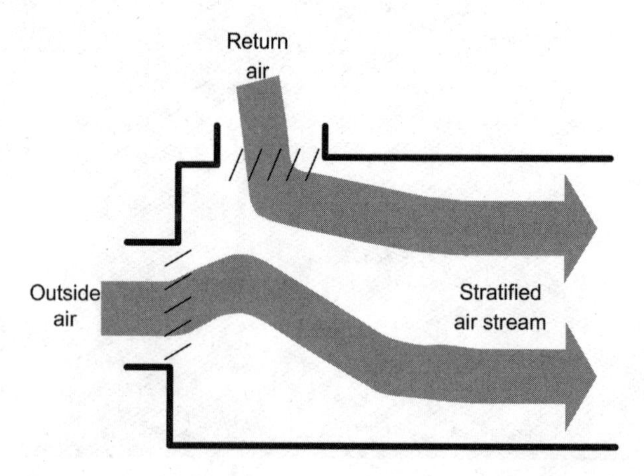

FIGURE 14.18 Stratification in air-handling unit air stream.

FIGURE 14.19 Upstream view of air mixers in mixed air plenum.

Most fan motors are three-phase 230/460 VAC. Always properly ground the motor chassis and air-handling unit body to prevent accidental electrocution in case of a line short. The action of the motor and fan causes the air-handling unit body to vibrate. This can cause electrical connections to work themselves loose and can even, over time, cause enough abrasion on wire insulation to create direct electric shorts to the conduit or motor body.

Most air-handling units have at least a hot water coil and many have steam coils. These pipes can be uninsulated around the valves and unions that connect the pipes to the coils. Temperatures of the exposed metals can run in excess of 120°F and can cause severe and immediate burns.

Outside air enters the air-handling unit all year round. This is necessary to maintain the air quality inside the building. If you are working in the mixed air or outdoor air intake plenums, be aware that you may be exposed to a constant wind of cold temperature air. Under the right conditions, this can cause cold burns and frostbite.

fastfacts

The windmill effect is not only dangerous to personnel, but can actually cause equipment failure. If the fan is turning fast enough in the wrong direction when the motor power is switched back on, the belts, pulleys, and motors can be damaged.

The outlet side of cooling coils usually has catch pans to contain any water that condenses on the coils or becomes entrained in the air stream and drops out away from the coil (Figure 14.20). These pans can fill up with water and if the conditions are right can end up looking like a biology experiment gone bad. The pans should be kept clean and possibly even treated to prevent growth of mold and mildew. Be sure to use relatively benign treatment since the pan is in full contact with the air that circulates throughout the building. Treatment similar to that used in evaporative cooling pans should work fine.

Heating and cooling coils consist of many thin fins of metal (Figure 14.21). These are generally sharp enough to cut like a knife. Unfortunately, there is often very little room to work in between these coils and it is very easy to slide your hand or arm along the face of the coil. Always wear long sleeve shirts and even gloves when cleaning or straightening coil fins.

The floor areas around large fans may by slippery due to any excess grease from the fan bearings that may have leaked out or sprayed out onto the floor. It seems like a bit of a cruel joke: make the floor slippery around a large, rapidly rotating fan. But it is not a joke—be very careful that you do not slip and, if you do, do not grab onto rotating shafts, fan cages or motors to keep from falling down. This problem is often compounded by the low-level or non-existent lighting in the fan rooms. You can't avoid the grease puddles if you can't see them.

On large air-handling units with doors or duct-access hatches, always be aware of the pressure effects on your ability to exit any spaces you may enter. One inch of water pressure is a little over 5 pounds per square foot. Each inch of water pressure difference on a standard size door can therefore exert more than 100 pounds of pressure keeping that door closed. It is entirely possible to have fan cabinet doors slam shut due to negative or positive pressure that is

FIGURE 14.20 Catch basin for water drops entrained in cooling coil leaving air.

FIGURE 14.21 Coil fins.

fastfacts

The pressure inside a mixed air plenum may be enough to pull in bees, wasps, and other unpleasant surprises without letting them leave. They tend to get angry when this happens.

so strong you can be trapped inside. Combine that with poor lighting levels, sharp coil fins, and strong winds and you could quickly make the local news for reasons that you would prefer to keep private. Keep in mind also that the air velocity increases dramatically as you approach the fan inlet. Airflow at 500 feet per minute (about 6 miles per hour) may suddenly become a few thousand feet per minute at the inlet vanes or ductwork. While this is usually not enough to drag a person, it is enough to grab pens, glasses, and small hand-held devices. If these encounter the rotating fan blades they can shatter explosively in your face.

TROUBLESHOOTING

The majority of problems with air-handling units are associated with the control of the unit. Failure of individual components (for example, valves, fans, and dampers) are discussed in other chapters. This section talks more about the failure of controls on the air-handling unit itself. The key point to remember is that many control problems with an air-handling unit may not be the fault of the control system. If an AHU has been operating for years without problems and suddenly has a major "control" problem, it could very well be that the solution is simple. A clogged filter, loose belt, motor phase loss, or bearing problem is most likely the cause. Remember that the control system only controls the air-handling unit and does not in any way heat or cool in itself. If other parts of the HVAC system such as piping, boilers, chillers, pumps, and electric power are not functioning correctly, then the air-handling unit will not function correctly.

It is not unusual for a well-designed air-handling unit to operate for many years with normal maintenance and care. Yet for some unknown reason in many instances when a problem does occur with an air-handling unit after many years of faithful service, the first thing that seems to happen is the need to dig into (and sometimes rip out) the control system to "fix" the problem. Many dollars and days later the problem is still there or new problems have been created. As always, with any problem, you should check the most obvious first. If your car quits running you would automatically check the gas gauge before working on the engine. The same logic follows here. A systematic approach to solving HVAC problems is usually the fastest and least costly in the long run. Many problems occur because of sloppy or incomplete initial installation practices.

Mixed Air Temperature (Too Hot or Too Cold)

It is pretty common to find that the mixed air bulb was placed in the most convenient location and not the best location in the original installation. Most control prints show the placement of the bulb at or downstream of the mixed air streams. If an air-handling unit has difficulty controlling its mixed air temperature and the controls are functioning correctly, then investigate the location of the mixed air transmitter or sensor bulb. Relocate the bulb if needed to reflect the average temperatures of the two mixed streams of air. In some cases the ductwork or dampers may need to be modified or baffles installed to prevent stratification of airflows or other improper air mixing.

fastfacts

A brand new AHU with 100 percent outside air and pumped hot water coils was still within the warranty period when the fan began tripping off in cold weather. The control company technician told the building owner that the only way to fix the unit was to replace the existing pneumatic three-way hot water control valve with an expensive electronic control valve with its own controller. The cost for this replacement was quoted at $4000. A simple relocation of the existing temperature transmitter on the coil and a retuning of the existing controller fixed the problem. This solution cost $100. The AHU has since worked without any similar problems for over 12 years.

fastfacts

There are many ways to test the air flow through louvers such as pitot tubes and hand-held anemometers. In a pinch, however, you can get a quick indication of the air flow direction by holding a tissue up to the damper itself and watching which way it blows.

If the bulb appears to be in the correct location, the problem may be due to incorrect pressurization of the mixed air plenum. If there is an exhaust fan that is running too high, it is possible that outdoor air is entering through the louvers and then heading backwards through the return air dampers and right back out of the building. A similar circumstance can occur with a return fan that is blowing too hard: the air can flow the right way through the return air dampers and then backwards through the outdoor air louvers. In both cases, outside air never enters the building, possibly leading to indoor air quality problems.

The direction and speed of wind can affect the amount of outside air entering the building. If the wind is blowing directly against the louvers it forces more air into the building than you would expect for a given damper position. If the louvers are on the lee side of the building then a negative pressure can be created and the effect is as if the return fan is running too high: air is drawn out of the mixed air plenum through the outdoor air louvers.

AHU Supply Air Temperature is Not at Setpoint or Varies Too Much

There are three places to look if the supply air temperature is not at the setpoint: the sensors, the coil capacity, and the controller. The first thing to check is the leaving air temperature sensor. Make sure this sensor is properly located and is responding as it should. If the sensor is electronic, check the power supply and output signal. Also check the signal at the controller to make sure there is no break in the control signal between the sensor and controller.

Another easy check is to see if the coil is at capacity. The first thing to look for is an isolation valve or balancing valve that has been closed—perhaps for recent maintenance—and is preventing water from reaching the coil. Also make sure the control valves are opening as they should. If the valves appear okay, you need to check the performance of the coils. Measure the inlet and outlet air temperature and the air flow rate. Use the method described in Chapter 3 to determine if the coil is working properly or if the capacity is lower than what it should be. If the coil is at capacity, consider increasing the hot water temperature or decreasing the cold water temperature. If the coil does not seem to be giving as much heating or cooling as it should, the coil could be fouled on the air side or water side. Check the filters to make sure they are clean and then check the coil fins for dirt or grease. If the fins appear okay, then you probably have a fouled water side which requires dismantling of the coil for cleanout.

Before dismantling the coils, however, you can do a final check to see if the controllers are not too sluggish or causing too much overshoot. If the coil is too sluggish, try doubling the proportional gain to see if the coil reacts faster. If the coil outlet temperature cycles too much, try increasing (slightly) the integral gain. Keep in mind that coils in larger air-handling units that do not have coil pumps may be hard to control due to hot or cold areas in the coil face. The addition of a coil pump with a two-way valve or three-way valve (as shown in Figure 7.12) may solve the air temperature control problem. Sluggish control action in a large non-pumped coil may indicate the above problem and not actually be a control problem.

Airflow Too Low / Fan Doesn't Start

Low airflows through an air-handling unit are almost always caused by some kind of obstruction in the air stream. In many cases the cause is obvious: a piece of insulation or a filter panel has come loose and is covering up the face of the coils. Check the filters and coils to make sure they are not clogged with dirt. Don't just check on the upstream side even though this is the easiest place to look. Excess grease from the fan bearing may catch in the fan blades and coils and get clogged with dirt. This reduces the airflow through the unit.

In direct expansion systems, a failure of the compressor control circuit can cause the cooling coil to ice up. In some cases the ice can get so thick that no air flow through the coil is possible. Don't just let the ice melt and restart the unit: if there is a problem with the defrost cycle or the compressor control, the problem will just happen again.

A reduced airflow may also be caused by a problem in the motor. Use a clip-on ammeter to check the phases of a three-phase motor to make sure all have current. It is possible to have lost a fuse on just

fastfacts

If a fan is operating with an unusual sound or vibration, perform an inspection to determine the cause as soon as possible to avoid major damage to the fan bearings, shafts, belts, pulleys or drive motor. It is much cheaper to fix a minor problem before it becomes a major one.

one or two phases, reducing the motor power. Some motors have a starting winding and a running winding. The relay that switches between the two windings can fail, causing the motor to run in the start winding all the time. If the motor is multi-speed, make sure that the correct speed setting is used and that any automatic switching relays are working.

Finally, check the fan blades, belts and bearings for any accumulated grease, dirt, oil, belt wear, belt tension, broken blades, etc. Usually if the fan bearings or belt are starting to go you will hear it. If there is a remote fill port for the oil, in some cases the flexible tube may become disconnected from the bearing, resulting in bearing damage due to lack of lubrication.

If a fan simply refuses to start, you can run the power and fan diagnostics listed in Chapter 13. If the motor checks out okay, you probably have a safety control that is preventing the fan from starting. Typical safeties are high pressure limit (as in Figure 14.22), low temperature limit, and motor overcurrent cut-out. NEVER jumper out a safety circuit to get the fan started. If a limit control trips the fan out, investigate and repair the problem before returning the air-handling unit to service. A nuisance trip may indicate that a freeze protect sensor is badly located, or that the fan may be cycling too high on start up and causing dangerous duct pressures. The safeties are there for a reason and should never be defeated.

PRACTICAL CONSIDERATIONS

Installation Considerations

There are a number of practical installation considerations that ensure long life and proper operation of an air-handling unit.

FIGURE 14.22 AHU supply air transmitter (left) and high pressure limit switch (right).

- The heating and cooling coils in an air-handling unit should be properly sized to meet their maximum loads. Oversized coils have greater initial cost, greater maintenance cost, and can be more difficult to control.
- Multiple coils in series should be used when there is a need for a large temperature change.
- Coils should provide for uniform heating or cooling of the airstream to prevent stratification.
- Reheat should be considered in systems providing dehumidification.
- Reheat coils should have a maximum temperature rise of about 30°F.
- In systems without reheat, the smallest increment of available cooling should equal the minimum load requirements to prevent short cycling.
- For DX systems, the short cycling of compressors under low load conditions can be prevented by installing multiple compressors, possibly of different capacities. It is also beneficial to use compressors that have unloading (i.e., part load) capabilities.

- In the mixed air plenums of air-handling units, the outdoor and return air dampers should direct the air streams toward each other. This helps prevent stratification of air as it passes through the coils.
- The spacing between coils in an air-handling unit should be big enough to allow the installation of temperature and humidity sensors. Similarly, there should be provisions made that allow the installation of sensors downstream from the air-handling unit, preferably in places where the air is well mixed.

Bearing Lubrication

Many factory-built units have ball or roller bearings that are greased with a flexible tube connected to a grease zerk mounted to the outside of the fan cabinet. These ports look like that shown in Figure 14.23. This allows easy greasing of the bearings without removing the side of the fan cabinet. However, this may also lead to bearing failure due to improper inspection of the bearings. Over time, excess grease may be thrown from the bearing into the fan blades and coils of the fan unit, causing dirt to collect. This can clog the airside of the coil and can cause the fan blades to become clogged with dirt. This reduces the airflow through the unit. Note: in some cases the flexible tube may become disconnected from the bearing, resulting in bearing damage due to lack of lubrication.

Freeze Protection

Stratification of airflow may cause frozen coils or nuisance fan trips due to large amounts of cold or hot air flowing through one side of a coil and very little air flowing through the other side of the coil.

fastfacts

An over or under-greased bearing usually results in a short bearing life. Always follow the bearing manufacturer's instructions.

FIGURE 14.23 Fan bearing grease port on side of AHU.

Baffles, air mixers, or redesigning of the air-handling units may be required to correct the problem.

Some air-handling units that use return air use a separate outside air damper alongside the main outside air damper to meet the fresh air requirement for the building air system. This usually smaller damper is fixed at 100 percent open when the fan operates. In very cold weather this can cause cold air to enter the heating coil unmixed and can create a cold spot in the coil, causing a low limit lockout or a frozen coil. Mixers or baffles may be required to correct the condition. In extreme cold weather the damper may need to be closed automatically to avoid a frozen coil or fan shut-down.

Stratification of airflow may occur in fan discharges if the fan or coils are placed in an incorrect position in the fan cabinet. In some cases warmer air may be discharged out of one side of a squirrel cage fan and cooler air out of the other side. If the air temperature sensor is mounted on one side of the duct, faulty readings may result in poor control of the discharge air temperature. An averaging bulb won't necessarily correct this condition. In addition, supply air ducts that take off near such a fan discharge may draw in air that is too cold or too hot, causing heating or cooling problems due to the inconsistent, non-uniform temperature of the supply air.

When using glycol in heating and cooling coils for freeze protection, use a low limit freeze stat even though the glycol may not freeze.

If the air is cold enough due to a heating coil flow failure, other equipment downstream (for example, reheat coils and domestic water piping) may freeze. The freeze stat protects not only the air-handling unit coils but also everything else in the downstream ductwork.

A hot or chilled water coil subject to airflow at 32°F or lower can freeze within a few seconds if the liquid flow through the coil stops.

An air-handling unit may use coils that are on the same water distribution system as several other hot or chilled water coils. If one coil freezes then the resulting water system pressure drop can create a domino effect where other coils now have a loss of flow and will freeze. Proper attention should be paid to each unit's controls and safeties.

Chilled or hot water coils that are drained in the winter for freeze protection may still freeze due to a small amount of water trapped in the coils that does not drain out. Air blown through the coils may dry out the coils over time or glycol may be pumped through the coils before a coil seasonal lay-up.

Additional freeze protection may be obtained by the use of a pressure switch on the coil steam line and wired to the fan motor controls. If the steam pressure falls below a predetermined safe pressure, the fan shuts down to help protect the coil from freezing.

fastfacts

A low temperature limit device is usually the last line of defense against having a frozen coil, but it does not correct an improper coil application, operation or design. If a low temperature device trips repeatedly and has to be reset, investigate and repair or have the system redesigned to solve the problem.

chapter 15

AIR DISTRIBUTION SYSTEMS

HVAC air distribution systems move air around a building with a system of air-handling units, fans, ductwork and dampers. The main functions of air distribution systems are to provide fresh air to the occupants, remove stale air and indoor air contaminants from the building, and to maintain the proper building pressurization to prevent unwanted infiltration. This chapter concentrates on the ductwork, sensors, and control of the air distribution system. For information about air-handling units, see Chapter 14. For information about zone terminal boxes, see Chapter 16.

THEORY OF OPERATION

In *all-air systems,* air is circulated via a series of ducts located throughout the building. These ducts supply air to terminal boxes that maintain temperature control for each zone (see Chapter 16 for more detail about terminal boxes). The air is then recirculated back into the building or exhausted from the building via the return air duct and exhaust dampers. Figure 15.1 shows a typical schematic of an air distribution system.

There are many variations of air distribution systems. *Constant air volume* (CAV) systems provide a constant flow rate of air into the ductwork. The air is then diverted to zones as necessary. *Variable air*

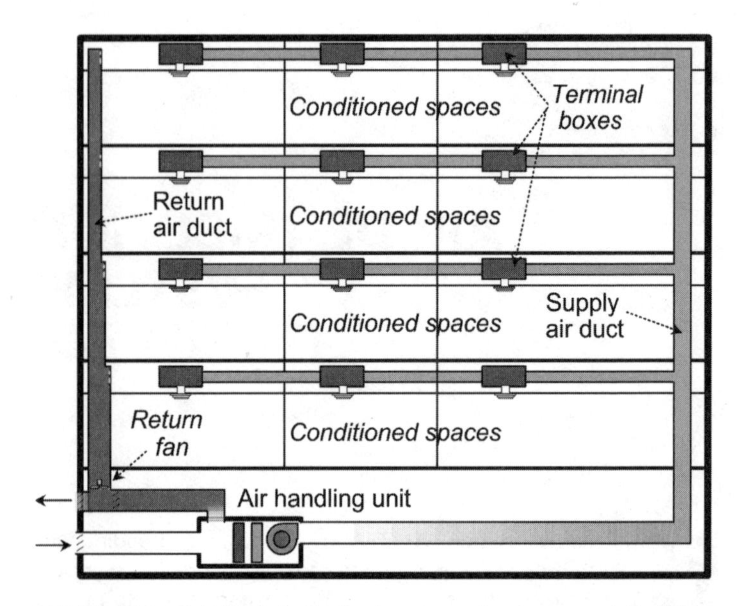

FIGURE 15.1 Simplified air distribution system for a commercial building.

volume (VAV) systems decrease the amount of air provided to the zones during times of low air demand. This reduction of airflow can lead to significant energy savings. *Single-duct* systems use a single supply air duct that sends air to the zones. The supply air is usually cooled to about 55°F. *Dual-duct* systems use two supply ducts, one with warm air at about 90°F and the other with cool air. A *multizone* system refers to a single air-handling unit that services many different zones with individual duct runs to each zone. Air systems are usually broken down into three categories:

- Low pressure systems have velocities less than 2000 FPM and duct static pressure less than 2 inches W.G. These are found in relatively small systems; for example, buildings less than 20,000 square feet.
- *Medium pressure* systems have velocities greater than 2000 FPM and duct static pressure less than 6 inches W.G.
- *High pressure* systems have velocities greater than 2000 FPM and duct static pressure greater than 6 inches W.G. These are found only in the largest buildings and manufacturing facilities.

The amount of air required is determined by the building heating or cooling load. As the load on the building increases, the dampers in the terminal boxes begin to open and the total resistance to air flow through the building begins to decrease. Figure 15.2 shows how the pressure drop across the air distribution system increases as the demand for air decreases (like trying to force air through a duct that is blocked). Also plotted in this graph is the fan curve (see Chapter 13). It is convenient to superimpose these curves so that the operating point of the system can be identified—it is where the fan and system curves intersect. This tells you how much air flows through the building for a certain air demand and fan rotation speed.

Return Fans, Exhaust Fans, and Relief Fans

While air is forced into the air distribution system by the main supply fan, there may be a number of other fans in the system that help circulate air and maintain building pressure.

- A *return fan* is located in the return duct downstream from the zones. It is used to maintain a slightly negative pressure in the return duct. Since the return fan does not do any conditioning of the air, it is located by itself (Figure 15.3) without any coils.

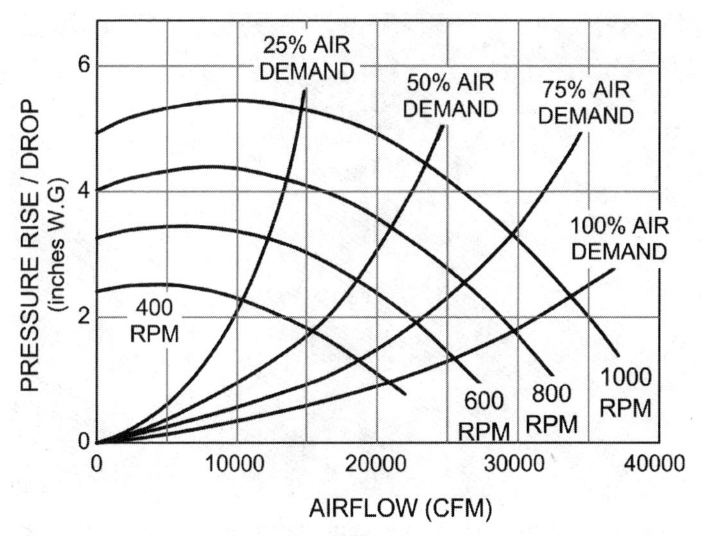

FIGURE 15.2 Hypothetical system and fan curves.

- An *exhaust fan* can be located just upstream from the exhaust dampers to force air out of the building. In fact, exhaust fans can be located anywhere in the building and are used for local building pressure control and smoke evacuation.
- *Relief fans* can also be located anywhere in the building. These fans usually do not stand alone but are incorporated into small air-handling units or fan-coil units and are then referred to as *make-up air units*. These are also used for building pressure control and local air conditioning.

Run-Around Loops

Run-around loops are used to help save money by recovering some of the heating or cooling energy that would normally leave the building. Run-around loops consist of two air-to-water heat exchangers connected by a continuously pumped closed-loop piping system as shown in Figure 15.4. As the exhaust air leaves the building, it passes through one heat exchanger and warms up the water in the loop. The water then passes through the heat exchanger in the outside air intake where it warms up the air and reduces the load on down-

FIGURE 15.3 Return fan showing upstream and downstream duct access doors.

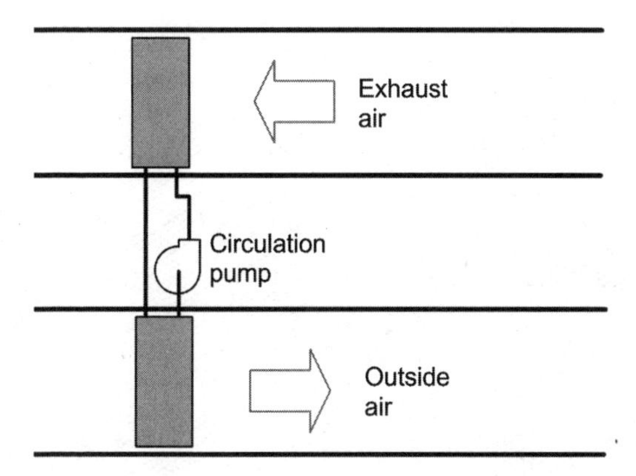

FIGURE 15.4 Inlet and exhaust air ducts showing run-around loop.

stream heating coils. Of course, the water must be treated with glycol to prevent it from freezing in the winter.

Run-around loops can also be used in the summer months when it is exceptionally hot outside. In this case, the exhaust air is used to cool the incoming outside air when the outside air is in excess of 90° or so.

Humidification and Dehumidification

In particularly dry climates, steam humidification can be used to increase the amount of water in the air supplied to the occupied zones. The steam is injected through multiple port injectors located within the air duct as shown in Figure 15.5. It is important to have dry steam in these applications because you do not want water droplets collecting on the bottom of the duct—this can lead to rusting ducts and microbial growth.

Steam humidifiers are controlled either by outdoor dry bulb temperature or by actual measurement of the duct air humidity. If measuring the duct air humidity, be sure to place the sensor far enough downstream to allow the steam to be fully dispersed. Otherwise you may have local "streams" of high humidity air that can throw off the control.

When sizing a steam humidifier, it may be better to purchase a unit that is slightly *undersized* rather than oversized. The traditional HVAC sizing technique is to calculate the capacity of your equipment based on design conditions and then multiply by some safety

FIGURE 15.5 In-duct steam humidifier bars (picture shown looking downstream).

factor. The inevitable consequence is that the equipment operates at part load for most of the time. In steam injection systems this can mean more chance for condensation of the steam and standing water in the duct.

Dehumidification of the air usually happens at the cooling coil in an air-handling unit. The process of using a chilled water coil to remove water from the air is described in Chapter 14. In particularly humid climates, however, you may encounter desiccant dryers used for dehumidification. In a desiccant dryer (Figure 15.6) a slowly rotating drum is used to transfer humidity from one air stream to another. The drum consists of a roll of paper with embedded water-

fastfacts

Using steam injection for humidity does not increase the air temperature very much.

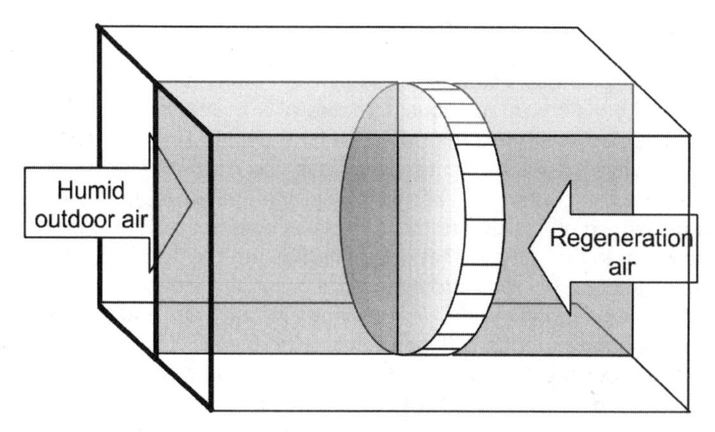

FIGURE 15.6 Schematic of desiccant dryer.

absorbing crystals (like the packets you find in salt shakers that say "Do not eat"). When the crystals are in the incoming outdoor air stream, they absorb the water and let drier air pass into the building. As the drum turns, the crystals then move into the *regeneration* air stream. On this side the air is extremely hot (>180°F) and forces the water to evaporate and leave the building.

The use of desiccant dehumidification depends on the building and climate. By removing water from the incoming air, you decrease the latent cooling load. However, you usually must use a lot of heating energy to create the necessary temperatures in the regeneration air stream. Even if you apply additional heating to the return air stream, it still takes a lot of energy to bring it up to the right temperature. However, if the regeneration heat is cheap (for example, as waste heat from another process), then desiccant dehumidification can provide a good supplement to the existing HVAC system.

CONTROLS

There are three conditions that must be controlled in air distribution systems: the air temperature, the air humidity, and the air pressure. The air temperature and humidity are controlled by the air-handling unit and any duct-based humidification or dehumidification systems. See Chapter 14 for more information about air-handling units.

Duct Pressure Control

The duct pressure is controlled to anywhere from about 1 inch W.G. in small buildings up to about 10 inches W.G. in very large buildings. Positive duct pressure is important for two reasons. First, the terminal boxes want to see inlet pressures above a certain level so that the dampers can control the zone air flow properly. Second, the building must remain pressurized so that outside air does not infiltrate into the building. In fact, any air that does infiltrate into the building will not be conditioned as well and can create occupant comfort problems. Unconditioned infiltrating air also brings in duct, dirt, pollen, and a host of other potential indoor air quality problems.

The duct pressure is typically controlled using a pressure sensor located some distance downstream of the fan (see Figure 15.7). The pressure signal commands a variable speed drive or dampers on the fan to modulate and change the duct pressure. You should not use very high gains on the controller since the duct pressure responds quickly to changes in the fan speed or damper positions. The pressure sensor is usually located downstream about two-thirds of the longest duct length, but this number is subject to strong debate.

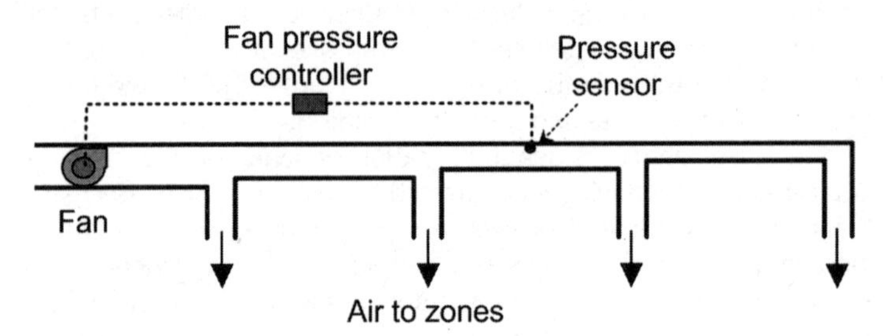

FIGURE 15.7 Location of duct pressure sensor.

fastfacts

If you pressurize the building too much (>1 inch W.G.) you will have trouble opening outside doors (or keeping them closed if they swing outward).

Some people insist it should be closer to the fan and others think it should be farther away. You may have to experiment with different locations to get the best control.

On the opposite side of the zones, a return fan draws air through the return duct back to the air-handler or expels it from the building. Return fans can be controlled in a number of ways:

- *Supply fan speed tracking* runs the return fan off the same control signal as the supply fan. This is a very simple technique but it ignores return duct static pressure.
- A variation of supply fan speed tracking is *airflow tracking control.* This requires duct airflow measurements and has the return fan control to about 90 to 95 percent of the supply duct airflow rate. Airflow tracking control is useful on the first floor (entry floor) of large buildings. A pressure-based controller would see sudden and significant changes in the static pressure each time a door opened or closed, and the fan would spend most of its time hunting for the right speed.
- *Return duct static pressure control* maintains the return duct as some slightly negative pressure (–0.1 to –1.0 inches W.G.) so that a proper air balance on the zones is guaranteed even if windows are open in the zone. In some cases the return fans are controlled to maintain the total building static pressure, not the duct pressure. This ensures proper pressurization of the entire building, provided that you are measuring the building static pressure measurements in several places.

In some cases, zones must be depressurized. For example, a laboratory zone should have a negative pressure with respect to the rest of the building so that chemical leaks do not contaminate other zones. Usually these special zones have their own, devoted air-handling unit, but the static pressure control must measure the pressure in surrounding zones to create the proper pressure differential.

An air distribution system should include both flow switches and pressure cut-out safeties. Flow switches are used to protect the fan and are discussed in Chapter 13. The pressure cut-out is a duct static high limit control and is used to shut off the fan when the pressure gets too high. This keeps the duct and duct connections from bursting. In normal situations the high pressure cut-out should be set to about five or six inches W.G. The duct static high limit control should be connected in series with the fan start circuit as shown in Figure 15.8. You will often find it convenient to wire the high limit control in series with the freeze protection control.

fastfacts

Setting the high pressure cut-out too low can cause nuisance trips on fan start-up.

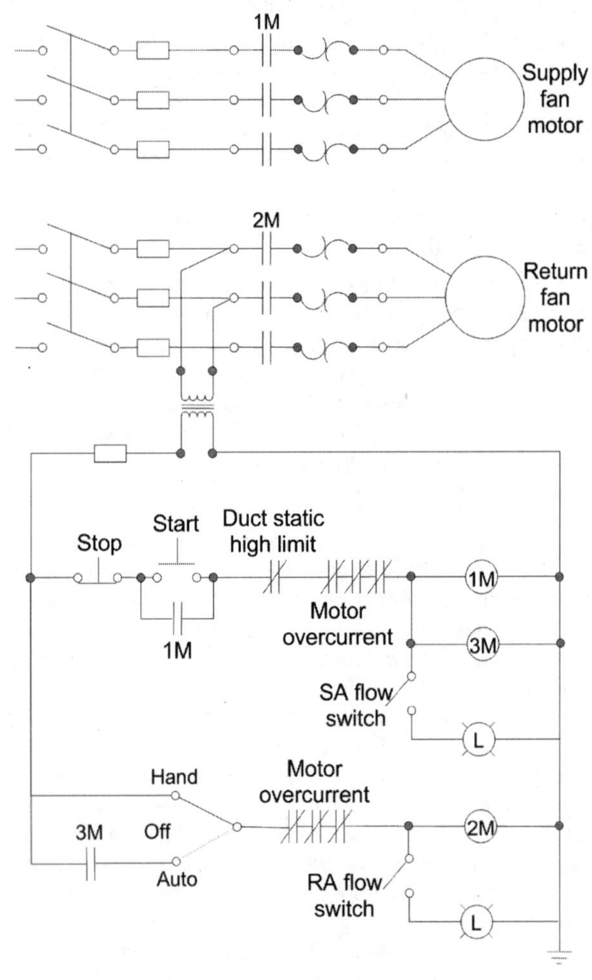

FIGURE 15.8 Fan start-up ladder diagram.

The duct static pressure controller is critical to the proper operation of VAV systems. If proper duct pressurization at the end of long runs is not present, flow starvation can occur at VAV terminals located far from the air-handler. The static pressure controller must be of PI design since a proportional-only controller would permit duct pressure to drift upward as cooling loads drop due to the unavoidable offset in P-type controllers. In addition, the control system should position inlet vanes closed during fan shutdown to avoid overloading on restart.

One useful technique if you have a complete data acquisition system in the building is to use a static pressure reset so that at least one mixing box damper is always greater than 90 percent open. The idea is to provide exactly the amount of air required by the zones, so if the mixing box dampers start to close you can reduce the duct static pressure. Of course, you should not let the pressure go below the minimum required for the boxes to operate, so have a lower limit of at least 1.0 to 1.5 inches W.G.

Building Warm-Up/Cool-Down

Many buildings are allowed to "float" at night, meaning that the HVAC equipment is turned off and the building temperature can drift up or down depending on the difference between the inside and outside temperatures. Of course, the building must be at a comfortable temperature when the occupants arrive so it is necessary to start the HVAC system beforehand. Starting the fans too soon (dotted line in Figure 15.9) wastes energy and money, while starting the boiler too late results in a cold building when the occupants arrive (dashed line in Figure 15.9). You have to experiment to know the proper start-up time, but it is usually around one to three hours before the building occupants start arriving.

During the warm-up or cool-down mode, any exhaust fans and relief fans are turned off, the building pressure control and the airflow tracking differential are reset to zero (if used), and the thermostat control is overridden or reset. The building essentially operates like a big wind tunnel in an effort to condition it as fast as possible.

Economizer and Night Purging

If the outdoor air temperature is close to the desired supply air temperature then the exhaust and outdoor air dampers are fully opened and the return air damper is completely closed. Under these conditions, 100 percent outside air enters the building and the chiller can

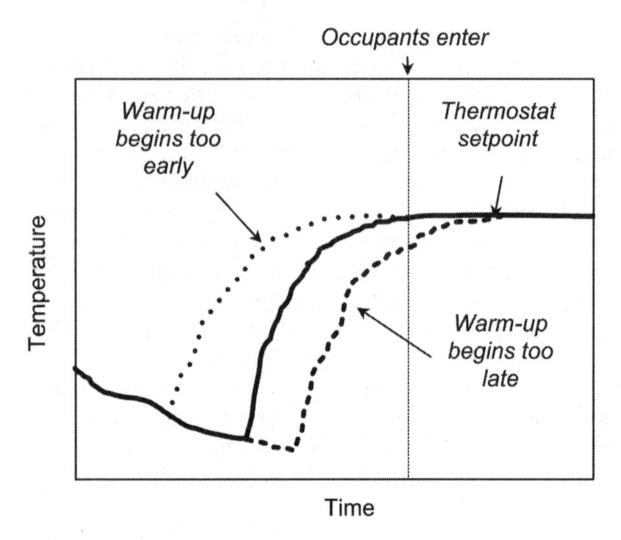

FIGURE 15.9 Effects of different warm-up start times.

be shut off because there is no need for cooling the air. This is called *economizer* mode.

Strictly speaking, the economizer mode should be used whenever the enthalpy difference between the outdoor air and the supply air is less than the enthalpy difference between the return air and the supply air. However, it is difficult to measure enthalpy so usually just the air temperatures are used. The economizer mode is sometimes used at night to cool off a building mass in preparation for the next day's cooling load. This is called *night purging*.

Figure 15.10 shows the typical controls and sensors used in an economizer system. This system is able to provide the minimum out-

fastfacts

Temperature-only economizer control in humid climates can actually end up using more energy than no economizer. You should really calculate the enthalpy differences between the various air streams.

side air during occupied periods when it is warm outside, to use outdoor air for cooling when appropriate by means of a temperature-based economizer cycle and to operate fans and dampers under all conditions. The numbering system used in the figure indicates the sequence of events as the air-handling system begins operation after:

1. The fan control system turns on when the fan is turned on. This may be by a clock signal or a low temperature space condition.
2. The space temperature signal determines if the space is above or below the setpoint. If above, the economizer feature will be activated to control the outdoor and mixed air dampers. If below, the outside air damper is set to its minimum position.
3. The mixed air PI controller controls both sets of dampers (OA/RA and EA) to provide the desired mixed air temperature.
4. The outdoor temperature rises above the cutoff point for economizer operation, the outdoor air damper is returned to its minimum setting.
5. The supply fan is off the outdoor air damper returns to its NC position and the return air damper returns to its NO position.
6. The supply fan is off the exhaust damper also returns to its NC position.

CAV and VAV System Control

Figure 15.11 shows a constant volume air distribution system equipped with supply and return fans, heating and cooling coils, and economizer for a single zone application. If the system were to be used for multiple zones, the zone thermostat shown would be replaced by a discharge air temperature sensor. This fixed volume system operates as described below.

1. When the fan is energized the control system is activated.
2. The OA temperature sensor supplies a signal to the damper controller.
3. The RA temperature sensor supplies a signal to the damper controller.
4. The damper controller positions the dampers to use outdoor or return air depending on the economizer settings.

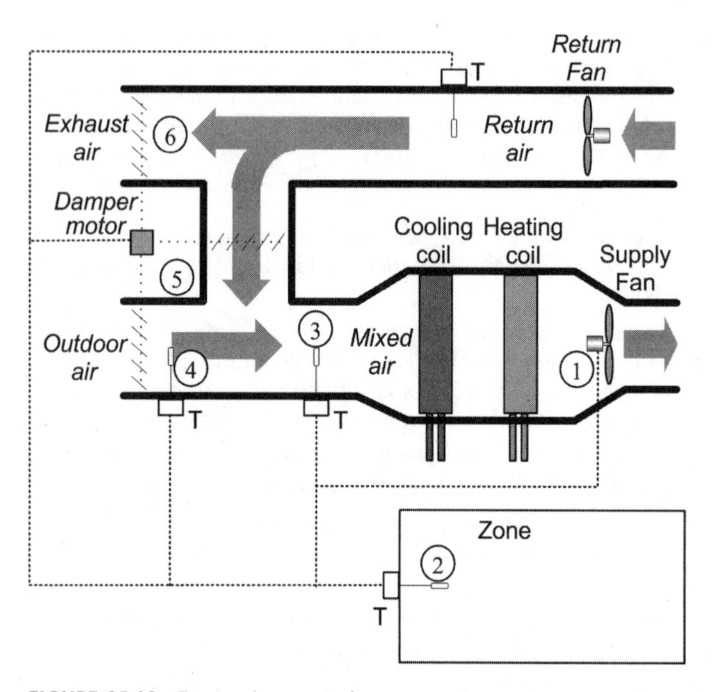

FIGURE 15.10 Economizer controls.

5. The mixed air low temperature controller controls the outside air dampers to avoid excessively low temperature air from entering the coils. If a preheat system were included, this sensor would control it.
6. Optionally, the space sensor could reset the coil discharge air PI controller.
7. The discharge air controller controls the heating coil valve, outdoor air damper, exhaust air damper, return air damper, and cooling coil valve after economizer cycle upper limit is reached.

A VAV system has additional control features including a motor speed (or inlet vane) control and a duct static pressure control. Figure 15.12 shows a VAV system serving both perimeter and interior zones. It is assumed that the core zones always require cooling during the occupied period. The system shown has a number of options and does not include every feature present in all VAV systems. However, it

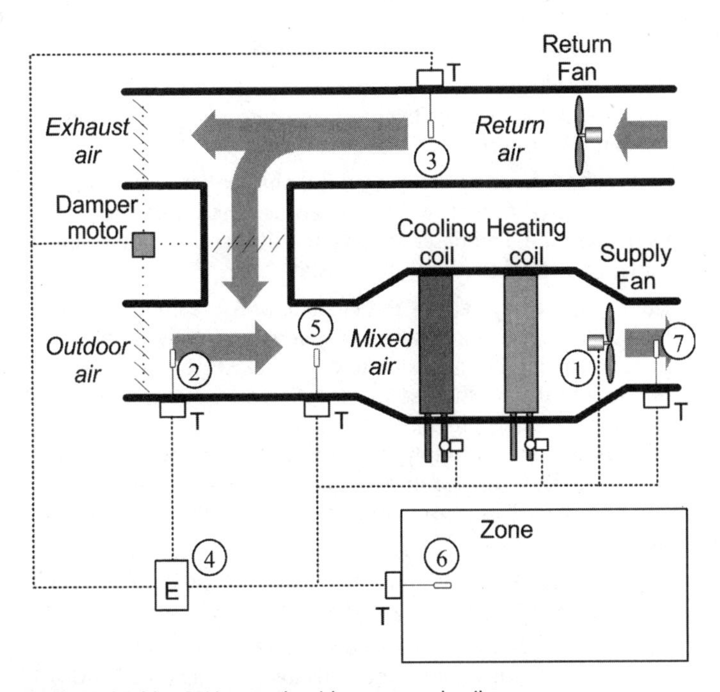

FIGURE 15.11 CAV controls with runaround coils.

is representative of VAV design practice. The sequence of operation is as follows:

1. When the fan is energized the control system is activated. Prior to activation during unoccupied periods the perimeter zone baseboard heating is under the control of room thermostats.
2. Return and supply fan interlocks are used to prevent pressure imbalances in the supply air ductwork.

   The mixed air sensor controls the outdoor air dampers (and/or preheat coil not shown) to provide proper coil air inlet temperature. The dampers will be at their minimum position when the outside air is around 40°F.
3. As the upper limit for economizer operation is reached, the OA dampers are returned to their minimum position. The return air temperature is used to control the morning warm-up after night setback. (Option present only if night setback is used).

4. The outdoor air damper is not permitted to open during morning warm-up.
5. Likewise, the cooling coil valve is de-energized (NC) during morning warm-up.
6. All VAV box dampers are moved full open during morning warm-up by action of the relay override. This minimizes warm-up time. Perimeter zone coils and baseboard units are under control of the local thermostat.
7. During operating periods the PI static pressure controller controls both supply and return fan speeds (or inlet vane positions) to maintain approximately 2.0 in. W.G. of static pressure at the pressure sensor location (or optionally to maintain building pressure). An additional pressure sensor (not shown) at the supply fan outlet will shut down the fan if fire dampers or other dampers should close completely and block air flow. This sensor overrides the duct static pressure sensor shown.
8. At each zone, room thermostats control VAV boxes (and fans if present); as zone temperature rises the boxes open more.
9. At each perimeter zone, room thermostats close VAV dampers to their minimum settings and activate zone heat (coil and/or perimeter baseboard) as zone temperature falls.
10. The controller, using temperature information for all zones (or at least for enough zones to represent the characteristics of all zones), modulates outdoor air dampers (during economizer operation) and the cooling control valve (above economizer cycle cutoff) to provide air sufficiently cooled to meet the load of the warmest zone.

SAFETY

Air distribution systems contain high velocity air at relatively high pressure differentials between the inside and outside of the duct. While an inch of water pressure may not sound like much, it is equivalent to about 5.2 pounds per square foot. This kind of pressure across a 3 foot x 7 foot door exerts a force of over 100 pounds, enough to injure the careless. There are basic safety rules for operating with ductwork:

- Never over pressurize ductwork as it could rupture or explode.

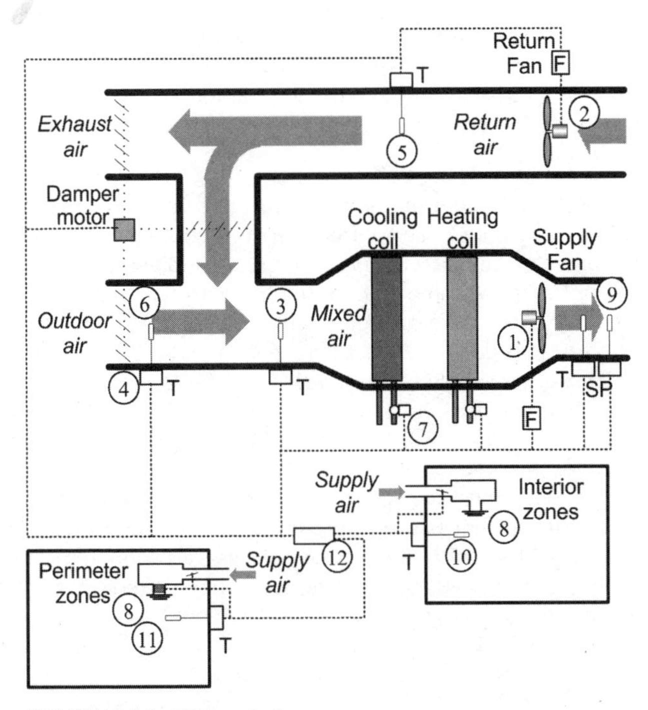

FIGURE 15.12 VAV controls.

- Extreme negative pressures may collapse the ductwork. This condition is known in the trade as "suck a duct" and should be avoided at all costs.
- Never enter ductwork alone.
- Ductwork may contain insects, dirt, and dust, standing water, loose insulation, chemicals or biological hazards. Always be cautious of where you place your hands and feet.
- Ductwork may have sharp edges that can cut like a knife. This usually occurs where ductwork has been joined or where there are penetrations into a duct.
- Never stand on unsupported ductwork. Despite its sturdy appearance, ductwork is usually just thin sheet metal and it could suddenly collapse.
- Glued-on nail type insulation attachments (Figure 15.13) will cause puncture wounds if stepped on.

- Extreme care must be exercised when working inside ductwork. Unlike in the movies, there is usually little or no light. Ducts are filled with turning vanes, transitions, dampers, drop-offs, splitters and other obstructions that may block your path, leaving you trapped.

Fire and smoke control is important for life safety in large buildings. The design of smoke control systems is controlled by national and local codes. Some components of space conditioning systems—for example, fans—can be used for smoke control but HVAC systems are generally not smoke control systems by design. When an air distribution system goes into *smoke evacuation mode*, the exhaust fan tracks above the supply fan to guarantee a net loss of conditioned air from a zone. Other air distribution systems in the building have static pressure reset upward to prevent smoke from infiltrating into other zones.

Any building operating with chemical or cook-line hoods must be controlled to prevent backpressure from bringing fumes back into the building. This is often a pretty complicated task during the commissioning of the building, but it must not be ignored since it represents a serious health risk.

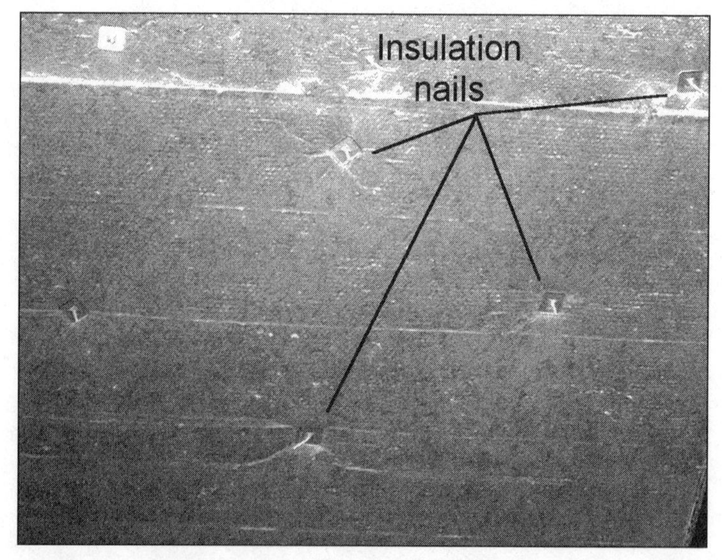

FIGURE 15.13 Glued-on nail type insulation attachments.

fastfacts

A zone in smoke evacuation mode does not get fresh air. This presents a suffocation threat to anyone still in the zones. Proper fire control modes in a building must include instructions for all occupants to leave the zones.

TROUBLESHOOTING

There are many problems that can occur in the ductwork. Many of these are related to the various components that exist inside the ductwork or the sensors located along the air distribution system. The problems generally cannot be quantified by the flow charts as listed in other chapters of this book.

Supply Air Flow Rate/Static Pressure Too Low

Some causes of low supply air rate can be easily found. For example, the airflow in a VAV system is often controlled by the duct static pressure. If the pressure sensor is located too near to the fan, or immediately after a turning section that forces pressure on to the sensor, the fan will slow down to relieve what it believes to be too high a duct pressure. It may even be that the duct pressure setting is too low or that a pressure reset control has inadvertently lowered the setpoint. In some cases, a pitot tube is used to measure the static pressure and the pressure sensor is accidentally connected to the dynamic pressure tap. Finally, it may simply be that the overall demand for air is so high that the fan does not have the capacity to provide enough airflow for all the zones.

However, these kinds of problems are relatively easy to solve. A loss of airflow can also occur when insulation comes off the inside of the duct (for example, the glue comes loose or insulation nails come loose) and the insulation hangs down, gets caught up in turning vanes, or clogs up reheat coils. All of these conditions result in reduced air flow. The insulation glue can just come loose after a period of time, but this problem tends to be more prevalent in vibrating ducts or in areas where people can walk around on top of the duct. In some cases, loose insulation can block the whole ductwork, where in other cases there are just little chunks that get caught in coils. The former problem is easy to identify and fix while the latter requires an investigation of each coil, although it's fair to say that if you find insulation in one coil there is a good chance it exists in all coils on that particular air duct branch.

Another reason for restricted airflow in ducts is a dropped fire damper. If the ductwork goes through floor or wall penetrations, there will be fire dampers to isolate the floors. The fire dampers are spring-loaded and are held open by small links made of a soft metal such as lead. The links are usually sized to release the damper at 140°, 160°, or 195°F. It is not uncommon in a double-duct system to have the dampers release on the hot deck, especially on a hot day or on a very cold day if the system has an air temperature reset.

If there is no or low flow in a particular branch of ductwork and the insulation and coils appear fine, check to see where this branch connects to the main supply duct. If the connection is very near the supply fan and has a side takeoff, the fast-moving air in the supply duct creates a venturi effect and can actually suck air out of the room through the diffusers. This usually happens in high load situations when there is high airflow. It is particularly prevalent in cooling VAV system where the chiller is out of capacity and the fan ramps up.

Supply Air Flow Rate/Static Pressure Too High

High duct static pressure, if not caught, can manifest itself by blowing the flex duct off of the main duct and can even blow the sides off the main supply duct. The key in maintaining the proper supply air flow and duct pressure is the placement and care of the static pressure sensor. If the duct appears to be bulging, or there are temperature or noise complaints, check to make sure that the duct static pressure is within the proper range. A clogged static pressure sensor can cause the fan to ramp up and quickly overpressurize the ductwork. Keep in mind that each inch of water pressure is equivalent to a little over 5 pounds per square foot, or close to 100 pounds of pressure on a typical duct access door. Overpressurization is dangerous and can lead to extremely costly repairs.

In the previous section, it was mentioned that a dropped fire damper can reduce static pressure, especially if the static pressure sensor is located between the fan and the dropped damper. In that case, the ductwork downstream of the fire damper has little or no static pressure. However, it is also possible to drop a fire damper and overpressurize the duct work. This occurs if the dropped damper is between the fan and the static pressure transmitter.

Supply Air Too Cold or Too Hot

In a single duct VAV system, the wrong air temperature is usually caused by improper controls at the air-handling unit. Refer to Chapter 14 for more troubleshooting tips on air-handlers. If the air-handler checks out okay, then see if the ductwork passes along an exterior wall or through an unconditioned space. If the outdoor temperature is cold, this can affect the supply air temperature, particularly if the ductwork is not insulated.

Also, in single duct systems, some air-handling units use a squirrel cage fan where the fan outlet is divided with the heating coil on one side and the cooling coil on the other side. Often, the coils only heat their side of the air stream so that hot air goes down one side of the duct and cool air on the other side. After some distance the two air streams mix, but take-offs located just down from the fan can draw in just hot or cold air. You can check for this with a duct temperature probe. If there is such stratification in the duct, you may need to install air mixers before the first duct take-offs.

In dual-duct VAV systems, a cause of cold air temperatures may be related to a dropped fire damper. Since the dampers are more prone to accidentally drop on hot air ducts, the hot air flow may be impeded and the terminal boxes only receive cold air so that you can have cooling and no heating. If the dual-duct system has recently had work done, make sure that the cooling coil is located in the cold deck and the heating coil in the hot deck. As ridiculous as it sounds, such mistakes have been made.

Supply Air is Too Humid or Too Dry

The humidity control in ductwork almost always occurs at the air-handling unit. Possible causes of poor humidity control include a humidifier not working, a cooling coil removing too much moisture from the air, or an evaporative cooler that is adding too much moisture.

One caution, if the air is too humid, is that mold can build up inside the ductwork, especially downstream of an air wash unit. Once mold and other organics have taken hold in the insulation, it is very difficult to get rid of them. Generally one must remove the insulation, clean the duct with a chorine-based cleanser or copper sulfate, and reinstall all new insulation. In the worst cases, mold can lead to severe occupant problems such as Legionnaire's disease and respiratory problems.

Too Much Noise from Supply Air Duct

Uninsulated ductwork is an excellent carrier of noise. Excessive fan noise can travel great lengths from the fan to the mixing boxes and into the occupied zones. If there are noise complaints, you may wish to insulate some sections of the duct solely for noise suppression.

Transitions can be a particular problem, especially if there is a constriction or expansion immediately after the fan. This can cause a lot of noise along with poor airflow. This is usually done in places where there is not much space between the fan discharge and where the duct take-offs start. Unfortunately, this is a basic design problem and is very expensive to fix. It usually requires reworking the entire ductwork near the fan.

Another, easier-to-cure source of noise is simply high duct pressure or velocities traveling through turning vanes, dampers, and coils. If may be possible to reset the duct pressure so that the noise goes away. In particular, it can be hard to keep the ducts from leaking in a VAV systems where the air flow varies all the time. Generally speaking, if the duct pressure is greater than 1 inch water, it can get noisy around the duct joints, seals, and flanges. A possible solution is applying hardcast to these sections. It helps seal the duct but is a

fastfacts

A technician was called to a building where the occupants like to do their own "maintenance." There had been several complaints of ducts leaking and excessive duct noise, to the point where the people in the building had been wrapping duct tape around everything. The technician began digging around in the mechanical room and found that the fan was running full open. When he investigated the fan pressure controller, he discovered that someone had put the duct static pressure setpoint up to 6 inches W.G. After making sure that the ductwork was not in immediate danger of splitting open, he went back to the occupants and asked why the fan pressure was set so high. They responded by telling him that "six inches sounded like a good number." The technician lowered the setpoint to about 2 inches W.G. and the leak and noise complaints stopped completely.

relatively expensive solution and makes future duct maintenance more difficult.

Finally, rectangular ductwork tends to vibrate more readily than oblong or circular ductwork. If the noise is unmanageable and cannot be fixed with duct pressure adjustments, consider replacing large rectangular duct sections with circular or oval sections.

PRACTICAL CONSIDERATIONS

In an effective air distribution system, care must be given to the design of the mechanical systems and to the selection of the fans, ductwork, dampers, and other components.

- If the ductwork runs through a space that is at a considerably different temperature than the air in the duct, the ductwork should be insulated.
- Temperature stratification in ductwork can lead to uneven control, occupant discomfort, and, in extreme cases, to equipment damage (for example, from freezing coils). To reduce stratification, mixing devices should be used. These can be mixing chambers, vortex blades, etc. The system fan can also be used as an effective way to mix the air. A single inlet fan typically mixes air more effectively than a double inlet fan.
- If possible, steam coils should be arranged so that the supply header is on the longest dimension. This helps reduce the temperature gradient across the coil and ensure more even heating of the air.
- Duct pressure, temperature, humidity and air flow rate sensors should be located where they accurately measure the values used in control loops. Temperature sensors should be located away from heat sources, point flow meters should be located in the center of the duct, etc.
- Humidifiers normally should be located downstream from a source of heat. If the air is too cold, the water introduced into the airstream may immediately condense and result in water damage to the duct.
- If freezing air temperatures are expected, an air-handling system must have low-temperature protection to prevent freezing of coils. It may also be worth considering coil pumps to force flow through the coils during periods of subfreezing temperature.

- Steam coil installations should ensure that the tubes are vertical and that the coil itself is correctly angled to allow for proper condensate drainage to the steam traps. The coils should also have vacuum breakers and steam traps of the type and size appropriate for the coil.
- You may want to use the interior space temperature (instead of the outdoor air temperature) to perform hot water reset. For example, use the output of relay logic or load analyzers connected to space sensors to reset multizone unit hot and cold deck discharge controller set points.
- Air-handlers should not introduce outdoor air to a building area that is unoccupied or during the warm-up period. The only exception to this is if you want to "flush" the building of stale air right before the occupants arrive.
- Flex duct is nice, but care should be taken in its use because it can flatten out after a while, particularly if it gets covered with conduit or cables. This can happen if the duct is located in a plenum along with electrical and communication utilities. Many brands of flexduct have nylon or stainless steel bands to keep it open, although sometimes hardcast must be applied to keep the flexduct open.

Fire dampers, while absolutely necessary, can be the source of a lot of headaches and maintenance costs. On hot days, an overactive hot deck can drop a fire damper, but on cold days, a hot deck temperature reset control can also lead to temperatures high enough to melt out the fire damper links. This can also happen if there are failures in the building management system. In fact, sometimes the dampers just decide to drop just from the repeated stress of daily heating. To avoid such problems, you may want to use the highest temperature links (for example, 190°F) in your fire dampers. In any case, fire damper problems can be very hard to track down and time consuming to find. Many new fire dampers are actuated by pneumatic actuators that fail safe (closed). These have a higher capital cost and have a relatively high maintenance expense but are usually cheaper in long run, particularly when trying to identify ones that have failed.

chapter 16

ZONE TERMINAL SYSTEMS

Air distribution systems are used to introduce conditioned air into the ductwork of a building. The ductwork then distributes this air throughout the building. As the ducts pass by occupied zones, some of the air is diverted into the zone to maintain the zone temperature. This chapter describes the operation of the zone terminal systems that do this.

THEORY OF OPERATION

Commercial and industrial buildings are broken up into zones. Each zone represents an individual thermal or functional area separate from those around it. For example, Figure 16.1 shows the typical zoning pattern for a rectangular commercial office building. The pattern would be repeated for each floor of the building. In the winter, the *perimeter zones* (that is, the zones around the outside edge of the building) might require heating while the *core zone* would require cooling. Certain zones of a building, such as elevator shafts and other mechanical pathways, almost always require cooling because of the heat build-up through the operation of the mechanical systems. The loads in the perimeter spaces also depend not only on the internal load factors such as occupants, lighting, and equipment, but also on the outdoor air temperature and the amount of sunlight hitting the building surface and entering through windows. As a result, the east and south zones may require cooling in the morning while the west and north zones require heating.

The conditions in an individual HVAC zone are controlled by a thermostat located in a (hopefully) representative area in the zone. Building zones like those shown in Figure 16.1 are often conveniently separated by walls and doors, allowing for easy separation of spaces and dedication of each thermostat to each zone. However, in some cases, a single thermostat is used to control the conditions in multiple rooms. This can work if the loads and occupancy schedules are more or less the same for each room, but can create problems if not.

In larger commercial buildings, the zoning patterns can be quite complex, depending on the function of the various areas within the building. Figure 16.2 shows the different regions within a supermarket. The regions with the refrigerated and freezer cases have significantly different air-conditioning requirements than the produce section and aisles. The problem is compounded by the generally open floor plan used by most supermarkets. In this case, the walls separating the zones are imaginary, and the thermostats are located on ceiling supports, aisle ends, and any other location deemed suitable for proper control of each individual region.

Once the zones have been defined, there are many different ways that the air can be conditioned to counter the different and changing loads in each. In *all air systems*, there are one or more central air-handling units that distribute air through ductwork to the terminal boxes at each zone. Figure 16.3 shows such an arrangement. The air that enters the terminal box can be reheated using electric or hot water coils, and it can be mixed with air that is recirculated from the

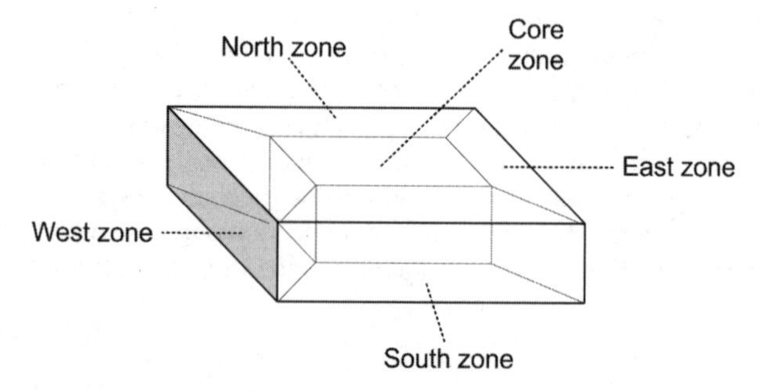

FIGURE 16.1 Typical zoning arrangement for an office building.

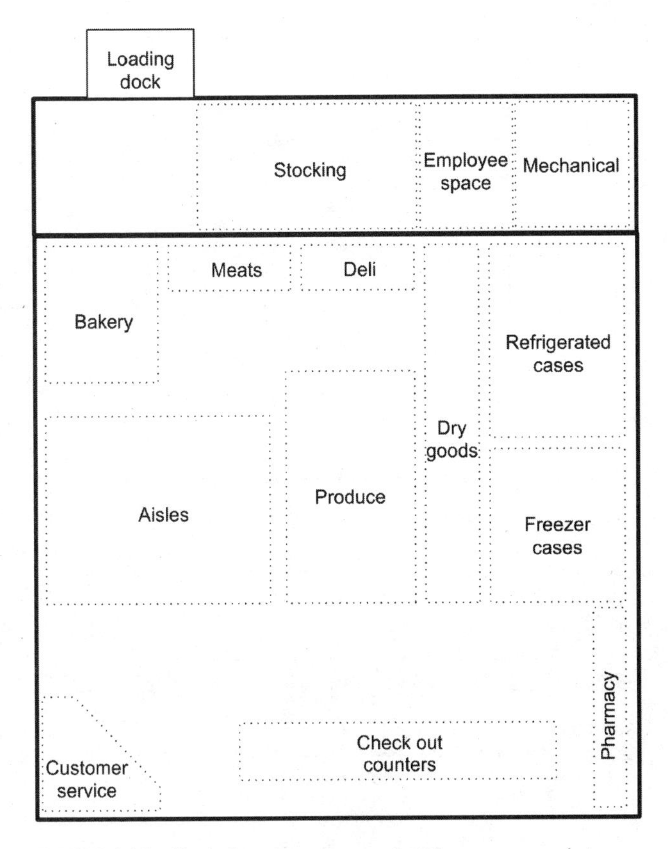

FIGURE 16.2 Typical zoning arrangement for a supermarket.

zone. The amount of supply air from the air-handling unit can also be modulated. In this kind of system, a thermostat on the wall controls the amount of air and the amount of reheat used in order to maintain a zone temperature setpoint.

There are many different variations of terminal boxes used in commercial buildings. The type of box used is a function of the type of air distribution system and of the characteristics of the zone. Figure 16.4 shows the components of a typical installation. The box is located in the space above the suspended ceiling in a zone. The air then travels from the box through flexible ductwork to the diffuser (Figure 16.5). This space is called the *return air plenum* because air from the zone passes through the space before it returns to the air distribution system.

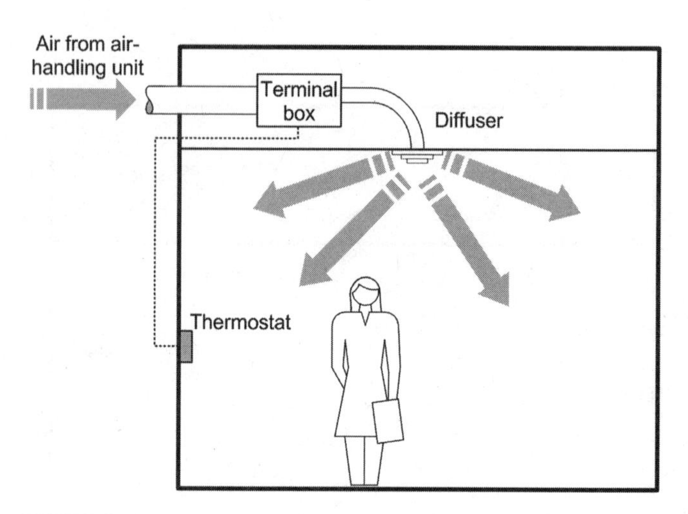

FIGURE 16.3 Zone terminal box.

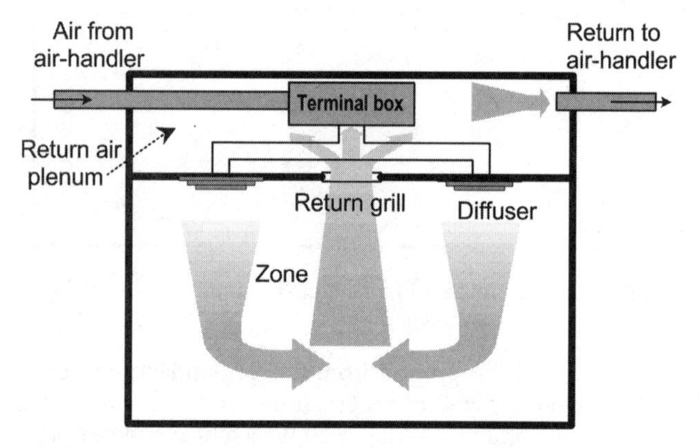

FIGURE 16.4 Schematic of zone terminal box and room air flows.

In some cases the return grill is connected directly to the return air duct—this is called a *ducted return*. It's much easier to control the amount of return air in a room with ducted return; these kinds of systems are common in laboratories and chemical rooms that use variable air volume systems. However, a nice feature of using an open return plenum is that the return grills can be placed where they are needed, such as over a heat source (a refrigerator, com-

FIGURE 16.5 Flex duct from terminal box outlet to zone diffuser.

puter, etc.). This is much more difficult on a ducted return, since rigid duct is usually used.

Unless you have a specialized application, it is much easier to use open plenum return. Easy movement of the return air grills is useful during building commissioning. For example, if there is a good distance between the return grill and the main return duct, there can be problems with the zone pressurization, leading to doors that won't close or that are hard to open. Another problem can occur when cold air in the plenum sinks through the return grills onto occupants. This usually occurs if there are static pressure control problems in the building and may indicate a problem not with the controls, but with the location of instrumentation.

A *pressure dependent* terminal box does not measure the supply air duct pressure or flow rate. The supply air damper moves to a preset position depending on the thermostat signal. If the duct pressure is too high, there may be too much air flow at a given damper position and you can easily overcool the room. If the duct pressure is not high enough, then not enough air will flow for a given damper position and you will have indoor air quality problems. *Pressure independent* terminal boxes measure either the duct pressure or the airflow and adjust the damper position based on that. Pressure independent boxes use a pitot tube, a flow cross, or a hot-wire anemometer to get the flow measurement.

Flow controllers are not always needed on smaller systems. A variable air volume system without flow controllers can work pretty well just using manual positions for low and high flow rates. These are set

fastfacts

A technician was always having trouble with one particular zone. It was a double duct VAV system with a flow controller that showed cool air flow even though the zone was too hot. He finally traced the problem to the hot wire anemometer used to measure the flow. Because of pressure differences in the ducts, air was actually flowing backwards through the cold air duct, and since the hot wire does not care what direction the air is flowing, it reported no problem. This device was replaced with a flow cross (that only senses flow in one direction) and the problem was solved.

using manual stops on the damper motor. You may find that some VAV boxes farther down on long runs operate towards the higher range of damper positions while ones closer to the fans see higher pressures and operate more closed to flow the same amount of air. This is not usually a problem as long as flow controllers are set up right. If the duct runs are long enough, you may need to put in flow controllers.

Single Duct VAV Boxes

In a single duct VAV box (Figure 16.6), the air from the air handler is usually cool, somewhere around 55°F. The amount of air entering the box is controlled by the damper located at the inlet. In a *fan-powered mixing box*, such as shown in the figure, there is a fan that forces air from the plenum to mix with the air from the air-handling unit. These units come with or without reheat coils. Reheat is typically only used in zones along the perimeter of the building. The reheat may be electric, steam or hot water coils in the terminal box, but can also come in the form of baseboard heaters in the zone itself. In either case, the box and the heaters should be controlled by the same thermostat so that there is synchronization between the damper and the reheat to prevent localize hot spots and cold spots in the room and to prevent simultaneous heating and cooling.

These types of boxes have a minimum position on the damper so that there is always some fresh air entering the zone. The exception to this occurs during morning warm-up when the damper may close and full reheat is used to rapidly heat the zone. Keep in mind that

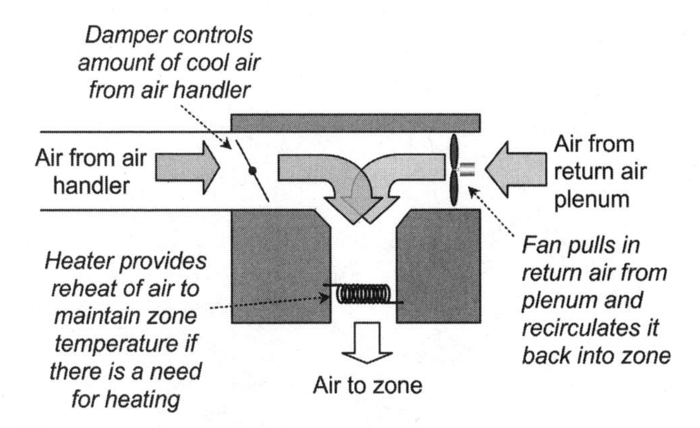

FIGURE 16.6 Components of a single-duct fan-powered terminal box.

full recirculation with reheat can cause excessive temperature set-point overshoot in the zone, which can fry computers and other equipment. In older buildings you may encounter *bypass VAV* boxes that control the amount of cold air entering the zone by bypassing this air back into the return plenum. These are not used very much any more because they waste energy.

Double-Duct Boxes

In a double-duct system, the terminal box has two inlets and two dampers (Figure 16.7). Since there is always a supply of hot air, no reheat is necessary. In this kind of system, the thermostat controls the position of the two dampers. It is important to have the dampers controlled so that they don't overlap (that is, so there is not simultaneous heating and cooling), although in practice this does happen. From a comfort standpoint, the dual-duct system is very nice and in the summer you can use outside air to heat with (even in summer, some rooms need a little heat).

In a double-duct VAV system, you need to pay close attention to the distance between the box and the diffuser. If there's a little bit of overlap in the box and the VAV air handler uses a single fan, one deck could be at a higher pressure, allowing air to flow backwards down the other deck. Usually the hot deck is at a lower pressure, so you have no heating capacity in the zone. If the air goes backwards through the supply duct, the flow cross or pitot tube will register zero air flow and the damper will open all the way. This does not

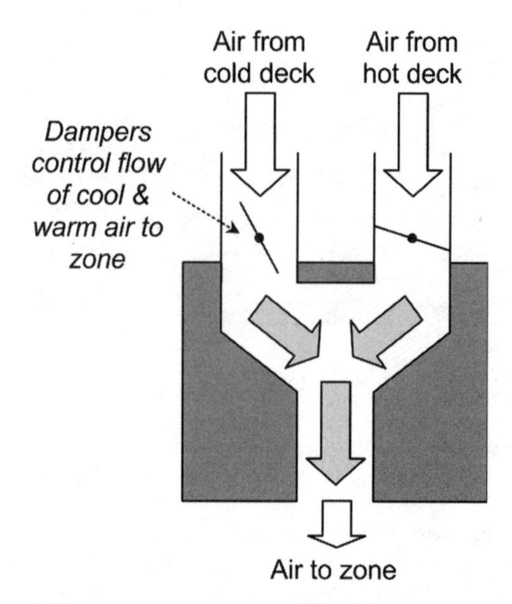

FIGURE 16.7 Components of dual-duct terminal box.

help things. For example, cold air might end up traveling back up the hot side and entering the hot deck supply and exiting in another zone that is calling for heat. Obviously, this can be a real problem.

On the other hand, single-duct systems can create a big problem in interior rooms with no baseboard and no reheat. If you set a minimum airflow rate that is too high and the zone is not fully occupied, it can get too cold. While the single duct system has lower initial cost and easier control, its advantages over double-duct VAV are not always clear from a comfort standpoint. For example, in the summer you have "free" heating of the hot deck, while with a VAV system you have to heat water or steam to provide the reheat.

Induction Boxes

A variation on the single-duct system is the induction box (Figure 16.8) where the supply air passes through a series of conical passages that create a mild suction and induce the secondary flow from the room. These kinds of units are usually found in perimeter zones under windows but can also be located in the return air plenum. A constant supply of air at about 72°F and high pressure (~6 inches

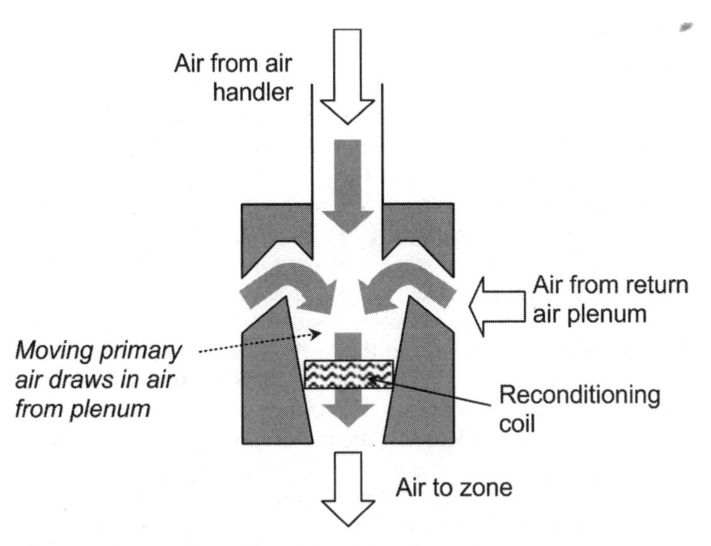

FIGURE 16.8 Components of induction terminal box.

W.G.) is supplied to the box. Hot water and chilled water valves modulate the flow of water in the reconditioning coil to maintain the room temperature. The supply air to an induction unit is usually 100 percent outdoor air.

It is very important to keep the static pressure up in the induction primary air line. These units also have a tendency to pull dirt and dust into the unit because it recirculates a lot of air back into the room. These are relatively high-maintenance units because the coils and induction nozzles must be kept clean. However, they are great for indoor air quality because the entering air is 100 percent outside air whether the zone is in heating or cooling mode.

fastfacts

A contractor connected a VAV box to the primary air for an induction system and then couldn't figure out why there was no cooling and why the box would not shut off. He spent two weeks trying to make it work before being informed that system was not VAV.

Diffusers

Diffusers spread air from the terminal box into the zone. Most diffusers are designed to fan the air out across the top of the ceiling so that cool zone air can slowly drop down into the zone without creating drafts. This works well in the cooling season but has obvious disadvantages in the heating season when the air just stays near the top of the room. Usual square diffusers (Figure 16.9) work well except that if air comes down across walls and across the floor it can cause a draft. The big challenge is to locate the diffuser so that it effectively distributes air in the zone without short-circuiting air back into the return air grill, and so that it does not create drafts. Sometimes a baseboard is required not for heating but simply to keep air traveling up the walls and windows.

Perception is a problem. If there is a group of offices where some diffusers are small and some are big, it is not uncommon for people to complain because they think they don't get the same amount of air out of each. This occurs even when it is easy to feel the air coming from the small diffuser because it is at a higher velocity. In high pressure systems, small diffusers are noisier than larger ones.

FIGURE 16.9 Zone air diffuser.

Fan-Coil Units and Radiant Heating

Air-water systems are another method of controlling the zone temperature. These systems usually involve a *fan-coil unit* that contains its own fan and hot and/or chilled water coils. The air flow is set by the occupants through a multiple stage switch, and the thermostat controls the amount of water circulated through the coils. The air in such a system often comes directly from the outside such as shown in Figure 16.10. In some cases, the fan-coil unit can use recirculated air from the zone (Figure 16.11) to help offset the load on the coils that would occur with 100 percent outside air.

Radiant heating systems use piping incorporated into the physical structure of the building, as shown in Figure 16.12. The pipes can be in the floor, walls, or ceiling, and hot or cold water flows through the piping in response to a thermostat signal. As the building surface changes temperature, so does the air in the space. Other variations of radiant surfaces use ductwork in the surface (usually the floor) for closed air circulation. These kinds of systems are fairly common in northern Europe but rare in the United States. The floor panels in a radiant heating system are often controlled by the outside air temperature. The maximum allowable temperatures are about 85ºF since people are walking on the floors. Wall and ceiling panels have less mass and can operate at temperatures up to 100ºF and 120ºF, respectively. Temperature sensors for radiant heating should be located away from the panels so that the control maintains the proper air temperature and is not biased by direct heating from the panels.

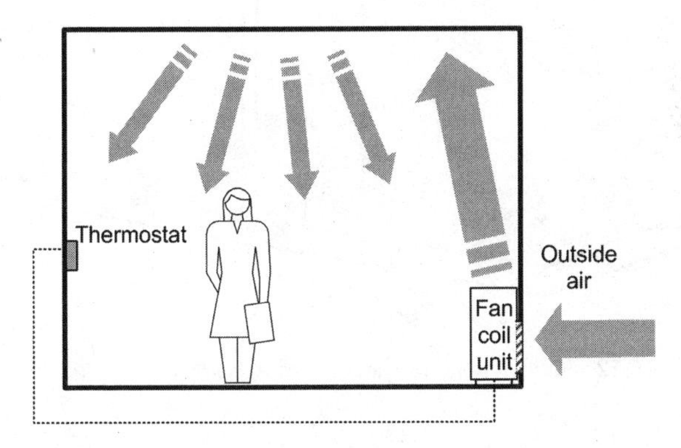

FIGURE 16.10 Fan-coil unit used for local control of space temperature.

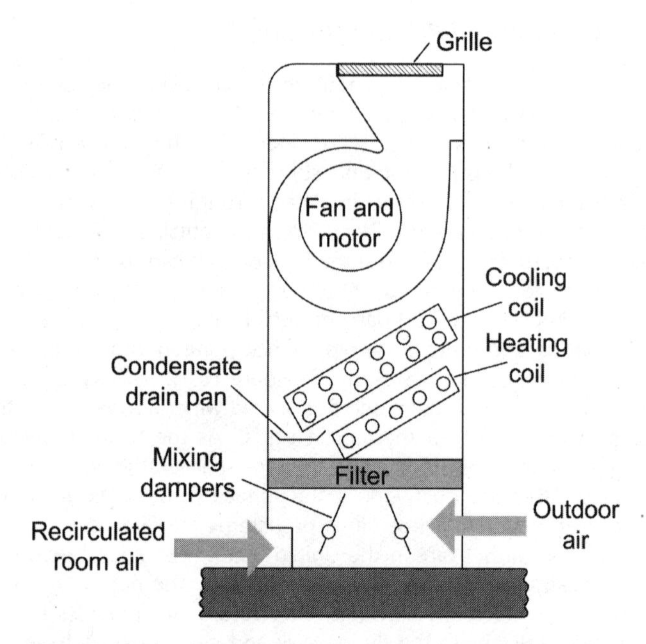

FIGURE 16.11 Cross-section of four-pipe fan-coil unit. *(From Kreider et al., with permission)*

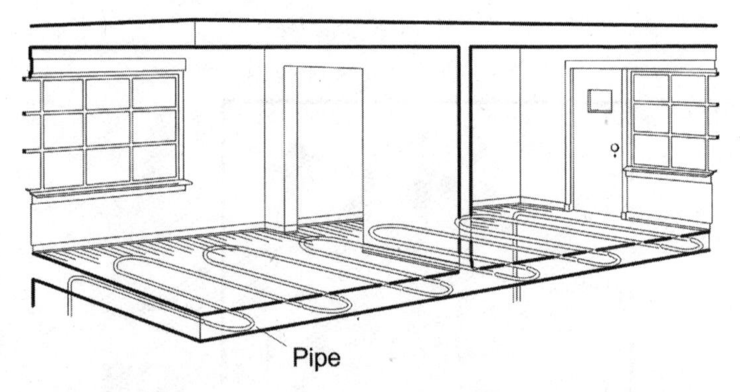

FIGURE 16.12 Radiant heating system. *(From Kreider et al., with permission)*

Thermostats

Thermostats measure the temperature in a zone, compare the temperature to the setpoint, and send a control signal to the zone terminal system. Most zone thermostats use some form of coiled bimetallic strip to either swivel a mercury-filled bulb to complete electric circuits on electric thermostats or to change the pressure control bellows on pneumatic thermostats. Figure 16.13 shows two pneumatic zone thermostats, both in good working condition, that show how the state of the art has not evolved much over the past century. The thermostat on the left was manufactured in the mid 1920s while the one on the right was manufactured in 2001. Both use the same pneumatic principles and even accommodate the same size tubing.

The primary function of the thermostat is to tell the zone terminal system that the zone is too warm or too cold and to take the appropriate action. There can be several other features of thermostats commonly found in HVAC systems:

- The anticipator on an electric thermostat tries to anticipate the residual heating that occurs after the heating system turns off. The anticipator is a small coil of wire that sits underneath the temperature-sensing element and is heated with a small electric

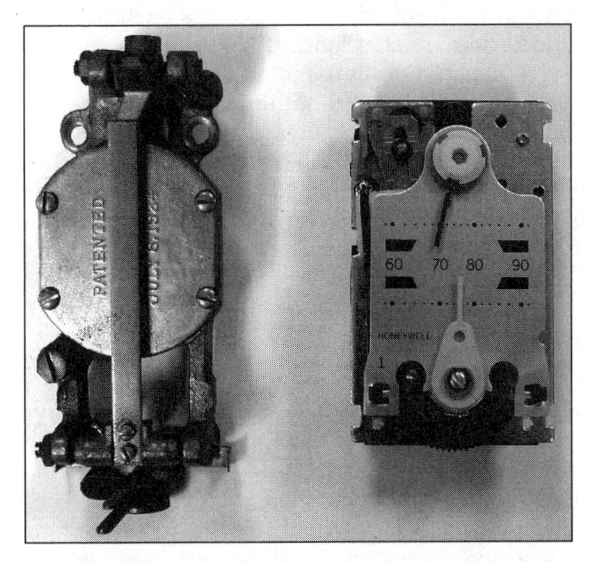

FIGURE 16.13 Old and new pneumatic zone thermostats.

current whenever the heating system is on. The longer the heating system is on, the hotter the anticipator gets, causing the temperature-sensing element to record a warmer zone temperature than actually exists. This causes the thermostat to turn off before the zone reaches the setpoint temperature. The anticipator is adjustable in the range of 2° to 10°F anticipation to account for the building dynamics and the distance of the thermostat from the diffuser.

- Dead band is used to prevent the heating or cooling system from cycling on and off too rapidly. The deadband allows you to program in some hysteresis so that the temperature at which the system turns on is a few degrees different from the temperature at which it turns off.
- Dual-temperature thermostats are essentially programmable devices that have an internal clock and the ability to use different setpoints for day/night schedules, weekend/weekday, summer/winter, etc.

Sometimes there are three to five rooms on a single VAV box and a single corresponding thermostat. This works well if all of the rooms have the same loads and occupancy schedules, but it creates problems when there are load differences in different rooms. The classic example is the interior receptionist office controlled by the thermostat in the boss's office that gets full afternoon sun. As the cooling load goes up after lunch, the boss's office is comfortable while the receptionist needs to put on a sweater, turn on the space heater, etc. In remodels, it is not unusual to find that a wall has been erected between the diffuser for a mixing box and the thermostat that controls that box. In both of these cases, there are no easy solutions to the comfort problems—there was a mistake made by the designing engineer that can only be fixed by a significant reworking of the terminal systems.

CONTROL

Control of the induction and fan-coil units is fairly straightforward. These operate like smaller air-handling units (see Chapter 14) and control water flow to maintain zone temperature. On the single and dual ducts, however, the controls are different.

In the VAV single-duct system, the thermostat controls three things: how much primary air enters the zone, how much air is recirculated back into the zone, and how much reheat is supplied. Figure 16.14 shows a typical control strategy for such a terminal box with reheat. If the zone is too hot (right side of the figure), the primary air damper is full open to allow as much cold primary air as possible into the zone. As the zone begins to cool off, the damper begins to close. If the zone gets too cold, then the fan may come on, recirculating warm air back into the zone. If the zone continues to cool off, then the primary air damper is kept closed to allow only the minimum amount of air necessary to maintain the indoor air quality, and the reheat control begins to ramp up.

The reheat system in single deck boxes can be water or steam coils, or staged electric heaters. In the case of water or steam coils, the increase of the reheat signal represents an opening of the valve. On pneumatic reheat valves, there is usually a spring range that can be adjusted so that there is no overlap with cooling. You don't want the damper opening up until the reheat valve is fully closed. On pneumatically controlled baseboard heating connected to the VAV box, the baseboard valve coil is usually 3 to 6 psi and the VAV damper starts out around 8 to 13 psi, giving a 1 or 2 psi deadband.

In the case of electric coils, the coils can be adjusted to come on at different temperatures. There are usually thermostats in the VAV box controls to set the zone temperature at which the heater stage comes on.

The control of dual-duct systems is similar except that instead of reheat you have the hot deck damper opening up. Figure 16.15 shows two different strategies for dual-duct terminal box operation.

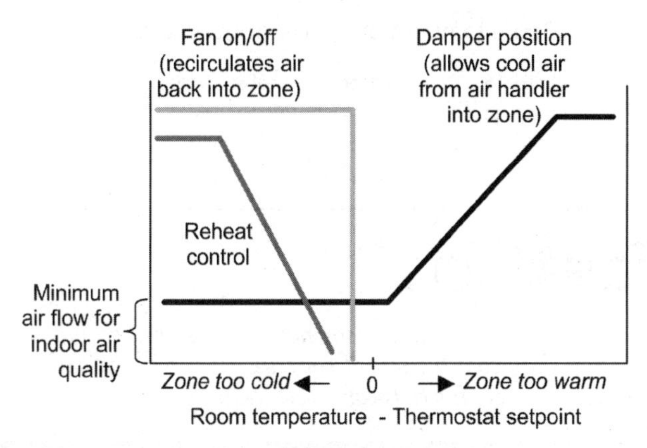

FIGURE 16.14 Reheat box control strategy.

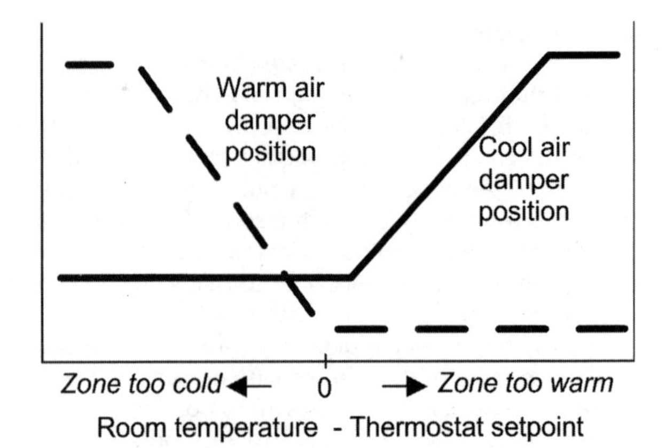

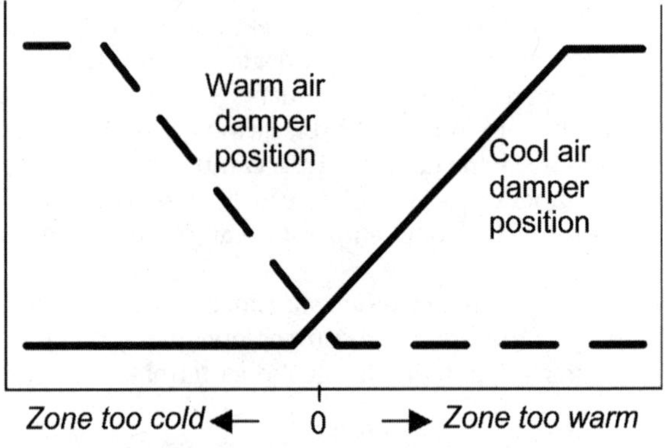

FIGURE 16.15 Dual-duct box control strategies.

fastfacts

If using electric reheat in the terminal box, you must ensure that the fan is on before the reheat coils are energized. Too little air flow across the electric coils can cause them to overheat.

Both of these allow for some minimum air flow necessary for indoor air quality control, but vary in having overlapping heating and cooling modes.

Most single-duct and dual-duct terminal boxes are meant to work with upstream duct static pressures of 1 to 1.5 inches W.G. If the static pressure is too low, the controller response will be sluggish. If the pressure is too high there will be trouble controlling the box air flow and the damper will be at the minimum position most of the time, causing noise problems. Inlet duct static pressures over about 2 inches W.G. generally create noise problems.

SAFETY

When working on terminal boxes, one is usually in an inconvenient, confined space like a zone return plenum. There are a number of hazards in such spaces, including building structural elements, suspended ceiling support wires, high voltage wires, and possibly steam and hot water piping.

- Many plenums are filled with wire-ways and trapezes used to support piping. The supports for these are dark and can be quite sharp. It is a good idea to wear head protection whenever you are inspecting or working in a plenum.
- The wire used to support the frames for suspended ceilings is rigid and the ends can be razor sharp. These can be very difficult to see, yet they can produce spectacular injuries. Always be wary when working above suspended ceilings. Eye protection is a must, and head and hand protection is recommended.
- Do not place heavy tools on suspended ceiling panels. A heavy drill or hammer can cause the panel to collapse, dropping the tools onto the heads of anyone below. Even if the panel appears to be sturdy, it may have been weakened by exposure to water or steam leaks. Keep all tools in a tool belt or on dependable surfaces.
- Even if vibration isolation duct sections are used, there can still be an electrical connection between the terminal box and any flex duct and supporting structure. A high voltage short-circuit can suddenly (and hopefully temporarily) energize the ductwork.

One of the big problems in a plenum space can be dust that has accumulated on the ceiling panels. If you pick up a panel and drop it down onto another panel, you can raise a pretty good size plume of dust and ceiling panel fibers. This is not so good to breathe and it can have other unintended consequences. For example, some smoke detectors (Figure 16.16) pick up particles in the air, not just heat. Creating too much dust in the vicinity of one of these may activate the alarm, so instead of simply fixing a terminal box problem, you have also managed to evacuate the building.

FIGURE 16.16 Smoke detector located on suspended ceiling.

fastfacts

During work on a VAV box, a drop-in light fixture had to be moved out of the way. As it was being moved, a large spark shot from the light to the suspended ceiling grid. It seems that the light fixture was operating through the grounded ceiling grid to complete the circuit.

TROUBLESHOOTING

A lot of maintenance problems with terminal systems are actually design problems. Sometimes the occupants take the HVAC systems for granted and are not willing to assume responsibility for zone comfort problems. This section describes some of the common troubleshooting techniques you can use for terminal systems. In most cases, the suggestions also apply to induction units and fan-coil systems.

Zone Too Hot

A zone that is too hot can indicate an overactive reheat system or a lack of cool air getting to the zone. A general flow chart is given in Figure 16.17. If this troubleshooting guide does not help, you will need to check to see if the zone loads are simply too great for the terminal system. You may need to use a capture box on the vents to check the outlet air flow rate, or use a portable pressure/flow meter to check the flow rate and compare this to the chart on the side of the VAV box. This may also need to be checked against the prints from the initial testing and air balance at commissioning. If the room has too much cooling load, the problem goes back to the engineer, who may specify that you need to put in a bigger box. However, this solution works only if you have enough capacity in the air system.

Note 1: The thermostat may not be in the room that is uncomfortable. This is both good and bad—it is easy to identify what the problem is, but it may not be so easy to fix. Find the thermostat

> **fast**facts
>
> *If many zones in the building are too hot, you probably have a systematic problem that can be cured by lowering the supply air temperature (single deck) or cold deck temperature (double deck). Take care, however, not to lower the temperature too much in a single deck system or you might throw the systems terminal boxes into minimum air flow. This can lead to indoor air quality problems and complaints.*

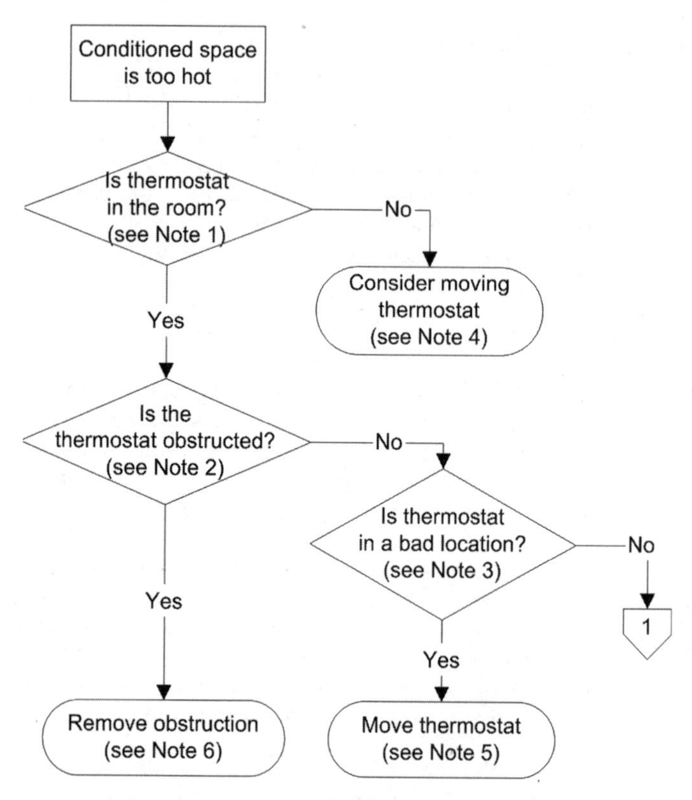

FIGURE 16.17 Troubleshooting flow chart for a zone that is too hot.

that serves the room. It also helps to identify any other zones that are served from this thermostat. See Note 4 for more information.

Note 2: Too often the location of the thermostat does not match with the occupant's desired furniture location. If the thermostat is located behind a bookcase, for example, then it will not accurately register the room temperature.

Note 3: If the thermostat is placed near a window or even on an exterior wall, it can pick up the ambient temperature effects more than the room temperature. If the thermostat is located directly underneath the diffusers, it may reflect the supply air temperature and not the actual zone temperature. Make sure that the thermostat is positioned to get a good, reliable measurement of the air

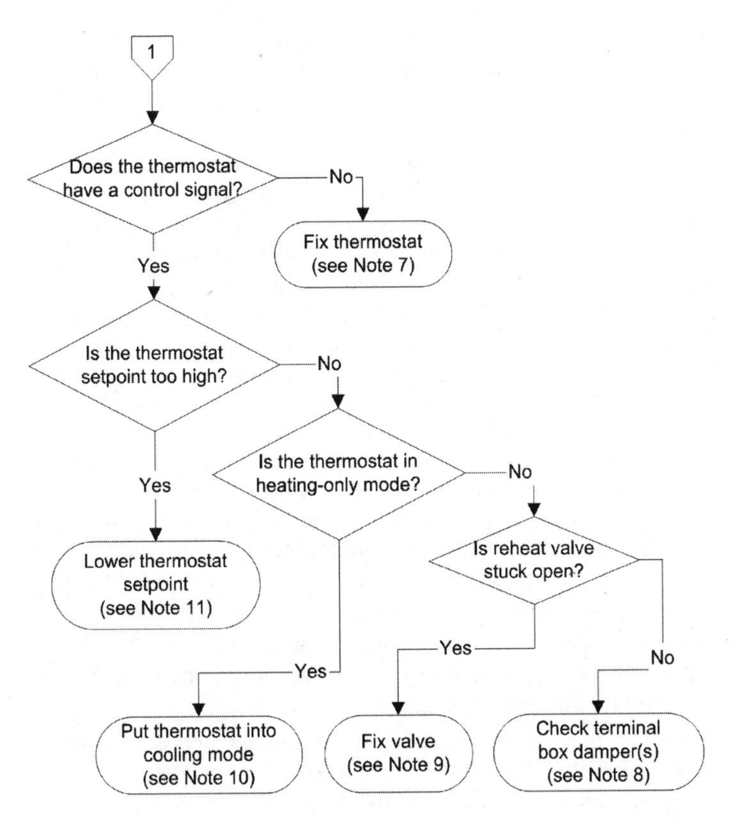

FIGURE 16.17 *(continued)* Troubleshooting flow chart for a zone that is too hot.

fastfacts

A "too cold" call came from the same office at the same time each day. No problem with the HVAC system could be found. On further investigation it was discovered that a copying machine near the thermostat was used for only one hour each day because it was too noisy. During this hour, heat from the machine reached the thermostat, causing the zone to be overcooled.

temperature. Note that some older buildings have walls that open up in the attic. This is common with older plaster and lath walls. If this happens, cold air can drop down the wall and cool interior walls behind the thermostats.

Note 4: Often several smaller offices are considered a single HVAC zone even though the loads in each may be very different. For example, a corner executive office may have a thermostat that controls the terminal box in charge of the corner office and the reception area to that office. If there is a lot of sunlight in the corner office, the solar heating of the office forces the terminal box to let a lot of cool air into the zones. Consequently, the reception area is too cold. A similar situation can occur when there are two similar rooms next to each other and connected to the same terminal box, but the one with the thermostat is unoccupied. This usually causes the other room to be too warm. In such a situation you are dealing with an inherent flaw in the system design. This is more difficult to fix than an obvious mechanical failure. If the thermostat cannot be moved to a more representative location, then one option is to reduce the air flow to the zone that is being overheated or overcooled.

Note 5: Moving a pneumatic thermostat is not a trivial task. This should be avoided if necessary. Try to identify any other problems with the thermostat before taking this step.

Note 6: It may not be possible to rearrange the occupant's furniture to give the thermostat better exposure to the actual room temperature. If this is the case, make sure that the occupant

fastfacts

A technician responded to an overheating complaint. He checked out the VAV box, reheat coil, and baseboard valve but could not find anything wrong. Even the thermostat seemed to be working fine. But when he finally took the thermostat off the wall to check the connections, he felt a cool draft. Cold air was infiltrating down inside the wall and was cooling the thermostat from behind. A little bit of spray foam behind the thermostat mounting plate cured the problem.

understands that the thermostat cannot properly control the space temperature unless it is treated with the same level of concern as the light switches, windows, etc.

Note 7: It doesn't matter if the thermostat is electronic or pneumatic: it should provide some kind of signal to the reheat coil. If electronic, check to make sure that the thermostat has power and that the wire connections are secure. If pneumatic, make sure there is air pressure and that the tubing is firmly attached on both the inlet and the outlet.

Note 8: It is not unusual to find the damper linkage bound up or loose. If the terminal box is a dual duct unit it may be that the heating damper is stuck open. If it's a single-deck system, it is possible that the primary air damper is stuck closed. Most manufacturers put the linkage on the outside of the box, but some are unfortunately inside and it's almost impossible to see what they're doing without taking the box apart. You should also look at the pitot tube or flow cross in pressure independent boxes to make sure the ports are not clogged with dirt or dust. Since the damper position is corrected for air flow, a sensor that registers no air flow can force a damper wide open. Some problems with the dampers can also include:

- Damper linkage can fall apart or fall off the end of the damper shaft.
- Gears in the electric damper actuator can jam or strip.
- Linkages can slip due to a loose set screw—the motor operates but the damper doesn't move.
- Sometimes electric motors go farther than they should, ripping the shaft out.
- The damper can be stripped off the shaft and be lying on the bottom of the terminal box supply duct.
- Something can get inside the damper and jam it shut or keep it open.

Once you take a box apart, it is not working under normal conditions and it may be difficult to troubleshoot duct pressure problems.

Note 9: If the reheat valve is stuck open, there is no way to decrease the amount of heating in the space. See the troubleshooting of valves in Chapter 6. A steam valve with even a small leak can provide considerable heating. Check the outlet pipe of the coil to see if it is warm. If it is, you probably have a leak.

Note 10: Some thermostats (mostly residential) have a heating mode and cooling mode switch. Make sure the switch is in the right position.

Note 11: Wouldn't this be nice? Turn the thermostat down and tell the occupants not to play with it.

Zone Too Cold

If the zone is too cold, first follow the flow chart for the thermostat diagnostics in the previous section. For example, if a heat-producing piece of equipment (photocopier, lamp, computer screen, etc.) is placed under or near the thermostat, the thermostat will be tricked into thinking the room is too warm and will consequently make the room too cold.

Another easy check is to see if the cold feeling in the zone is caused by drafts. The diffusers in a zone should direct the air horizontally across the ceiling of the zone so that the air can slowly drop without creating drafts. Some diffusers, however, are meant to stir the air in the zone or are non-directional grills. The solution in this case may be as simple as replacing the diffuser or, better yet, the placement of a vane insert on top of the existing grill.

If the box is connected to a single-duct VAV system with reheat and the thermostat is working normally, there is a good chance that the problem is with the reheat coils in the box or in the baseboard heating. If it's a pneumatic control system, the first thing to look for is the spring range on the reheat valve to make sure there's no overlap: you don't want the damper opening up until the reheat valve is fully closed. While you are looking at the valve in the terminal box, make sure that the reheat coil itself is not clogged up with dirt, dust, or

fastfacts

A technician responded to a cold complaint in a zone and found that the dual-deck box had been mounted upside down, so that the hot deck was hooked to the cold deck port and vice versa. The ports were well marked but something obviously went wrong during the installation.

fastfacts

A technician was troubleshooting a lack-of-heat complaint in a VAV box with a single-row reheat coil. He increased the hot water temperature until it was over 200°F but still had trouble meeting the room temperature. He finally had to turn up the supply air temperature, but this created problems in areas with high heating loads. He checked the specifications of the terminal box, but the performance simply did not match the manufacturer's rated specifications. The only thing to do was replace the box.

pieces of insulation from the duct and that any pitot tubes or flow crosses on pressure independent boxes are not clogged. If there is air flow through the coil and the spring is okay, then check the water or steam flow through the coil, making sure the piping is hot on both sides. If it is not hot on the outlet and the airflow is at rated flow, then:

- coil may be clogged,
- valve may not be opening fully,
- balancing cock may be closed, or
- hot water pressure or temperature may be too low.

It is also possible that the coils on a VAV box are sized incorrectly and that there's too much air running through the coil to be heated properly. In a box that does not have a reheat coil, the problem is probably that the minimum air flow into the zone is too high. The only recourse is to lower the minimum flow setting, but doing this may violate indoor air quality standards.

If the problem is occurring in a single-duct VAV terminal unit without reheat, the zone may be over-cooled if the proper minimum airflow requirements for the space are followed. This usually leads to setting the minimum airflow to the off position to increase the space temperature, causing IAQ problems due to low or no minimum air. Heat loads in the room cannot be counted on to allow the minimum VAV terminal unit airflow. It is not unusual to find the use of electric space heaters in these rooms, even in summer months. This wastes energy and makes room temperature control difficult. The problem in that zone may be eliminated by raising

fastfacts

Cold air in the hot decks of dual-duct systems can also happen if there is a dropped fire damper somewhere in the hot deck. The cold air shoots back down the duct with the dropped damper and creates problems everywhere. This happens relatively frequently.

the main supply air temperature, but this may lead to insufficient cooling in other spaces. A correct design of the original HVAC system would have prevented these problems.

An unusual situation can happen with dual-duct systems that can make the zone too cold. If the air-handling unit has only one fan, and there is a little bit of damper position overlap in the box, one deck can be at a higher pressure and the air ends up going down the other deck. If the air in the lower pressure duct is going backwards, then the flow cross or pitot tube picks up a zero signal and the damper will open all the way. Usually the hot deck is at a lower pressure, and cold air might end up traveling back up the hot side and entering the hot deck supply. This can make the zone too cold, and it can also make other nearby zones too cold because the cold air in the hot deck ends up exiting in zones that are calling for heat.

PRACTICAL CONSIDERATIONS

There are many different "tricks" that can be used to ensure that even simple terminal box control systems can still properly maintain the setpoint temperature in a space.

- Return air grills in zones should be located so as to eliminate short-circuiting of the supply air. Ideally, the return air vents to the plenum are located on the opposite side of the room from the supply air diffusers. It is relatively easy to change the location of the return air vents if necessary.
- Diffusers in zones with low ceilings should not blow directly downward. This tends to make the occupants uncomfortable.
- Sometimes it is better to use several small diffusers than one large one.

- Temperature and humidity sensors should be located on an interior wall where they can measure values that are representative of the whole space. If necessary, these sensors can be located in the return air duct as close to the space as possible. This is sometimes required if an appropriate location in the zone cannot be found.
- When performing an air balance on a VAV system, always use a capture box to measure flow. Do not rely on data from the VAV flow controller to balance a VAV system since there's no guarantee this is correct.

Pay close attention to the installation and always inspect the installer's work. A VAV terminal unit that is mounted upside down may not operate correctly. An improper amount of straight duct run to or elbows directly before the VAV terminal unit inlet may cause erratic operation, or may not allow the unit to flow the correct amount of air under some operating conditions. Always follow the manufacture's instructions when installing a VAV terminal unit. Make sure that the connections on any differential pressure pick-up probes or flow crosses are correct. If necessary, use a U-tube manometer to identify the high and low connection taps. Note, however, that airflow controllers may not be required if the distance from the fan to the terminal unit is short and the duct static pressure is constant. A simple damper operator with a manual high and low airflow stop controlled by a thermostat works well in these conditions. In the case of long duct runs or varying duct pressure due to other operating terminal units, the use of flow controllers is required.

Maintenance is unfortunately overlooked too often for terminal boxes. Fan-powered VAV terminal units require regular maintenance to operate properly. If a regular maintenance schedule is not followed, filters may become clogged, fan motor shafts may seize up, and reheat coils or electric heating elements could become nonfunctional. One quick way to check for problems in a terminal box is to compare the air flow rates and reheat status as you change the thermostat setpoint temperature.

- A properly operating cooling only VAV terminal box controller maintains the preset minimum flow at a high thermostat setting and the preset maximum flow at the low thermostat setting. At the setpoint of the thermostat the airflow should be at a point between the maximum and minimum flow setting.
- A properly operating VAV terminal box with reheat maintains the preset minimum flow and provides maximum heat at a

high thermostat setting and the preset maximum flow with no added heat at the low thermostat setting.

- A properly operating double-duct VAV terminal box controller supplies maximum preset hot air flow at a high thermostat setting and maximum preset cold airflow at a low thermostat setting. At the thermostat setpoint both the hot and cold airflows should be at the preset minimum.

Be careful when making adjustments to the mixing boxes, and make sure you know what you are adjusting. A properly operating flow controller should maintain a constant, correct flow to the room in accordance with the demands of the space thermostat under all supply duct static pressure conditions. Always check the VAV fan system and ducting first before adjusting the thermostat or VAV box controller. A change in one VAV box controller may affect another VAV box controller. It is easy to throw the entire VAV system out of balance if enough changes are made to enough boxes. The only way to put a system back in proper balance is to rebalance the complete VAV system.

Zone Pressurization

In open plenum returns, some areas farther away from the main return duct don't get very good circulation. This can create a balancing problem unless you find a good place to put the space static pressure sensor. This can be a complicated problem in zones with fume hoods or special pressurization needs.

Space static (not duct static) can be a challenge to maintain. Sometimes if there are several air distribution systems serving a single building, the systems can work against each other. For example, the supply air from one system leaves a room and travels down the hall to the return of another system. This can create a wind-tunnel effect, particularly if you have one fan system on one side of the building and another on the opposite side of the building.

Another problem occurs if a space goes negative (pulling out more air than allowing in) during cold weather. Subfreezing air can infiltrate into the zone, enter the pipe chases and end up freezing pipes. If you are having regular trouble with frozen pipes in a building, check the building static pressure control. This can freeze pipes in exterior walls.

chapter 17

EVAPORATIVE COOLING

Evaporative cooling can be used in dry climates quite effectively. Such systems are gaining in popularity for a number of reasons: CFCs are not used in the cooling process, energy consumption is significantly less than with conventional cooling, and high ventilation rates are a natural by-product. This chapter describes the basic functions and control algorithms for evaporative cooling.

THEORY OF OPERATION

Evaporative coolers (sometimes called *swamp coolers* or *air washers*) take advantage of the cooling induced by the evaporation of water into an airstream. A paper, cellulose, fiberglass, or wood fiber media is soaked with a continuous supply of water. Air then passes through the media, and as the water evaporates it draws energy from and cools the airstream.

The process takes advantage of the psychrometric path of evaporating air. Figure 17.1 shows the psychrometric chart with an example. This chart shows air entering at 95°F and about 25 percent relative humidity. As water evaporates into this airstream, the evaporation process decreases the air temperature and increases the air humidity along the wet-bulb line, with the leaving air condition dependent on how long the air remains within the evaporative media. The chart here shows the air leaving at about 71°F and 85 percent relative humidity. The only energy requirements are for the fan and the water pump.

Direct evaporative coolers don't actually reduce the energy content (enthalpy) of the air. The air is sensibly cooled (the dry bulb temperature drops), but the latent load is actually increased (moisture is added to the air), so that there is no net increase or decrease in the enthalpy of the air.

The *effectiveness* of an evaporative cooler provides an indication of how close the leaving air temperature approaches the wet-bulb temperature. In other words, the effectiveness tells how far the process moves along the wet-bulb line towards the saturation curve. For example, if the air leaving the evaporative cooler is equal to the entering wet bulb temperature (that is, it is at saturation), then the effectiveness is 100 percent. If the temperature of the air leaving the evaporative cooler is the same as the entering dry bulb temperature then the effectiveness is zero. The effectiveness (sometimes called the *saturation efficiency*) is given as

$$\varepsilon = \frac{T_{db,\,in} - T_{db,\,out}}{T_{db,\,in} - T_{wb,\,in}}$$

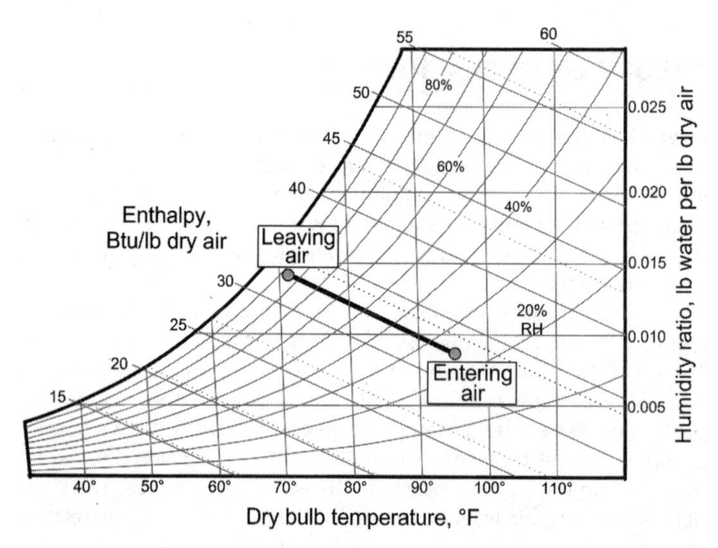

FIGURE 17.1 Psychrometric chart showing path of evaporative cooling.

where T_{db} and T_{wb} are the dry-bulb and wet-bulb temperatures. The effectiveness depends on the thickness of the evaporative media, the amount of water in the media, and the face velocity of the air passing through the media. Figure 17.2 shows one manufacturer's cited effectiveness for different media thicknesses. As the media thickness increases, the effectiveness also increases. This makes sense, given that there is more surface area available for evaporation. However, note that as the media thickness increases, so does the pressure drop across the media. In fact, the pressure drop increases as the square of the airflow rate, so that roughly speaking if you double the airflow rate the pressure drop quadruples. This quickly leads to significant fan power increases necessary to achieve the desired airflow.

Direct evaporative cooling circulates the air that passes through the media directly into the conditioned space. Figure 17.3 shows the components of a typical rooftop direct evaporative cooler with side discharge. In this system, the media is constantly soaked with water using a small pump that draws water up from the sump to the top of the pads. Depending on the type of media used, the direct systems can have airflows anywhere from a couple thousand to a couple hundred thousand CFM per unit.

In air-handling units, a direct evaporative cooler has the same configuration as a cooling coil: it stretches across the face of the air-handling unit and has a catch pan for water that drips off the media. Figure 17.4 shows the downstream side of a large evaporative cooler with the circulation pump situated above the pan. In these kinds of systems, the evaporative cooling media comes in modular panels (Figure 17.5) that may be arranged several layers deep across the width of the coil.

Indirect evaporative cooling systems cool an airstream without humidification. Figure 17.6 shows the basic components of an indirect system where outside air is cooled by evaporation and then passed through an air-to-air heat exchanger to cool the building supply air. Indirect systems can be used to precool air supplied to cooling coils. This helps reduce the electrical load associated with a chiller or DX unit. Indirect systems can also be successfully used in recirculated return air streams where the air is already warmed by occupants and lighting.

Combined *direct and indirect* systems can produce cooling equivalent to conventional mechanical cooling systems. The indirect system is used to drop the inlet air temperature, and then the direct system provides additional cooling and humidification. Figure 17.7 shows how the process would appear on a psychrometric chart. The

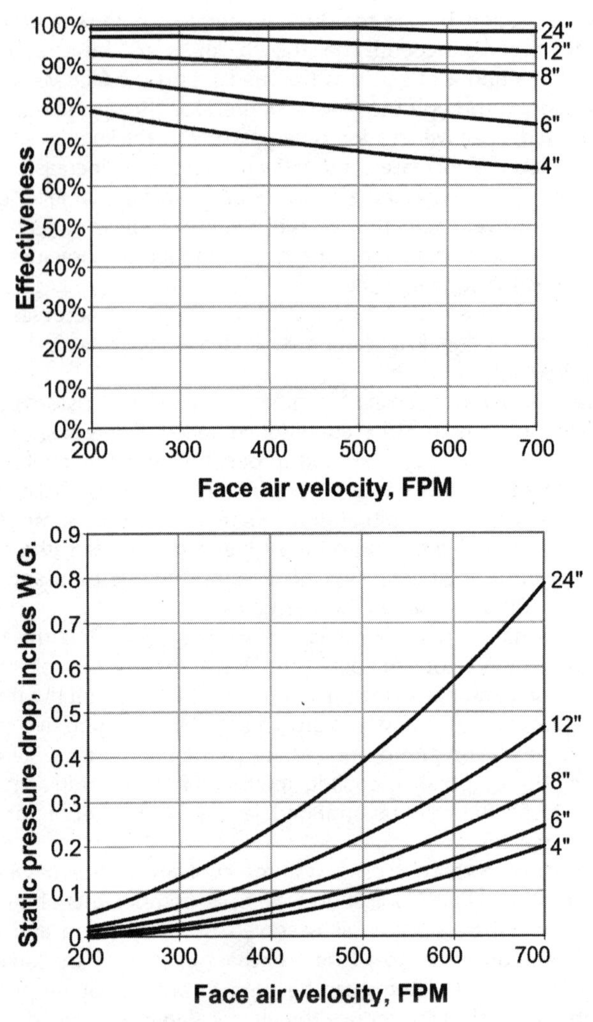

FIGURE 17.2 Evaporative cooling effectiveness and pressure drop for different media thicknesses.

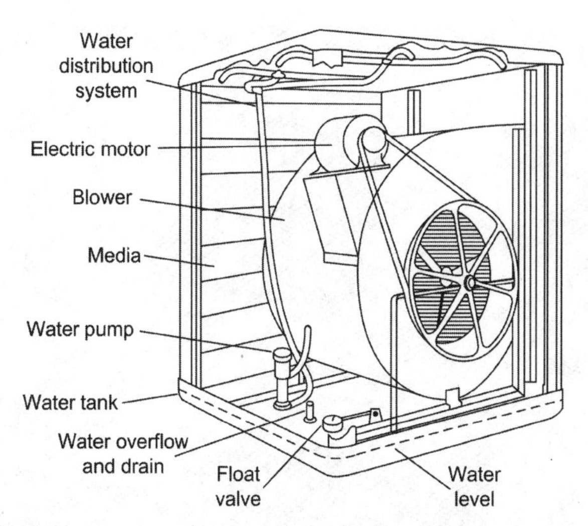

FIGURE 17.3 Components of direct evaporative cooler. *(From Kreider et al., with permission)*

FIGURE 17.4 Swamp cooler circulation pump situated above sump.

FIGURE 17.5 Evaporative cooling media. *(Courtesy of Munters, Inc.)*

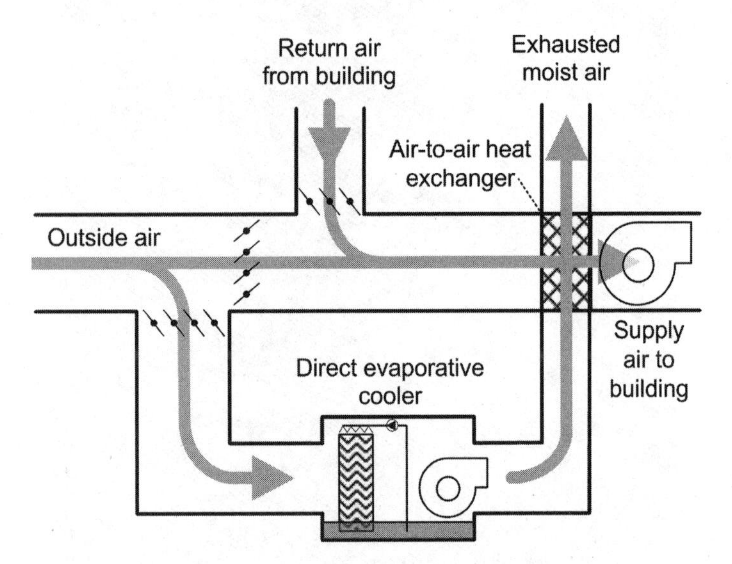

FIGURE 17.6 Schematic of an indirect evaporative cooling system.

indirect portion of the cooling occurs at a constant humidity ratio and the direct portion at a constant wet-bulb temperature. In direct-indirect systems, the equivalent COP can be on the order of 10 to 15. This compares very favorably with even the best mechanical cooling systems.

Another trick with evaporative cooling is to use a small chiller or other cooling device to cool the water before it is sprayed onto the media. The cooler water produces cooler air temperatures and lower humidities in the outlet air stream. However, the water must be cooled below the wet bulb temperature for this to work.

Some new air-handling units come with evaporative coolers in which the circulator pump sits up above the water level in the sump (as in Figure 17.4). These can lose their prime and then fail to operate.

The outlet dry bulb temperature of an evaporative cooler can be no less than the inlet wet bulb temperature. For this reason, evaporative coolers are more effective in dry climates. However, the potential for economic evaporative cooling exists in many regions of the United States.

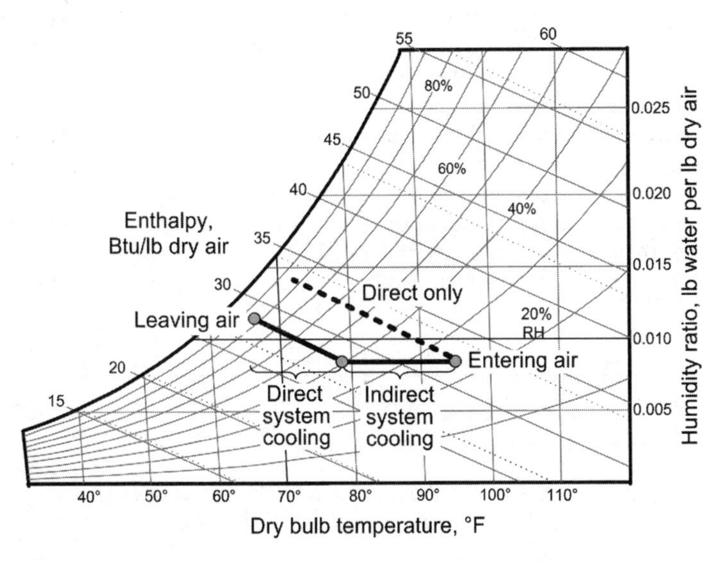

FIGURE 17.7 Psychrometric chart showing direct-indirect evaporative cooling.

CONTROLS

Most direct evaporative systems have multiple winding motors. The control is limited to a simple on/off and by stepping the motor rotation speed (which in turn varies the air flow rate). In large buildings with multiple evaporative coolers serving the same space, each individual cooler is connected to a separate thermostat that turns the fan on and off.

Control of the direct and indirect cooling systems is similar to that of a conventional DX cooling system. When there is a call for cooling, the dampers that allow air to pass through the air-to-air heat exchanger in Figure 17.6 begin to open and the evaporative cooler fan turns on. In a direct system these components are within the main air stream. Connected to the fan on/off circuitry is a circulator pump that keeps water flowing across the media. In direct systems, the start-up sequence may be staggered so that the circulator pump comes on first, allowing the media to be wet before the air starts flowing. This can help prevent an initial rush of warm air into the space.

Control of the direct-indirect evaporator is illustrated in Figure 17.8. Figure 17.8a shows the operation under economizer mode where the cooling coils are bypassed to avoid the associated pressure drop to save fan energy. Figure 17.8b shows the air path on the first call for cooling, where the indirect evaporative coil is activated. As the cooling load increases, the direct system is activated to supplement the indirect system as shown in Figure 17.8c. If additional cooling is required, the mechanical system is turned on. In any case, make sure that the outside air dampers are 100 percent open before you turn on the circulator pump.

Using a bypass around the evaporative cooler for temperature control can be problematic. Usually in a face and bypass situation, the circulation pump keeps running. This can cause a swampy odor to build up; when the face damper is opened you may have indoor

fastfacts

Evaporative coolers have about a half-hour carry-over as the pads dry out once the pumps are turned off. They are definitely not off/on devices.

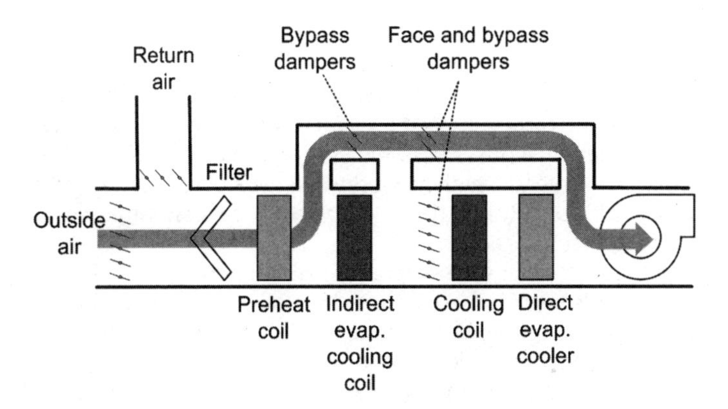

a. economizer mode

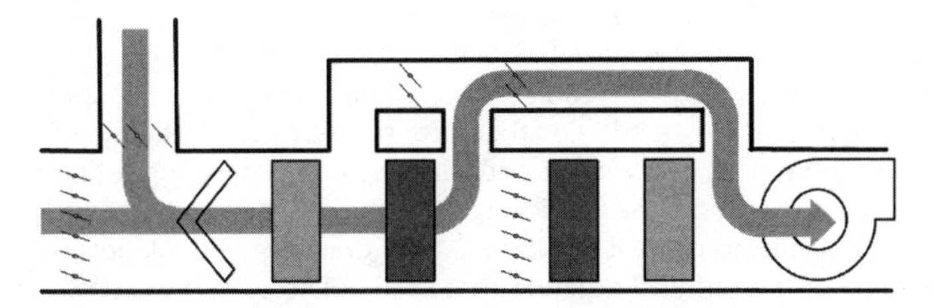

b. indirect evaporative cooling only

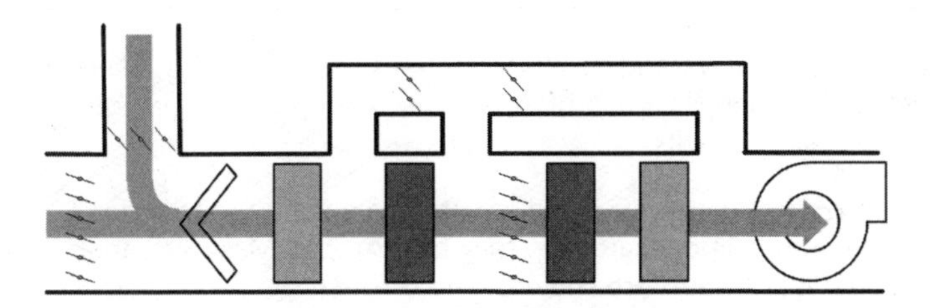

c. direct-indirect evaporative cooling

FIGURE 17.8 Mechanical cooling system with direct-indirect evaporative cooling and economizer.

air quality complaints. If this problem occurs, you should cycle the circulator pump on and off for temperature control.

In some instances, evaporative cooling is used for humidity control. If this is the case, you need a steam or hot water coil for reheat at the zone terminal boxes or at the evaporative cooler itself. This is necessary for winter when the air is dry and the air leaving the media gets very cold. Usually you have better control success if the reheat coil is right after the cooling coil.

Don't use return air with evaporative cooling, especially if the circulator pump control is used for both humidity control and cooling. Recirculating the return air back into the building can increase the humidity to uncomfortable levels in the occupied spaces. This can also lead to mold growth in papers and books. If necessary, use a high-limit humidity cut-out and chilled water or DX cooling. If you are using the cooler for humidity control, also consider some form of reheat coil. Do not try to stage evaporative coolers with different pumps or one pump with different speeds. You can turn it on or off, but if you try to stage the coolers you will most likely have hot spots in the pad and create additional control problems downstream.

Drain the pans every night and turn the pump off using a timer if necessary. Letting the media dry out for at least an hour each evening seems to make it last longer. Plus, draining the tank gets rid of some of the dirt and organics that accumulate during the day.

SAFETY

As with all motors and electrical systems, care must be taken to ensure that loose hair, clothing, and jewelry do not get caught in moving parts or belts.

- Most direct evaporative systems stand alone on rooftops, far from where immediate help may be available. Make sure that your whereabouts are known.
- The fan belts on rooftop evaporative coolers are rarely in cowlings or other protective enclosures. The fans can start at any time, so be careful where you place your fingers. Never hold on to the fan pulley for leverage or support.
- Many direct systems operate on relatively low voltage (120 VAC). This can mean that there is no local electrical disconnect. The remoteness of rooftop units can also prevent a direct line of site between the circuit breaker and the equipment. When servicing

such a device, make sure that the power is off to the circuit breaker and that the circuit breaker is properly tagged.

- By nature of their operation, evaporative coolers have electricity and water in close proximity. When working on water-related items such as the pump or water lines, take care to not have water spray on the motor or electrical connections.

Regularly inspect the catch pans under the media. These can rust out easily if you use galvanized or painted steel. Once the pan begins leaking, it can weaken the flooring of the cooler and cause shifting of the fan or damage to the rooms below.

Use stainless steel for the evaporative cooler pan. They may be more expensive initially but they will be cheaper in the long run.

TROUBLESHOOTING

This section discusses some of the potential problems and solutions involving evaporative cooling systems. Note that the majority of problems have very simple solutions. For example, if there is no water in the pan (Figure 17.9) or the media itself is so old that it is falling apart and not able to hold water (Figure 17.10), there will be no cooling.

FIGURE 17.9 Swamp cooler sump refill level float not working (no water in pan).

FIGURE 17.10 Old evaporative media crumbles at the slightest touch.

For more difficult problems with an evaporative cooling system, refer to the following troubleshooting flow charts. These charts are for direct evaporative coolers but many of the solutions also apply to indirect systems. If there is an indoor air quality problem (either noise or odor), see Figure 17.13. If the conditioned space is too cold, see Figure 17.12. If the conditioned space is too hot, see Figure 7.13.

Note 1: The belts and bearings should be checked periodically to catch problems before they become critical. Many motor bearings are sealed and require no maintenance, but the fan bearings require annual lubrication.

Note 2: If the building is too cold and the system appears to be operating correctly, it may simply be that the cooling load in the zone is too small for the given capacity of the evaporative cooler. Check to see if the loads (computers, kitchen equipment, lights, number of occupants, etc.) have decreased significantly since the equipment was installed. What follows are questions to address.

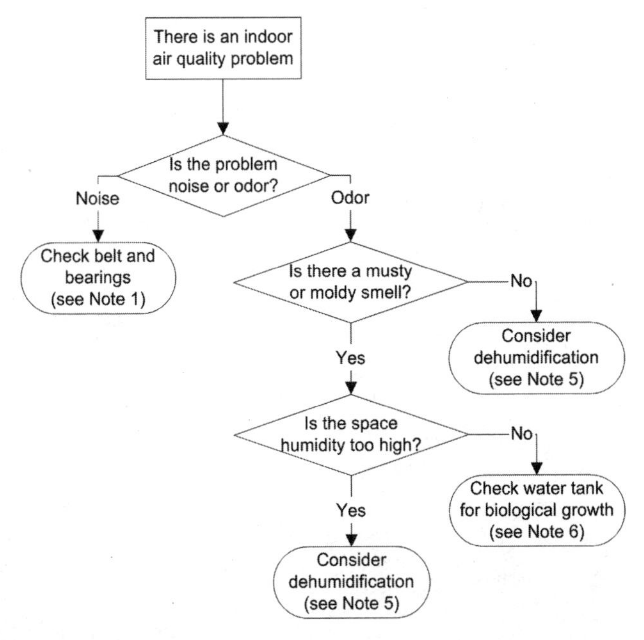

FIGURE 17.11 Troubleshooting evaporative cooler IAQ problems.

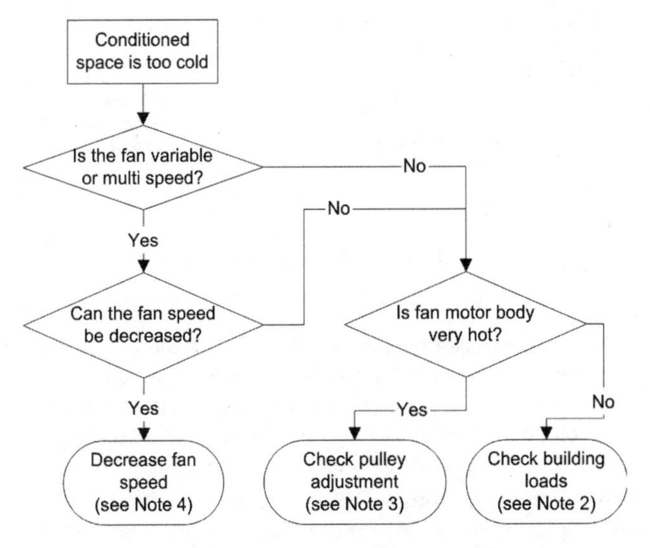

FIGURE 17.12 Troubleshooting evaporative cooler too cold.

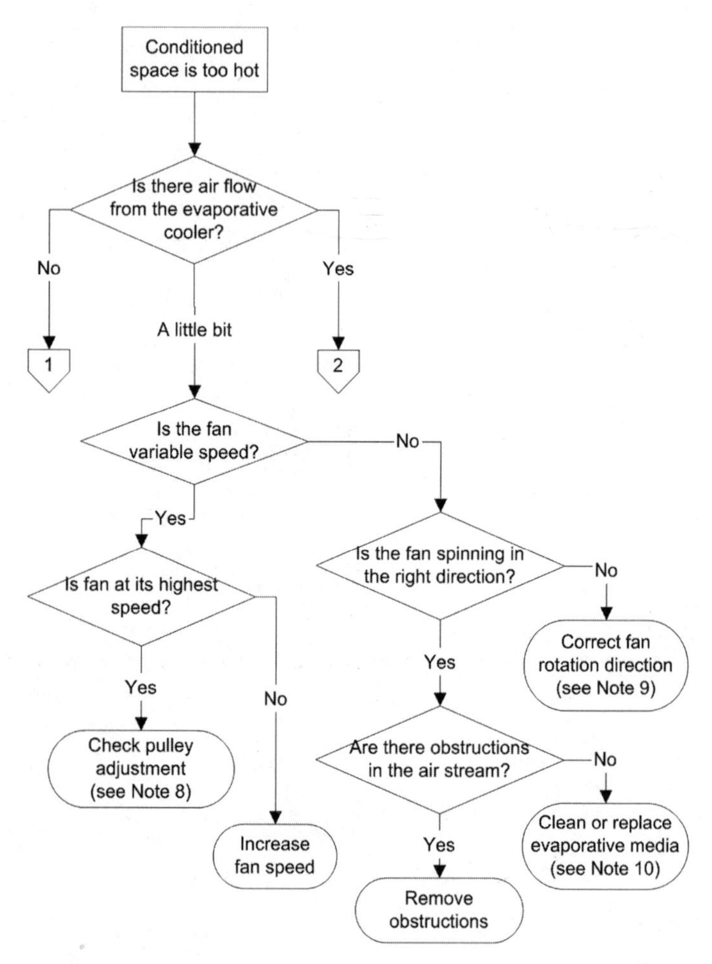

FIGURE 17.13 Troubleshooting evaporative cooling too hot.

- Is the unit controlled by a thermostat? If so, is the thermostat correctly placed so as to get a good sense of the zone temperature? If the thermostat is located far from the air stream coming from the cooler then the cooler may run too long and over-cool the zone before the cold temperature registers on the thermostat.
- Is the humidity or outdoor air temperature too low? If so, consider replacing the pulley on the motor to decrease the fan rotation speed.

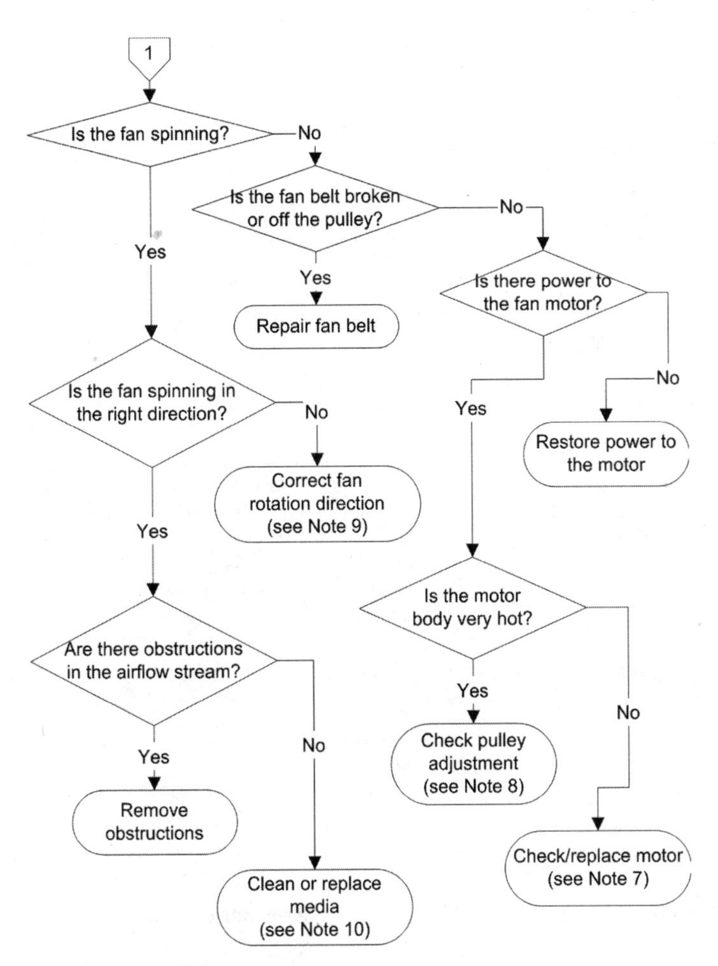

FIGURE 17.13 *(continued)* Troubleshooting evaporative cooling too hot.

Note 3: The pulley on the motor shaft is probably adjustable. That is, the width of the pulley can be modified by releasing a set screw and twisting the two halves of the pulley. Sometimes the set screw comes undone and the pulley can tighten, torqueing the motor shaft and causing undue strain on the motor. This can also cause the fan to spin too fast, putting a higher load on (and shortening the life of) the motor. Check the pulley for the proper setting and correct or replace if necessary.

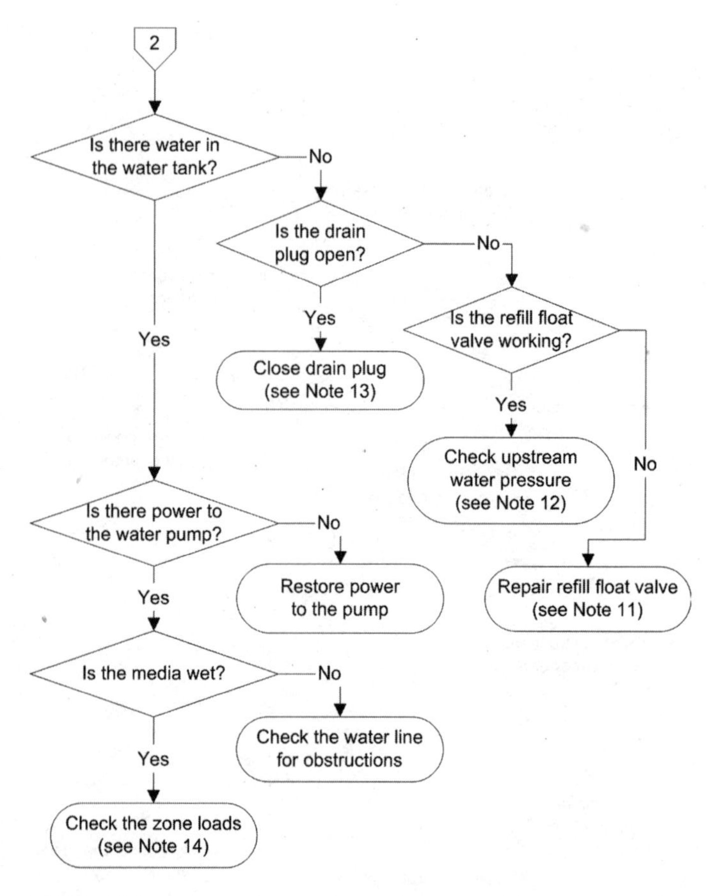

FIGURE 17.13 *(continued)* Troubleshooting evaporative cooling too hot.

Note 4: The fan speed may be set too high. Reduce the fan speed to lower the airflow rate. Also, if the fan speed is controlled by a thermostat, check to see if the thermostat is staging the fan correctly.

Note 5: The direct evaporative cooling process adds a considerable amount of water to the air. This can raise the indoor humidity to almost saturation and can make things uncomfortable for the occupants. A side effect of this indoor humidity is the potential for bacteria, fungus, and mold growth. If you suspect any of these occurring, you should consider a dehumidifier for the zone. This can be particularly bad if there is not enough exhaust from the zone, in which case this may not be a good application for evaporative cooling.

Note 6: The water tank should be treated to prevent biological growth both in the tank and on the evaporative media. Many companies make slow-dissolving tablets that can be added to the tank to prevent such growth. Depending on the usage of the cooler, these tablets can be added monthly, seasonally, or even annually. If possible, drain the pan once a day to let the media and the pan dry out completely.

Note 7: If the motor is not working despite the presence of power, it may need to be replaced. Disconnect the motor power, remove the fan belt, and reconnect the motor power. If the motor does not operate even under no load, it probably needs to be replaced.

Note 8: Most motors have a thermal breaker that stops the motor if it gets too hot. If the motor body is too hot to touch, the pulley probably needs adjustment or replacement. See Note 3 for more information.

Note 9: It is very unusual to have the motor turning in the wrong direction, but it is possible that the fan cage is mounted backwards. If the motor is a large, three-phase motor, it may also be wired incorrectly.

Note 10: Over time, the evaporative media becomes clogged with dust, pollen, and other airborne particles. If the media are noticeably obstructed, it may be possible to wash them with a hose, making sure to spray the water in the opposite direction of the normal airflow. Keep in mind that in large systems the media last only about 3 or 4 years before it needs replacing. After a while it gets mushy and just collapses. If there are several layers (for example, front, middle and back) the outside layers may look okay but the middle section may be squashed and restricting the air flow.

If the media disintegrates, pieces can pile up inside the pump. It can also clog the distribution pipe that goes across the top of the media, preventing water from reaching the media.

Note 11: With known water pressure on the upstream side of the float valve, depress the valve (i.e., drop the float) and see if water comes out the valve. If so, the float level may need to be adjusted. Otherwise, the valve may need to be replaced.

Note 12: Sometimes the easiest solutions are the simplest ones. Check to make sure that there is water pressure leading to the float valve. Often at the end of the cooling season the water is shut off.

Note 13: Winterization of evaporative coolers includes draining the water from the water tank. Verify that the drain plug is not allowing water to drain from the tank. Sometimes this plug doubles as an overflow drain.

Note 14: If the building is too warm and the system appears to be operating correctly, it may simply be that the cooling load in the building is too large to be satisfied by the cooler. Check to see if the loads (computers, kitchen equipment, lights, number of occupants, etc.) have increased significantly since the equipment was installed. Other things to check are:

- Is the humidity or outdoor air temperature too high? Remember that the minimum leaving air temperature from the evaporative cooler is the ambient wet bulb temperature. If it is too humid outside then there will be little if any evaporative effect.
- Is the unit controlled by a thermostat? If so, is the thermostat correctly placed so as to get a good sense of the zone temperature? If the thermostat is located directly in the air stream coming from the cooler then it will probably shut off before the rest of the zone is cool enough.

fastfacts

On the first warm day of the year, occupants of an evaporatively cooled building called the service technician complaining that there was no cooling. The technician showed up and noticed that the pump was running but the media pad was wet only about half-way down. He waited for a while, and soon it got wet all the way to the bottom and started cooling. The technician left, and within the hour received another call of no cooling. He returned and noticed the pad was wet only about two-thirds of the way down. Everything else seemed fine so he left, only to experience similar calls for the next several days. After a week, the frustrated technician entered the mechanical room without bothering to turn on the light but used his flashlight instead. This time he noticed that the pump wasn't running. As it turned out, electricians had recently remodeled and had wired the pump to the mechanical room lighting circuit.

PRACTICAL CONSIDERATIONS

In rooftop direct evaporation systems, the face velocity of the air stream should be no more than about 5 feet per second. Any higher than this and water droplets begin entering the building through the air stream. In direct systems, high airflow rates may cause these entrained water droplets to fall and gather in the ductwork. This can cause corrosion in the ductwork or in the air-to-air heat exchanger. Also remember that the higher the airflow rate, the more rigid the evaporative media must be to prevent deformation and breakage.

In general, evaporative cooling can be used when the wet bulb temperature is less than about 75°F during the cooling period of the day. The cooling effect of evaporative coolers can be enhanced by using low-temperature water, as long as the water is cooler than the air wet bulb temperature.

Evaporative units that have good exposure to the sun may actually lose water to the ambient air through direct evaporation of the water from solar heating. While it is assumed that most of the air surrounding the evaporative cooler is drawn through the media by the fan, it may not always be so. It is useful to periodically make sure that all the media is wet. Since the evaporation is the only cooling mechanism, any dry patches of the media mean that you are introducing warm outside air directly into the building. If the dry patches are predominantly on the sides exposed to the sun, you may want to consider some kind of shading.

Direct evaporative coolers can also be used to improve efficiency of air-cooled condensers on chiller systems. Since the fans already exist, one need provide only the media and a circulation pump.

fastfacts

Direct evaporative coolers do not actually decrease the building load. They only "hide" the load by decreasing the air temperature. The humidity increases and the enthalpy of the air actually stays about the same.

Chemical Treatment and Dirt Build-up

Chemical treatment can be added to the sump tank if necessary, although problems like Legionnaire's disease rarely occur because the tanks operate at low temperatures. If you do use chemical treatment in a direct evaporative system, remember that the water is in contact with the air that circulates throughout the building. Chlorine tablets or other biocides can create indoor air quality problems for the occupants. There is not really much need for chemical treatment of the evaporative cooling water if the sump is drained and dried out every night.

One of the other names for evaporative coolers is air washers. This is appropriate, since they really do clean out the air, but the dirt just ends up in the pan. Use pneumatically actuated ball valves where possible for flow control, since solenoid valves tend to get clogged easily. You want a lot of valve area that won't get clogged with dirt. Also make sure there is a bleed off in the system to remove the total dissolved solids. Once these start to coat the media, it will not last much longer.

The dissolved solids can also be a problem for the next coil downstream of the evaporative cooler (such as a reheat coil). If too much water circulates into the media or if the air flow rate is high, you can have carry-over where drops are blown into the next coil. This can cause calcium build-up on the downstream coil. In fact, you should avoid carry-over under any circumstance, since excess water can also miss the catch pan and end up leaking onto the floor below.

Carry-over usually happens when it is cold outside and the water is not evaporating fast enough.

RESIDENTIAL SYSTEMS

Residential systems are similar to the commercial systems discussed elsewhere in this book, except that they are much smaller. Furnaces are used in residential and light commercial applications to provide heated air to condition occupied spaces. These are often used in tandem with direct expansion cycles to cool the house in the summer. Heat pumps are unitary pieces of equipment that "move" heat to or from the house depending on the need.

PRINCIPLES OF OPERATION

Furnaces

Gas or oil-fired furnaces use a fan to pass air over the external surfaces of a heat exchanger and to circulate the resulting warm air throughout a house (see Figure 18.1). Most furnaces use a standing pilot light to provide combustion to the burner manifold when needed, although some newer furnaces have intermittent pilots where the pilot is lit first by a spark igniter. In natural gas furnaces, a *combination gas valve* provides fuel to the pilot (if used) and also sends fuel to the main burner. The more expensive gas valves have a pressure regulator to maintain a constant fire level. In an oil-fired furnace, oil is forced into the combustion chamber using a small pump. Usually a *thermocouple* is used to verify that the pilot light is working. The thermocouple produces electricity that goes to the combination gas valve and holds the pilot cutoff open. If the pilot goes out, gas flow ceases to the pilot and to the main burner. Within the combustion chamber, fuel is burned in the combustion air to create hot gases that percolate up through the heat exchanger and then pass through the flue to the outside.

A *natural draft* furnace has a draft hood to direct the flue gases that naturally flow up through the flue. These are common for residential use. Some furnaces may be *forced draft* or *induced draft* where a blower motor is used to force air into the combustion chamber. The absence of a draft hood reduces off-cycle losses, resulting in increased efficiency.

Many new furnaces also have totally sealed combustion chambers where no room air is used. These are referred to as *direct venting furnaces*. All combustion air comes in through a PVC pipe vented to the outside. A blower is used to pressurize the combustion chamber. In these kinds of systems there is no interaction between the room and burner/combustion chamber, so the room can be tightly sealed.

Furnaces can include DX coils connected to split systems or heat pumps. These coils are used for cooling in the summer. There can also be spray humidifiers for dry climates. Furnaces with cooling coils usually have two-speed motors: a low speed for heater operation and a high speed for cooling. The motors in furnaces are usually in the return air stream and rely on the air stream for cooling.

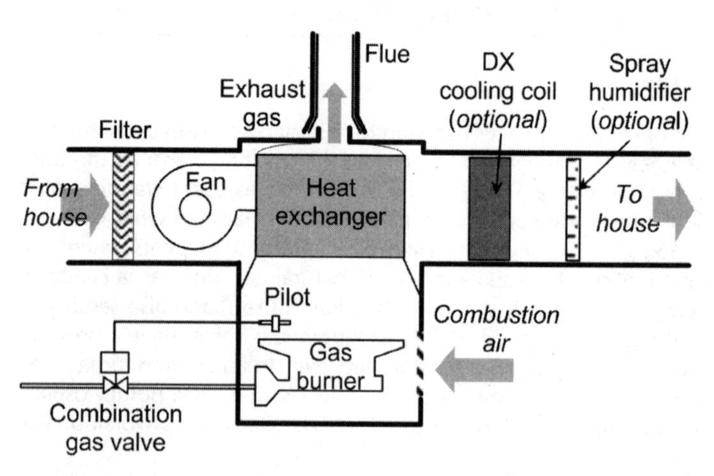

FIGURE 18.1 Typical residential furnace.

Gas-fired furnaces can be *condensing* or *non-condensing*, the differences being efficiency and venting requirements. The most basic and least expensive (and least energy efficient) is the non-condensing furnace. The name comes from whether or not water condenses out of the air (water being one of the principal products of natural gas combustion).

In a non-condensing furnace, air is drawn directly from the mechanical room into the combustion chamber. The fuel and combustion air burn less efficiently than in a condensing furnace. To avoid condensation, a higher temperature flue exhaust is maintained. The hot flue gases naturally draft upwards and out of the building through a Type B vent (rated at 550°F). The local codes set standards for the amount of combustion air required and vertical flue restrictions. Generally, all vent piping must be in a constant upward direction and must exit through the roof. Water vapor does not condense out of the flue gases, and the typical efficiency at sea level is about 80 percent.

In contrast, the condensing furnace is more efficient and may allow a more flexible design with fewer code restrictions. A secondary heat exchanger captures additional heat and increases the furnace efficiency to around 90 percent. These furnaces also cool the flue gases to the point where water begins to condense out of the gas, hence the name. The lower temperature flue gas allows the use of plastic vent piping and vent configurations unsuitable for non-condensing furnaces such as sidewall terminations and horizontal runs. Typically, condensing furnaces save space and materials through the reduced venting requirements. Never retrofit a non-condensing furnace into the same flue duct used for a condensing furnace.

Manufacturers typically certify their equipment for site elevations up to 2000 feet. Above this, the ratings are reduced 4 percent for each 1000 feet above sea level. For example, in Denver (at 5300 feet) the actual output of a furnace rated at 100 MMBtu/hour unit would be about 20 percent less, or 80 MMBtu/hr.

Oil-fired furnaces follow the same concepts as the standard draft furnace except for their burners and flue requirements. In order to utilize oil as a fuel, an oil pump pressurizes the fuel through atomizing nozzles. The pump pressure, orifice size and nozzles regulate the firing rate. The flue requirements are Type L (rated at 570°F). In place of a draft hood, oil-fired furnaces use a barometric draft regulator that controls the pressure above the combustion chamber to around –0.02 to –0.05 inches W.G., depending on the furnace size.

An *electric furnace* is similar to the combustion furnaces, except that electric resistance-type heating elements are used instead of fuel-fired components. The initial cost of electric furnaces is usually less than combustion furnaces, although their operating costs may be higher. They are usually used in applications where only electric power is allowed, such as remote sites with no gas service, or in buildings where architectural limitations limit the ability to vent the flue gases.

Furnaces come in a variety of configurations to fit in just about any mechanical space. Vertical discharge furnaces bring return air in at the bottom of the furnace and send conditioned air straight out through the top, while downflow discharge furnaces take air in from the top and push the air out the bottom. These can be useful for basement or attic installations, respectively. When working in narrow enclosures (such as crawl spaces underneath houses) you can use horizontal discharge furnaces where the air comes in one side and leaves through the other. You may also want to consider using a "low boy" (Figure 18.2) where the return air, supply air, and flue gases all enter or leave through the top of the low-profile furnace.

In applications where the furnace is in an unconditioned or semiconditioned space (such as a crawl space) you should insulate the duct work leading from the furnace to the floor registers. This saves you a lot of money in energy costs and may extend the life of the furnace since it does not cycle on as much.

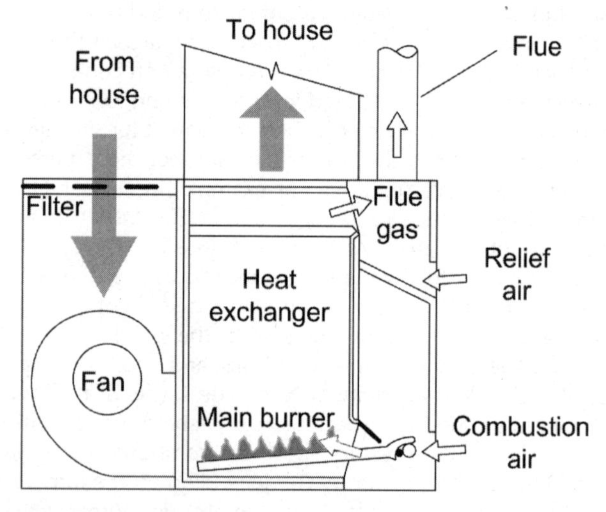

FIGURE 18.2 Basement ("low boy") residential furnace. *(From Kreider et al., with permission)*

Heat Pumps

Heat pumps are basically small chillers like those described in Chapter 8. The principles of operation are exactly the same, just on a much smaller scale. The big difference between a large chiller and a heat pump is the reversing valve that switches the direction of refrigerant flow through the coils. Figure 18.3 shows a schematic of the refrigerant flows when a heat pump is used in both heating and cooling mode. The flow of refrigerant through the compressor is always in the same direction (as it must be) but the flow through the coils reverses direction and the coils swap heat flows. In the winter the heat pump makes the outside even colder while in the summer it makes the outside even hotter. Of course, this only works up to a certain limit. If it is too cold outside, a heat pump cannot move very much heat to the house. For this reason, heat pumps used in cold climates should have electric or gas heating back-up.

When used in conjunction with a residential central air system, the indoor coil is incorporated into the residential ductwork, often as a DX coil as shown in Figure 18.1. While the majority of heat pumps use an air-to-air heat exchanger as the outdoor coil, in some places the outdoor coil consists of a long series of pipes buried under ground. These *ground source heat pumps* are popular in the southeast U.S. and essentially store the summertime heat in the ground to use in the winter. There are also heat pumps that use rivers or underground aquifers to collect or dump the heat.

CONTROLS

Residential system control is relatively simple. An electric thermostat inside the house determines if the inside temperature is below or above the setpoint and turns on the furnace or heat pump. On/off control is used almost exclusively since the houses tend to heat up or cool off fairly quickly. Normally, there is a 24 VAC control transformer that operates the combination gas valve solenoid, the thermostat, and other associated circuits (see Figure 18.4). The transformer supplies current directly to the thermostat; the thermostat has circuitry that allows the user to switch between heating and cooling mode or to shut off the system altogether.

The operation of thermostats is discussed in Chapter 16. Residential thermostats can be quite simple with just an on/off temperature sensor, an anticipator, and a heating mode/cooling mode switch. Programmable thermostats, on the other hand, can let you

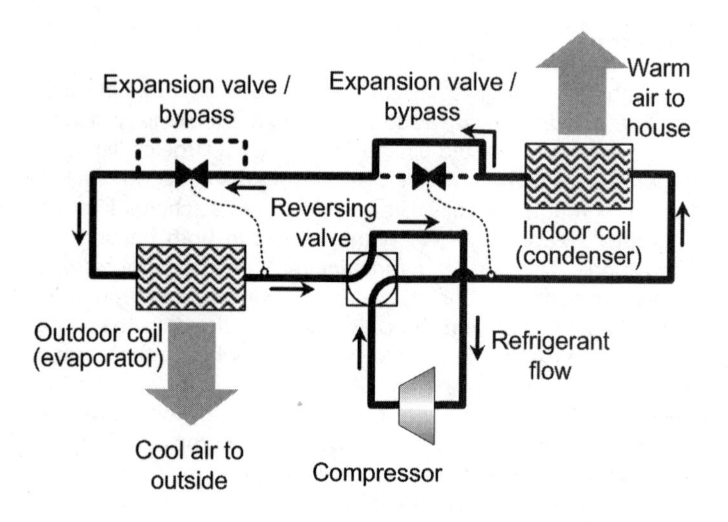

a. heating mode

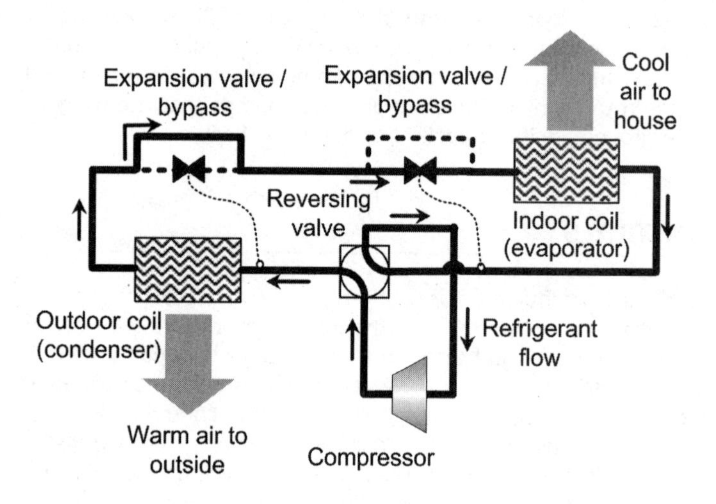

b. cooling mode

FIGURE 18.3 Heat pump schematic showing refrigerant flows.

schedule different heating and cooling setpoints. One problem with programmable thermostats is that people like to play with the buttons and can accidentally turn on heating in the summer or cooling in the winter. Another problem with programmable thermostats is that they use batteries to back up the programmed setpoints in case the main power goes out. In the summer it is not uncommon for the main 24 VAC power to be shut off, and by the time winter comes the batteries have drained and the thermostat won't work.

When a thermostat calls for heating, the combination gas valve determines if the pilot is lit (or sparks the electronic ignition if available) and then opens the gas flow to the burners. If there is a blower, it comes on first, and a pressure switch verifies combustion airflow before opening the gas valve. Once the burner is lit, the main fan comes on after a preset delay. This delay allows the gas valve to verify the flame and to let the heat exchanger warm up before blowing space air across it. Once the furnace is operating, a thermostatic high-limit switch monitors the heat exchanger temperature and shuts off the gas valve in the event of overheating. Once the thermostat setpoint is reached, the gas valve cuts off gas flow to the burners. The main fan continues to run for several minutes after the burner is off since the heat exchanger is still warm and can be used to heat the house.

The fan/high limit switch checks the temperature of the heat exchanger. This switch has two outputs: one turns on the fan when the heat exchanger is hot enough and the other is the high-temperature shutoff.

In cooling mode, the system operates in much the same way: the thermostat sends the signal to turn on the house air circulating fan and the heat pump compressor. There is no delay between the compressor coming on and the fan coming on because you do not want the DX coil to get too cold.

Combustion furnaces have a high-limit switch that cuts off the burners if the furnace temperature gets too high. There are also flame relays that cut off the gas flow if the flame goes out. The flame relays on gas furnaces are usually thermocouples, although some may use flame rods as discussed in Chapter 11. If the flame sensor does not see flame at the main burner after three to four seconds it shuts the main gas valve off. Some of these shut-offs are true lockouts that need resetting before the furnace starts again. Electric furnace safeties include electric overload protection and temperature limit switches. On heat pumps there are high and low pressure safety switches similar to those found on chillers.

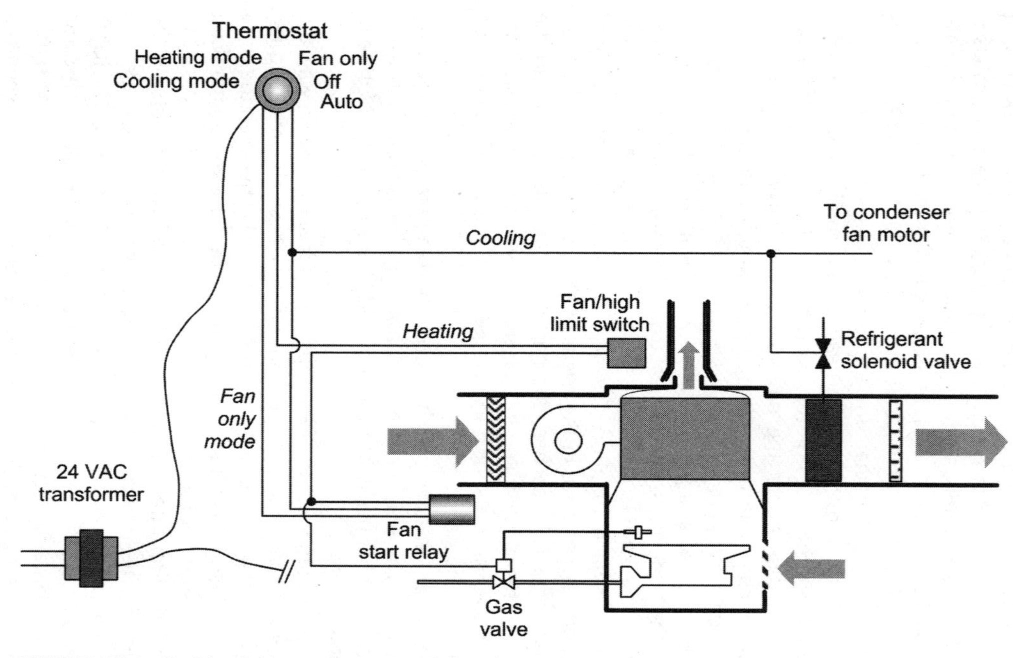

FIGURE 18.4 Residential control system schematic.

SAFETY

Furnaces have the same cautions as boilers. Be careful about back pressure forcing combustion gases back down the flue. This can cause flames to shoot out from the bottom of the furnace and can lead to smoke and carbon monoxide build-up in the house. It is always a good idea to have smoke and CO detectors in the house and near the furnace.

Whenever you are servicing a furnace or heat pump, turn off the power. The fans can start without warning and are generally in an enclosed space where you may not be able to get out of the way quickly. The belts on the furnace fans are rarely covered and can snag clothing and hair very easily. Keep in mind that once the power is off, the fan may start rotating backwards from a natural draft from the wind. If the fan is switched back on while it is rotating backwards, the belts or the fan motor can be damaged. However, resist the temptation to stick some kind of wedge between the fan pulley and housing to keep the fan from spinning. If you forget about it and then turn the fan back on, the results are immediate and depressing.

Make sure there is enough combustion air to the room with the furnace (no obstructions to vented air intakes). In most older houses this is not a problem, but in newer, tightly sealed houses with a natural draft furnace, there can be big indoor air quality problems. If the house is too tight, air can come backwards down the flue vent, and fire from the main burner can come out across the floor.

If you have a standing pilot with a thermocouple, make sure that the pilot flame is in close contact with the top of the burner. It should be arranged so that when the main burner opens the pilot ignites the gas just as it comes out of the burner. If not aligned correctly, the gas can build up before it ignites and you get a small explosion with a popping sound. Sometimes the explosion is big enough that it actually blows out both the flame and the pilot. In severe cases where the pilot is badly misadjusted or put back in the wrong place you can have a serious explosion.

With a natural draft furnace make sure there are no flammables in the area around the base of the furnace. They could get drawn into the flame area by the draft, leading to an uncontrolled fire or even an explosion.

Make sure the vent piping is in good condition. Unfortunately, many flues travel through unpleasant or inaccessible locations, like attics or serviceways. If possible, check to see that the vents are intact and have not come apart at the seams. You want all the flue gases to exit outside and not end up in the house. Flue gases can also enter the house if you have a cracked heat exchanger. While not that common, a break in a heat exchanger weld can lead to carbon monoxide building up in the house. You can check for this by using a CO sensor.

Use a CO sensor to check for flue gas leaks while the furnace is running. If you check while the furnace is not running, the gases may have already dissipated.

If a belt has come off the sheaves and is found to be lying loose around the pulleys with the fan drive motor running, leave the area and turn off the motor with the motor disconnect or at the breaker panel. Do not enter the furnace room until the motor has come to a complete stop. When the motor slows down the belt may become entangled in the motor pulley and whip around with great force. Always follow the manufacturer's instructions when replacing belts, and in particular do not force belts over the sheaves to install them since this can internally damage the belts. Never try to get more air flow out of the fan by changing the pulley size unless the manufacturer explicitly states it's okay to do so. A squirrel cage fan that is turning too fast can come apart and basically explode without warning. In the presence of gas and flame this can be pretty dangerous. If you suspect a gas leak, evacuate the house and call the utility company. Do not close doors on your path of exit and try not to create a static electric shock.

TROUBLESHOOTING

Furnace service calls are usually associated with burned-out motors, broken belts, seized fan shafts, or extinguished pilot lights. In some cases, the filters have not been changed in so long that there is almost no airflow through the furnace.

If the furnace does not start, the problem could be in the gas supply or electric control circuit. First check the obvious things, like the filter and the thermostat. Also make sure that the motor belt is intact and that the fan shaft hasn't seized (that is, you can turn it easily by hand). If you still have a problem, refer to Figure 18.5 for a troubleshooting flow chart that addresses this situation. You can also check Chapter 11 for troubleshooting boilers; many of the troubleshooting techniques (particularly for oil-fired boilers and furnaces) are the same.

Note 1: Check the spark igniter for cracks in the porcelain, dirt or grease in the spark gap, and for correct installation. No spark, no fire.

Note 2: Sometimes the wind can come back down the flue and blow out the pilot. Or if the furnace has been idled for the summer, it's possible that the pilot has been purposely extinguished. Light the pilot following the manufacturer's instructions that are probably attached to the furnace access door.

Pilots can blow themselves out under the right conditions. If there is too much distance between the pilot and the burner, gas can build up when the main gas valve opens. When it finally ignites, you get a gas pop that can be forceful enough to blow out the pilot.

Note 3: If the gas pressure is too low there may not be enough gas getting to the pilot or the burner. If the gas pressure is too high, the furnace may trip out on over-pressure (if available). The pressure should be between about 6 and 16 inches W.G. Check the nameplate pressure against the inlet gas pressure, both when the main burner is off and then when it is on. Sometimes just the act of having the main burner come on can drop the gas pressure

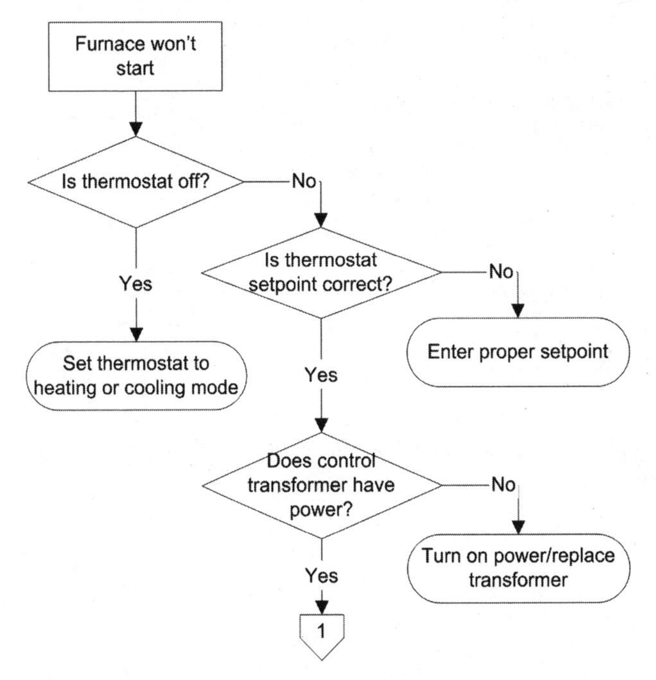

FIGURE 18.5 Troubleshooting flow chart for non-starting furnace.

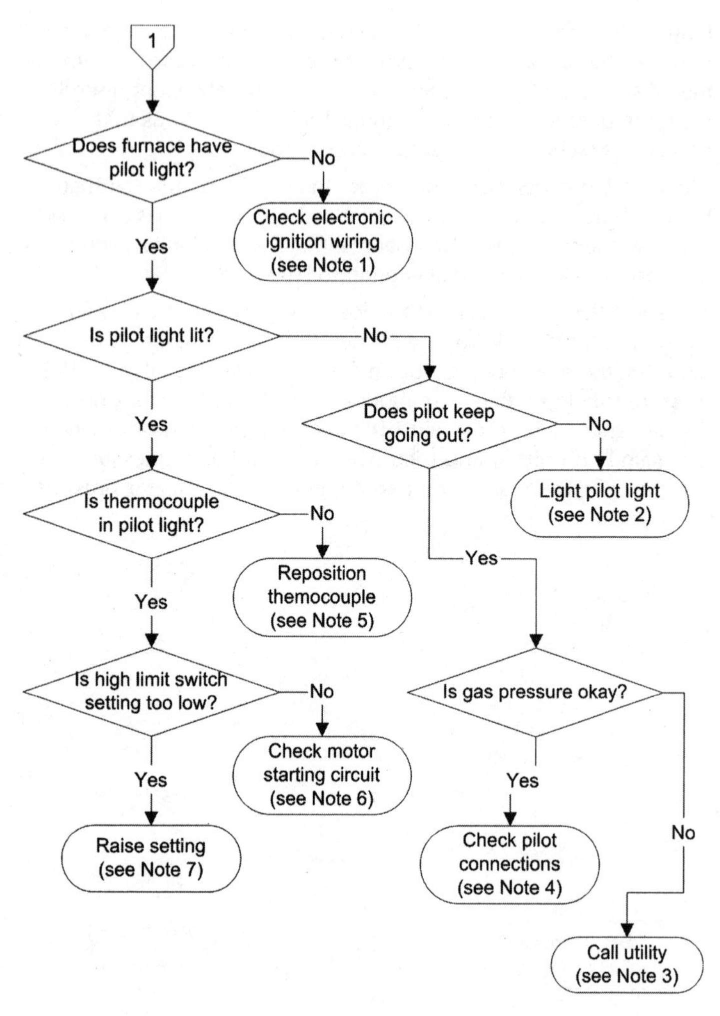

FIGURE 18.5 *(continued)* Troubleshooting flow chart for non-starting furnace.

below the minimum required to operate the furnace. In case of either high or low pressure, you have to contact the utility or gas supplier to upgrade your service.

Big fluctuations in the gas pressure may also cause the pilot to go out. You can adjust the pilot flame with the flame adjust screw, but when the main gas valve comes on it may drop the pressure enough so that the pilot goes out. Fluctuations can also occur when other gas equipment in the house (for example, the hot water heater, gas clothes dryer, etc.) turn on and off. You may need to put a regulator ahead of the combination gas valve.

Note 4: Check to make sure that the pilot nozzle is clean of dirt and grease and that the pilot gas tubing is clear. Also check the aluminum pipe connecting the nozzle to the combination gas valve. This can get clogged so that there is not enough gas pressure to keep the pilot lit. Check both the gas and electrical connections where the thermocouple goes into the combination gas valve. Since there is a very small voltage coming from the thermocouple, it is important to have a good electrical connection to prevent resistive losses and lots of nuisance trip-outs of the pilot.

When you work on a furnace, go ahead and replace the thermocouple. Generally they last four or five years before they need replacing anyway. It is usually not worth it to scrimp on costs. The more expensive ones usually last longer and end up costing less in the long run since they are less problematic and can also have slightly higher voltages that prevent nuisance trip-outs.

Note 5: Make sure the tip of the pilot flame is hitting the tip of the thermocouple—the thermocouple should be red on the end. The pilot flame should be blue and just surrounding the thermocouple. If there is too much flame then the flame can wrap up over the top of the thermocouple and blow itself out, while too little flame won't keep the thermocouple hot enough. There is a pilot flame adjustment screw on the combination gas valve that you can use to modify the flame size. If you cannot adjust the pilot size, there is a chance that the combination gas valve itself is defective.

Note 6: The fan starting circuit has several components that may go bad. Check the fan start relay that uses the 24 VAC control signal to send 120 VAC to the fan. If this relay is stuck then the fan cannot start. Also, some motors use starting capacitors that may need to be replaced.

Note 7: The high-limit switch is actually two thermal switches: one lets the fan turn on when the heat exchanger is hot enough and the other turns the fan off when the heat exchanger gets too hot. Make sure that the setpoints are not reversed.

Furnace Cycles Too Much

A furnace that cycles too much wastes energy and will burn out the fan motor too soon. Cycling too much generally means more than about 10 starts and stops per hour. It usually happens when there is not enough air flow across the heat exchanger and it is dropping out on the fan high limit switch. This can happen if the belt is slipping or if the filters are clogged. If these check out okay, then check the high limit switch. If the cut-out temperature is set too low, the furnace cycles every time the furnace begins to get up to the proper constant operating temperature (see Note 7). The last thing to check is the gas pressure as discussed in Note 3. Fluctuating gas pressure can lead to furnace cycling. If you do not have a gauge to check it with, just watch the pilot light for a while to see if it changes size.

If the furnace seems to be working okay, the problem is probably something with the thermostat. For example, too much anticipation (see Chapter 16) causes a furnace to shut off well before it has warmed the occupied spaces. Lower the anticipator to its smallest value and see if that helps. Also see if there are drafts near the thermostat (including in the wall behind the thermostat mounting plate). It could also be that the thermostat just needs replacing.

Space is Too Hot or Too Cold

If the furnace is unable to condition the space and is not cycling too much, the first thing to check is the thermostat. Not only check the setpoints but also see if it has been placed in some kind of manual setting that keeps the system from turning on or shutting off. Make sure the thermostat is not out of calibration. You can test this by slowly rotating the thermostat and seeing if it completes the control circuit when it passes through the current room temperature. Also make sure that the thermostat is not in a drafty location.

Thermostats with mercury bulbs must be mounted level. If not, the thermostat contacts may connect or disconnect before they're supposed to, leading to no heating or over-heating.

Equipment can get stuck in certain modes either through manual settings or through equipment failure. For example, the contactors on a heat pump control can weld, causing the compressor to stay

fastfacts

A technician responded to a house where the homeowner complained of a thermostat not working. When she investigated the furnace, she found that the gas valve had become stuck open. The heat exchanger was cherry red and the fan motor and ductwork were smoking. The homeowner had no idea how close he was to losing his house.

on after the house is cool enough. Similarly, the combination gas valve can become stuck in the open position. Both of these conditions are potentially dangerous, since it is then the sole duty of the pressure or temperature limit switches to shut the equipment off. Fortunately, most gas valves fail safe, meaning that they fail in the closed position. In fact, sometimes the gas valve can become stuck shut and prevent any heating at all.

Another common reason for a central air system to be unable to heat a house is inadequate air circulating through the system. As always, check the filters. They should be replaced once or twice a year, but more often than not they are replaced one or twice a decade. Check the air side of the heat exchanger for lint and dust build-up. If the system has an evaporator coil for cooling, check to make sure this coil is also clean. If there is a two-speed fan, check to make sure that the proper speed is being used for both cooling and heating. Usually the fan operates at higher speed for cooling mode, and it is not uncommon for the control wiring to get switched around.

Ducts can fill up with dust and dirt. Not only does this create an indoor air quality problem, but in some circumstances the dirt can restrict airflow from vents into the rooms.

Furnace is Too Noisy

Noisy furnaces should not be ignored. A squeak or a chirp or a growl can all mean that the furnace is trying to tell you something and that you should listen. Unusual noises usually become apparent as they echo up through the ductwork. A squeaking or dull roar says check the tension of the motor belt. There should be about one inch of slack when you press on the belt halfway between the motor and the fan cage. If the belt is too loose or too tight, then it may be rubbing against the pulleys and may fray soon. A chirping sound is asking you to check the bearings. When was the last time they were lubricated? It might be time to add a couple drops of oil into the bearing ports. A squealing noise says that the pulleys are out of alignment and are about to chew through the belts. A twisting metal sound means that the fan blades are loose and are about to make new vent holes in the furnace body. Replace or tighten all loose blades.

PRACTICAL CONSIDERATIONS

There are a number of things you can do both in installation and for maintenance that helps ensure a long life and proper operation of the furnace.

- The air flow rate for a furnace should be properly sized to meet the maximum heating loads for the house. Oversized fans have a greater initial cost, greater maintenance cost, and can be more difficult to control.
- If the system has a cooling coil, make sure the fan has two speeds and is sized to deliver the correct amount of air in cooling mode. Make sure that the fan is wired correctly so that the high speed is used for cooling mode and the low speed for heating. If the fan is wired incorrectly, the evaporator can get too cold and ice over. This also causes low temperature heating because too much room air flows across the heat exchanger.
- Change the filters at least once per year, twice if possible. By their very nature, filters get clogged up. They are supposed to. But when they do, there is not a lot of house air moving through the furnace and most of the combustion heat goes up the flue. Not only is this inefficient, but the exchanger and combustion chamber can overheat.
- Many squirrel-cage fans have oil-sleeve bearings—they are quieter than ball bearings—but they must be oiled every year. Three to four drops of household oil is enough. Check them every time you change the filters. In belt driven fans where the bearings have not been oiled for some time the shaft can grind through the pillow block bearings right on down to the fan housing and bearing supports.

- When you change the filters, check the belts for wear and fraying. Sometimes the belts come off completely, although you usually notice that from the lack of air conditioning in the house. If a belt is slipping, the heat exchanger can run hot and the furnace efficiency will drop. Also check the fan blades. Dirt can build up on the blades, either just from the airborne dirt or by clinging to excess oil that drips from the bearings.
- Make sure the pilot and burner gas flames are blue. If the flame is yellow then you have too much gas and not enough air. This causes soot to accumulate in the heat exchanger and flue. Use the gas valve adjustments to get the correct air-to-gas ratio. Use a flue gas analyzer if necessary.
- Some motors (whether they are belt or direct drive) sit in a rubber insert that holds the motor in place. Eventually these inserts become hard and brittle and fall apart and the motor can fall completely out of the brackets.
- When you change the filters, look for dirt build-up in the motor. Most motors are in the airstream and eventually fill up with dirt and dust. A dirty motor can overheat, causing the oil to evaporate out of the bearings and the motor to start running hot. The bearings can freeze up, or, in extreme cases, the motor lid can catch on fire.
- Be careful about putting too many gas appliances on a small or low-pressure gas line. This can cause frustrating intermittent problems with the furnace. It can be working fine until the hot water heater comes on and brings the pressure down far enough to knock the furnace off-line.
- Sealed combustion furnaces have a pressure switch that has factory settings for determining when the combustion chamber pressure is high enough to start. At high altitude installations (above 3500 feet), the factory setting may need to be adjusted so that they kick in at a lower pressure. Sometimes natural draft furnaces do not draw well at altitude.

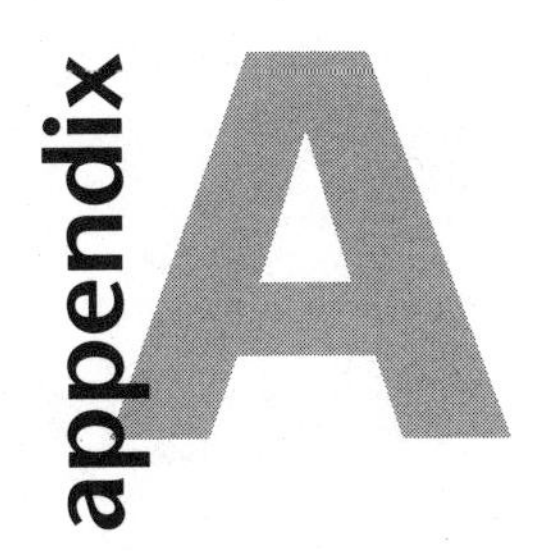

TOOLS FOR THE TECHNICIAN

As with most technical work, your life will be greatly simplified if you have and use the right tools for the job. You can spend an hour trying to jury-rig a system to remove a valve cover or debug an electric circuit, or you can get the right equipment and have the job done in a few minutes. This section lists some of the basic equipment you should keep with you. Remember that the cost of parts and special tools is usually much smaller than the labor costs for most service calls. If you use a service truck, try to keep it well stocked. A good variety of spare parts in the truck means less time spent chasing down the parts you need. Having the right parts and the proper tools available makes the service call go quicker and results in better customer relations.

COMMON HAND TOOLS:

- Set of screwdrivers in both Philips and straight blade.
- Set of Allen wrenches in both inch and metric sizes.
- Pliers, both standard and narrow-nose.
- Side cutters.
- Wire strippers.
- Nut drivers in both inch and metric sizes.

- Electrician's knife and electrical tape.
- Flashlight.
- First aid kit.
- Tool kit to put all this in.

CALIBRATION TOOLS

- Calibration (jeweler's) screwdrivers.
- Thermostat adjustment tools.
- Test gauges.
- Thermometers.

DIAGNOSTIC TOOLS

- Volt-ohm meters.
- Voltage tester.
- Clamp-on amp probe.
- A 4 to 20 MA power supply.
- Squeeze bulb tester.
- Equipment production information.

CONSTRUCTION TOOLS

- Tubing benders.
- Tubing cutter.
- Hack saw.
- Electric drill with bits.
- Socket set.
- Open and box wrench set.
- Knock out punches.
- Hole cutters.
- Soldering iron.

- Step ladder.
- Extension ladder.
- Drop light with ground fault protection.

MAINTENANCE TOOLS

- Pipe wrench.
- Strap wrench.
- Mapp gas torch and solder.
- Fire extinguisher.
- Specialty tools for different types of equipment maintenance.

THINGS TO CARRY WITH YOU AT ALL TIMES

- Identification tag.
- Cell phone.
- Notebook and pencil.
- Pocket flashlight.

Always remember that the most important tool on the job is attitude. "Safety first" is more than just words—it is a motto literally to live by. It only takes a few seconds to check for live power before working on equipment, but it is a few seconds well spent. Set up a lock-out/tag-out center (Figure A.1) and use it.

If something looks like a dangerous situation, it probably is, so don't give in to the temptation to get a job done quickly. Always let your supervisor know where you are so that if you become stuck somewhere, at least someone knows where to begin looking.

FIGURE A.1 Lock-out/tag-out center.

USING MULTIMETERS

When working with a multimeter, there are a few simple rules that ensure safe, proper operation and good readings. A multimeter measures current, voltage, and resistance. You need to select the proper range on the multimeter to have good resolution of the measured value. Figure B.1 shows a typical multimeter face with a range selector knob, AC/DC toggle, and the probe inputs.

Resistance measurements are used to test the electrical resistance of components. In most HVAC applications this means checking a resistor to see if it is burned out. A special kind of resistance measurement called a continuity check is used to see if a wire is broken or if there is a short circuit. For example, a continuity check can determine whether or not a fuse is blown. To test this, connect the voltmeter as shown in Figure B.2 and set the range to 2 Ω or 20 Ω. If the reading is low, for example, between 0.1Ω and 1 Ω, then the fuse is good. If the reading is off the scale (usually denoted by a blinking display or the letters "OL" appearing in the display) then the fuse is not good. You can also use continuity checks to identify short circuits, accidental groundings, etc.

Voltage measurements must be open-circuit. That is, they must not allow any current flow through the multimeter. To take a voltage measurement across any two points, connect the multimeter as shown in Figure B.3 and set the range knob to the appropriate value.

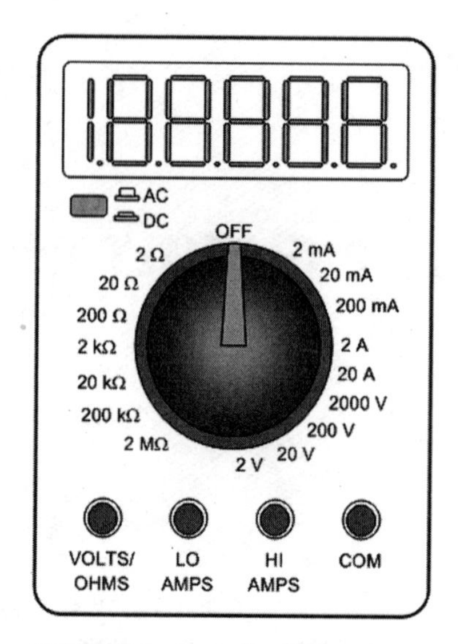

FIGURE B.1 Typical multimeter.

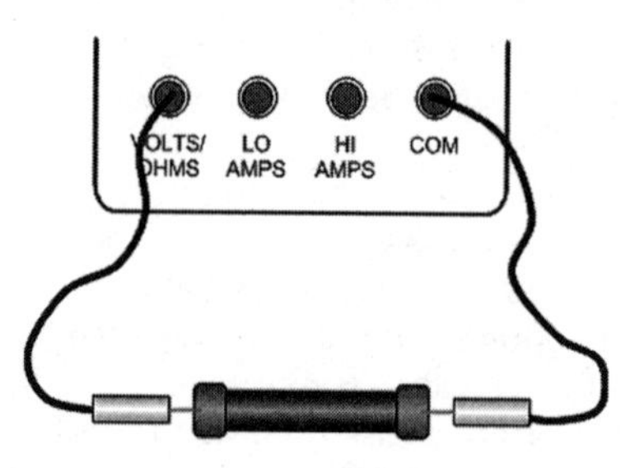

FIGURE B.2 Measuring resistance.

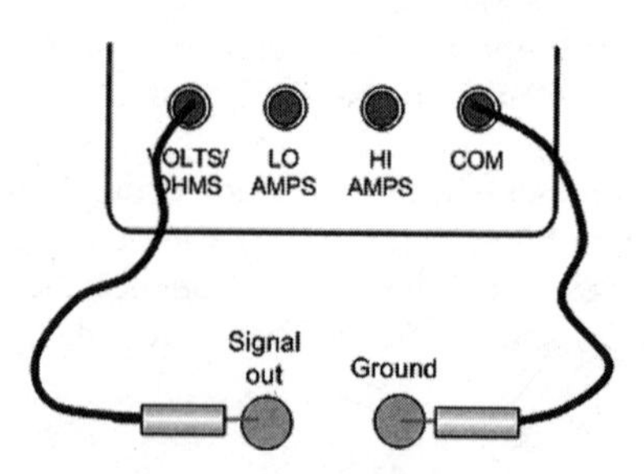

FIGURE B.3 Voltage measurements.

Most voltage sensors put out a 0 to 10 VDC signal, so use the 20 V range and have the AC/DC button set to DC.

You can also measure AC voltages with this exact same wiring by changing the AC/DC switch to AC. Be very careful when measuring high AC voltages (115 VAC and above). The plastic probe handles can insulate up to a point, but at higher voltages there is always the possibility of an electrical arc. Also, make sure your fingers are not accidentally touching the metal sections of the probes and that your fingers do not slip off the ends of the probes while you are taking the measurement.

If you measure AC voltages with the switch set to DC, you may not have any reading at all on the multimeter. Even though the *peak* AC voltage may be very high, it oscillates between a large positive number and a large negative number and the *average* voltage is close to zero.

In contrast to voltage and resistance measurements, current measurements must be made using a closed circuit. In other words, the multimeter must be in series with the current flow you are measuring. This requires breaking the existing circuit and re-connecting the multimeter in line as shown in Figure B.4. You cannot measure current across two lines (well, you can, but you will short out the two lines since the multimeter acts as a closed circuit when measuring current).

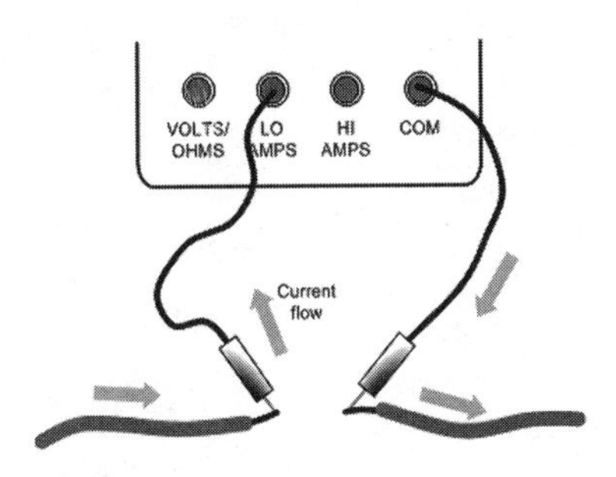

FIGURE B.4 Measuring current.

NOMENCLATURE

A	amperes
AC	alternating current
ACC	air-cooled condenser
AHU	air-handling unit
Btu	British thermal unit
CAV	constant air volume
CFM	cubic feet per minute
CHW	chilled water
COM	common
C_d	damper loss coefficient
c_p	specific heat
C_v	valve flow coefficient
CWR	condenser water return
CWS	condenser water supply
DA	discharge air (supply air from fan)
DC	direct current
DP	dew point
DPDT	double-pole, double throw
DHW	domestic hot water
DX	direct expansion
EA	exhaust air
E/P	electric-to-pneumatic
EPA	Environmental Protection Agency
ft	feet
GPM	gallons per minute
HOA	hand-off-auto

HP	horsepower
HVAC	heating, ventilation, and air-conditioning
HW	hot water
i	current
in. W.G.	inches of water gauge
I/P	current-to-pneumatic
kW	kilowatt
kWh	kilowatt-hour
LAT	leaving air temperature
lb, lbs	pounds
LT	low temperature limit sensor or switch
LTW	low temperature water
mA	milliamperes
$\dot{m}$	mass flow rate
MA	mixed air
MAT	mixed air temperature
MCC	motor control center
MMBtu	millions of Btu
MTW	medium temperature water
NC	normally closed
NO	normally open
OA	outside air
OAT	outside air temperature
P/E	pneumatic-electric
PF	power factor
psi	pounds per square inch
psig	pounds per square inch gauge
PI	proportional plus integral controller
PID	proportional plus integral plus derivative controller
P_t	total pressure
P_v	velocity pressure
$\dot{Q}$	energy flow rate
RA	return air
RPM	revolutions per minute
RTD	resistance temperature device
SA	supply air
SAT	supply air temperature

SP	static pressure
SPDT	single-pole, double throw
SPST	single-pole, single throw
T	temperature
TDS	total dissolved solids
TES	thermal energy storage
TEV	thermal expansion valve
UV	ultraviolet
V	volts
VAC	AC volts
VAV	variable air volume
VDC	DC volts
VFD	variable frequency drive
VSD	variable speed drive
W	work
W.G.	water gauge
ΔP	pressure difference
ΔT	temperature difference
η	efficiency
Ω	ohms

BIBLIOGRAPHY

ASHRAE (1981). *Standard 55-1981: Thermal Environmental Conditions for Human Occupancy.* American Society of Heating, Refrigeration, and Air-Conditioning Engineers, Atlanta, GA.

ASHRAE (1997). *Handbook of Fundamentals.* American Society of Heating, Refrigeration, and Air-Conditioning Engineers, Atlanta.

ASHRAE (2000). *Handbook of Systems and Equipment.* American Society of Heating, Refrigeration, and Air-Conditioning Engineers, Atlanta.

Avallone, E. and T. Baumeister, eds. (1996). *Marks' Standard Handbook for Mechanical Engineers.* McGraw-Hill, NY.

Honeywell (1988). *Engineering Manual of Automatic Control for Commercial Buildings.* Honeywell Inc., Minneapolis, MN.

Kreider, J.F., P. Curtiss, and A. Rabl (2001). *Heating and Cooling of Buildings: Design for Efficiency, 2nd Ed.* McGraw-Hill, NY.

Leatherman, K. M. (1981). *Automatic Controls for Heating and Air Conditioning.* Pergamon Press, Oxford.

Omega Engineering (2000). *The Temperature Handbook.* Omega Engineering, Stamford, CT.

Stein, B. and J. Reynolds (1992). *Mechanical and Electrical Equipment for Buildings.* John Wiley and Sons, Inc., NY.

INDEX

A

B

C

D

N

O

P

R

S

T

U

V

W

Z

NOTES

NOTES